PHIL EDMONSTON'S LEMON-AID

NEW CARS AND MINIVANS 2004

PHIL EDMONSTON'S

LEMON-AID NEW CARS AND MINIVANS 2004

PHIL EDMONSTON

PENGUIN
CANADA

PENGUIN CANADA

Penguin Group (Canada), a division of Pearson Penguin Canada Inc.,
10 Alcorn Avenue, Toronto, Ontario, Canada M4V 3B2

Penguin Group (U.K.), 80 Strand, London WC2R 0RL, England
Penguin Group (U.S.), 375 Hudson Street, New York, New York 10014, U.S.A.
Penguin Group (Australia) Inc., 250 Camberwell Road, Camberwell, Victoria 3124, Australia
Penguin Group (Ireland), 25 St. Stephen's Green, Dublin 2, Ireland
Penguin Books India (P) Ltd, 11, Community Centre, Panchsheel Park, New Delhi – 110 017, India
Penguin Group (New Zealand), cnr Rosedale and Airborne Roads, Albany, Auckland 1310, New Zealand
Penguin Books (South Africa) (Pty) Ltd, 24 Sturdee Avenue, Rosebank 2196, South Africa

Penguin Group, Registered Offices: 80 Strand, London WC2R 0RL, England

First published 2003

10 9 8 7 6 5 4 3 2 1

Manufactured in Canada.

National Library of Canada Cataloguing in Publication data available on request.

Visit the Penguin Books (Canada) website at **www.penguin.ca**

CONTENTS

PHIL EDMONSTON'S LEMON-AID

NEW CARS AND MINIVANS 2004

KEY DOCUMENTS

The following photos, charts, documents, memos, court filings and decisions, and service bulletins are included in this index so that you can easily find and photocopy whichever document will prove helpful in your dealings with automakers, government agencies, dealers, or service managers. Most of the service bulletins outline repairs or replacements that should be done for free.

INTRODUCTION I TOLD YOU SO

PART ONE DEALING WITH DEALERS

PART TWO LEGAL RIGHTS AND WRONGS

PART THREE NEW-VEHICLE RATINGS

APPENDIX I

I TOLD YOU SO

If you had read this book a few years ago you probably would have bought a better car for much less money. You would have known that Mercedes quality is on the decline and that Ford and Chrysler cars and minivans are quite unreliable. *Lemon-Aid New Cars and Minivans* would have also warned you about Toyota and Honda engine and transmission defects that are just coming to light.

Honda Service Bulletin 02-062

August 19, 2003

Applies to:

1999–2001 Odyssey - All
2000–01 Accord - All
2000–01 Prelude - All

Warranty Extension: Accord, Odyssey, and Prelude automatic transmission
(Supersees 02-062, dated February 25, 2003)

Background:
In certain vehicles, a higher than normal number of automatic transmissions may have defects in material or workmanship that could cause premature wear of failure. To ensure that customers have adequate warranty coverage, American Honda is increasing the warranty on the transmission and the torque converter to 7 years or 100,000 miles [160,000 km], whichever occurs first.

- If your diagnosis confirms a problem with the torque converter or an internal transmission failure, replace the affected part, and file a warranty claim using the warranty information in this service bulletin (02-062).

This September 2003 Honda bulletin admits that *Lemon-Aid* was right three years ago—Honda quality has declined.

Unlike any other auto book on the market, *Lemon-Aid* is a uniquely Canadian owner's manual that pulls no punches in disclosing what's a fair price for a new car or minivan, which vehicles are likely to break down or quickly lose their value, which repairs are the automakers' responsibility, and which scams you should avoid. We believe that paying a fair price is even more crucial this year, now that a whole slew of multi-thousand-dollar rebates make many vehicles once-in-a-lifetime bargains—if you know what to buy and how to use rebate and depreciation timing to save thousands of dollars.

This guide continues a 32-year tradition of publicizing abusive auto industry practices and providing hard-to-get information that may save your life, or at least make you a smarter consumer. *Lemon-Aid*'s information comes from owners who buy and drive these vehicles, not travel-junket-junkie, free-car-mongering car columnists. Our database includes Canadian and U.S. sources and is refined throughout the year with input from owner complaints, automaker whistle-blowers, lawsuits, judgments, confidential technical service bulletins (TSBs), and independent garages.

This gives us the facts to separate the good from the bad, the sizzle from the steak, the hype from the truth. For example, we have found that South Korean automaker Hyundai produces vehicles that are almost as reliable as Japanese makes, and can cost about one-third less. We like Toyota's 2004 Sienna minivan, but are worried about first-year glitches. We are happy to see Mitsubishi enter the Canadian market but feel the company's financing and DaimlerChrysler dealership network are shaky. And we are profoundly disappointed by Saturn's omnipresent factory-related defects.

Guess we won't get a Christmas card from GM this year, eh?

Saturn's promotion is much better than the product.

Lemon-Aid was the first book to blow the whistle on Ford Taurus, Sable, and Windstar transmission and coil spring failures, as well as the rising number of powertrain defects and safety-related problems found in Toyota and Honda vehicles since "de-contenting" came about in 1997. We were also first off the mark in exposing Ford and GM's failure-prone engine intake manifolds and Chrysler's defective paint and biodegradable automatic transmissions and brakes. Following *Lemon-Aid*'s intervention and printing of their internal service bulletins in this guide, Chrysler and Ford have paid hundreds of previously rejected owner claims and extended their warranties.

The 2004 edition combines test results with owner feedback to provide a critical comparison of 2003–04 cars and minivans. If improvements and additional safety features don't justify the higher costs of newer models (most don't), we say so. Safer, more reliable, and often cheaper alternatives are given for each vehicle. Front, offset, side, and rollover crash test results are also included, along with an exhaustive list of useful and useless accessories and safety features, and guidelines for how much profit dealers make on each vehicle.

So, get yourself a large cup of coffee, sit down, and peer into the future. And, for real fun, pass this book on to your relatives, friends, and colleagues and watch their jaws drop as they find out how their car or minivan rates in *Lemon-Aid*.

And, you too will be able to say, "I told you so."

Phil Edmonston
October 2003

PART ONE

DEALING WITH DEALERS

1

Luxury Lemons

Volkswagen was among the five worst brands in the industry, among 37 ranked [in the latest J.D. Power survey]. Ford's Volvo and Land Rover divisions also performed below average, with Land Rover ranking behind the beleaguered Daewoo brand and better only than Kia, made by the Korean automaker Hyundai. According to the study, Kia made the two least durable vehicles: the Sportage sport utility and the Sephia sedan, which has been discontinued.

At Mercedes-Benz, the report said, the most troublesome vehicles were the M-Class sport utility and the E-Class, one of its highest-volume sedans, which starts just under $50,000 and can cost considerably more.

The New York Times
July 8, 2003

Former Ford USA Prez Sees Yellow Over Poor Quality

In a shocking admission before getting sacked, Ford's chief executive officer Jacques Nasser said that a slew of quality problems in new and redesigned vehicles launched in the U.S. and Europe last year cost the company more than $1 billion in lost profits and hampered its growth plans.

"I'm totally pissed off with it," Mr. Nasser told a meeting of securities analysts also attended by about 350 of Ford's top executives. "That will never happen again," he added.

The Wall Street Journal
January 12, 2001

Ford's Windstar, now called the Freestar, has been one of the most problem-plagued minivans on the market. Forget name changes, Ford should up the warranty to 7 years/160,000 km.

It's still happening Jacques, and it's mostly your fault. You can't bully suppliers, cut their prices, and ask them to give you more for less. What you get is less—quality.

Prices down, quality down

Cars are an endangered species; last year's Detroit Big Three car sales of 3.4 million were down 44 percent from 6.1 million in 1990, even though overall Big Three car-truck sales were up about 400,000. Automakers are alarmed by these figures and are afraid to raise prices on their 2004 cars, though they'll sneak some higher prices into their truck and SUV lineup.

Industry insiders predict that new car prices will rise by about 2 percent on most models with few improvements to justify the higher prices. And watch out for price increases snuck through the back door; many standard safety and convenience features will become optional and freight and PDI charges will pass the $1,200 mark.

The Detroit Big Three use an interim pricing strategy where price increases are spread throughout the year; Asian and European makes generally announce one price in the fall. However, some Toyota dealers are trying to keep prices high by sticking to a "no-haggle" policy set up and then abandoned by Toyota, following its $2 million price-fixing settlement with Ottawa last year (as a result of complaints lodged by *Lemon-Aid*).

Currency exchange rates and high federal and provincial sales taxes are also putting a brake on higher prices. Afraid they'll price vehicles out of the reach of average Canadians, automobile manufacturers sell their vehicles for much less in Canada than in the States. In a 1999 study of new-vehicle prices in Canada and the United States, Dennis DesRosiers, a Toronto-based auto consultant, found that Canadians pay far less than Americans for the same car or truck. He pegged the difference to be $4,000 per car and $8,000 per light truck.

Leasing costs have also moderated, especially with high-end cars. Some of the luxury makes offering leasing bargains include Ford's Land Rover, and Volvo brands and the Mercedes E-Class (all models scoring poorly in J.D. Power's latest reliability survey). A notable exception to this price-cutting is

BMW, which is reporting record sales across the board, with chronic shortages of its new 5 Series and X5 SUV.

If you *must* buy a new car or minivan this year, seriously consider buying a cheaper 2003 model off the dealer's lot, particularly if it hasn't been substantially upgraded this year (check Part 3). It will likely offer more standard features than the 2004 version and be eligible for generous factory rebates as well. Of course, if you can afford to wait six months, most of the 2004 price increases will be wiped out by a new round of factory rebates and low-cost financing programs.

There's not much difference between a 2003 and 2004 Honda Accord—except for price. Smart shoppers will opt for a 2003 six months into the production schedule to benefit from second-series quality improvements.

The quality crisis

Ford, Chrysler, and (to a lesser extent) GM build poor quality, unreliable vehicles. Of the three, though, General Motors has made the most progress after phasing in more reliable components in its 1999 truck lineup and then carrying them over to its SUVs, vans, and cars. Not surprisingly, most of the product development money spent by automakers in recent years has been spent on pickups, vans, and SUVs, since they are the most profitable sellers.

Chrysler is still on a cost-cutting binge (dictated by Daimler-Benz in Germany) that isn't likely to bring up its low quality scores in the near future. However, its 7-year powertrain warranty will alleviate some of the pain.

And then there's Ford, Detroit's poster child for unreliable, low-quality vehicles. The company is rife with poor administration combined with unrealistic cost-cutting goals that are never met. Imagine, the company's newly redesigned 2004 F-150 has one of the highest suggested retail prices around, yet the base truck is sold at a loss. Program after program is begun with all the attendant hoopla, then dropped with nary a word of explanation. Dealers are treated like serfs and only consulted after key decisions have been made. Further, warranty payouts are Scrooge-like and owners are often forced to pay for the factory's mistakes.

In the winter of 2002, stung by owner complaints regarding powertrain and other major component failures, Ford USA convened an urgent meeting with most of its suppliers. After ripping through the supplier body, telling them how shoddy their products were, Ford was told that the parts suppliers were only supplying what Ford procurement had ordered. They reminded Ford that Toyota, Honda, and Nissan—companies that were at the top of J.D. Power's quality surveys—were their best customers. The culprit? Ford's cost-cutting edicts that called for constant changes in parts that weren't subsequently tested.

Ford adjourned the meeting and stopped blaming suppliers for its quality woes.

Buying a New Car or Minivan

There are about 4,000 vehicle dealerships in Canada and they all want your money—almost $30,000 from the average vehicle transaction, says Dennis DesRosiers, a Toronto-based auto consultant. And be prepared to pay almost $10,000 annually for the care and feeding of your new "wheels."

The Canadian Automobile Association's 2003 survey, based partially on research by Runzheimer Canada, a Toronto-based management consulting firm specializing in travel and living costs, says that the average vehicle is driven 20,247 km annually. Based on 18,000 km a year, maintenance costs hover around 2.9 cents per kilometre and insurance for a 2003 Cavalier would run $1,962 ($1,774 for a 2003 Chrysler Caravan). The CAA's estimate of annual ownership costs for that same Cavalier is $9,525, a Caravan would cost $9,986 (52.9 and 55.5 cents/km, respectively).

Before paying such big bucks, you should know what your real needs are and how much you can afford to spend. Don't confuse needs with styling (do you have a "bucket bottom" to conform to bucket seats?) or the trendy with the essential (will a cheaper wagon or hatchback suit you better than an SUV)? Visiting the showroom with your spouse, a relative, or a friend will help you steer a truer course through all the non-essential options you'll be offered.

Most sales agents admit that women shoppers are far more knowledgeable about what they want and more patient in negotiating the contract's details than men, who tend to be mesmerized by many of the techno-toys available and often skip over the fine print and bluff their way through the experience. In increasing numbers, women have discovered that minivans and SUVs are more versatile than passenger cars and station wagons. And, having spotted a profitable trend, automakers are offering increased versatility combined with unconventional styling in so-called "crossover" vehicles. These blended cars are part sedan and part station wagon with a touch of sport utility added for function and fun. For example, the Chrysler Pacifica is a smaller, sporty crossover vehicle that looks like a miniature SUV. Touted by Chrysler as the "sports tourer," the Pacifica is the modern, reiteration of traditional Chrysler minivans. Women drivers and young executives with new families are the targets for this vehicle, which starts at $43,395 for the front-drive model and $45,995 for the all-wheel-drive version.

Chrysler officials don't want you calling the Pacifica a mini-minivan, or a wagon; they prefer the term "sports tourer." I prefer to call it overpriced and predict large price cuts a few months after its launch.

What can I afford?

Determine how much money you can spend, and then decide on what vehicles in that price range interest you. Have several models in mind so that the overpriced one won't tempt you. As your benchmarks, use the ratings, alternative models, estimated purchase cost, and residual value figures shown in Part Three of this guide. Remember, logic and prudence are the first casualties of showroom hype, so carefully consider your actual requirements and how much you can budget to meet them before comparing models and prices at a dealership. Write down your first, second, and third choices relative to each model and the equipment offered. Browse the automaker websites and *www.carcostcanada.com/en/* for the manufacturer's suggested retail price (MSRP), promotions, and package discounts. Look for special low prices that may only apply to Internet-generated referrals. Once you get a good idea of the price variations, get out the fax machine at home or work (a company letterhead always impresses) and make the dealers bid against each other (see page 63). Then call the lowest-bidding dealership and ask for an appointment to be assured of getting a sales agent's complete attention, and take along the downloaded info from the automaker's website to avoid arguments.

Vehicle ownership in Canada is astoundingly expensive and is directly affected by what vehicle you choose and where you live. Montreal is the most expensive metropolitan area in the nation to drive a car, with Toronto and St. John's ranked second and third, according to the latest analysis by Runzheimer Canada.

Costs are also very dependent upon which model you choose. Runzheimer's analysis of 25 passenger cars, light trucks, vans, and sport-utility vehicles showed the 2002 Audi Quattro costs three times as much to own and operate on an annual basis as the Saturn, and both the Lincoln Town Car and Cadillac DeVille are twice as expensive. The Saturn SL2 shows annual operating or running costs of $3,408 and annual fixed or ownership costs of $7,214, for total annual costs of just over $10,600. Annual operating costs for the Audi A8

Quattro are $5,744 and fixed costs are $28,291, for total annual costs of slightly more than $34,000. The Lincoln Town Car Executive and Cadillac DeVille luxury sedans show annual operating costs of $4,752 and $5,296 respectively, and annual fixed costs of $17,031 and $16,413, for total annual costs of slightly more than $21,700 each (see table below).

PROJECTED OWNERSHIP AND OPERATING COSTS FOR SELECTED 2002 MODEL CARS—CANADIAN

		(Litres)	Annual Costs		
Make & Model	Cylinder	Displace	Operating	Fixed	Total
Audi A8 Quatro	8	4.2	$5,744	$28,291	$34,035
Lincoln Town Car Exec.	8	4.6	$4,752	$17,031	$21,783
Cadillac Deville	8	4.6	$5,296	$16,413	$21,709
Buick LeSabre Ltd.	6	3.8	$4,736	$12,069	$16,805
Mercury Grand Marquis GS	8	4.6	$4,688	$11,922	$16,610
Chrysler Intrepid ES	6	3.5	$4,608	$9,575	$14,183
Ford Taurus SE	6	3.0	$4,576	$9,008	$13,584
Mazda 626 LX	6	2.5	$4,272	$8,884	$13,156
Nissan Maxima GXE	6	3.0	$4,304	$8,289	$12,593
Buick Century Custom	6	3.1	$4,064	$8,400	$12,464
Pontiac Grand Am SE	4	2.2	$4,064	$8,000	$12,064
Toyota Camry LE	4	2.4	$3,856	$8,160	$12,016
Chevrolet Cavalier LS	4	2.4	$4,016	$7,595	$11,611
Saturn S Series SL2	4	1.9	$3,408	$7,214	$10,622

Costs include operating expenses (fuel, oil, maintenance and tires), fixed expenses (insurance, depreciation, financing, taxes and licencing) and are based on a 36-month/96,000-kilometre retention cycle.

All values expressed in Canadian dollars.

Source: *www.runzheimer.com*

Other relatively expensive vehicles in the analysis, all with annual costs exceeding $16,000, include the GMC Yukon light truck, Ford Explorer, Chevy Blazer, Buick LeSabre, and Mercury Grand Marquis. At the other end of the scale, cars with annual costs in the $11,000–$12,000 range include the Chevy Cavalier, Toyota Camry, and Pontiac Grand Am.

What are my driving needs?

The ideal car should be easy to drive, with a minimum of high-tech features to distract and annoy you. In the city, a small wagon or hatchback is more practical and less expensive than a sport-utility. However, if you're going to be doing a lot of highway driving, transporting small groups of people, or loading up on accessories, a medium-sized wagon or small sport-utility could be the more logical choice for price, comfort, and reliability.

If you travel less than 20,000 km per year, mostly in the city, choose a small car equipped with a 4-cylinder engine that produces between 120 and 140 hp to get the best fuel economy without sacrificing performance. Anything more powerful is just a waste. On the other hand, extensive highway driving demands the cruising performance, extra power for additional accessories, and durability of a larger 6-cylinder engine. Believe me, fuel savings will be the last thing on your mind if you buy an underpowered vehicle.

Be especially wary of towing capabilities bandied about by automakers. They routinely exaggerate towing capability and seldom mention the need for expensive optional equipment or that the top safe towing speed may be only 72.4 km/h (45 mph), as is the case with some Japanese minivans.

Remember, you may have to change your driving habits to accommodate the type of vehicle you purchase. Front-drive braking is quite different from braking with a rear drive, and braking efficiency on ABS-equipped vehicles is compromised if you pump the brakes. Also, rear-drive minivans and vans handle like trucks, and will scrub the right rear tire during sharp right-hand turns until you get the hang of making wider turns. Limited rear visibility is another problem with larger vans, forcing drivers to carefully survey side and rear traffic before changing lanes or merging with traffic.

Do I feel comfortable in this vehicle?

The advantages of many sports cars and minivans quickly pale in direct proportion to your tolerance for a harsh ride, noise, a claustrophobic interior, and limited visibility. Minivan and van owners often have to deal with a high step-up, a cold interior, lots of buffeting from wind and passing trucks, and poor rear visibility. With these drawbacks, many buyers find that after falling in love with the showroom image, they end up hating their purchase—all the more reason to test-drive your choice over a period of several days to get a real feel for its positive and negative characteristics.

Check to see if the vehicle's interior is user-friendly. For example, can you reach the sound system and AC controls without straining or taking your eyes off the road? Are the controls just as easy to operate by feel as by sight? What about dash glare onto the front windshield and headlight aim and brightness? Do rear-seat passengers have to be contortionists to enter or exit, as is the case with many two-door sport-utility vehicles?

To answer these questions you need to drive the vehicle over a period of time to test how well it responds to the diversity of your driving needs, without having some impatient sales agent yapping in your ear. If this isn't possible, you may find out too late that the handling is more trucklike than you'd wanted. But you can conduct the following showroom test. Adjust the seat to a comfortable setting, buckle up, and settle in. Can you sit a foot away from the steering wheel and still reach the accelerator and brake pedals? When you look out the windshield and use the rear- and side-view mirrors, do you detect any serious blind spots? Will optional mirrors give you an unobstructed view? Does the seat feel comfortable enough for long trips? Can you reach important controls without moving off the seatback? If not, shop for something that better suits your requirements.

What safety features are best?

Automakers are loading 2003–04 models with features that wouldn't have been imagined several decades ago because safety devices appeal to families and boost profits almost 500 percent. Yet some safety innovations, like antilock brakes and full-powered airbags, don't deliver the safety payoffs promised by automakers and may create additional dangers. Some of the more effective safety features are head-protecting side and de-powered airbags, adjustable brake and accelerator pedals, standard integrated child safety seats, seat belt pretensioners, innovative head restraints, backup object sensors, sophisticated communication systems, and stylish designs that provide tanklike crash protection.

Seat belts provide the best means of reducing the severity of injury arising from both low- and high-speed frontal collisions. In order to be effective, though, seat belts must be adjusted properly and feel comfortably tight without undue slack. But owners often complain that seat belts don't retract enough for a snug fit, are too tight, chafe the neck, and don't properly fit children. Some automakers have corrected these problems with adjustable shoulder-belt anchors that allow both tall and short drivers to raise or lower the belt for a snug, more comfortable fit. Another important seat belt innovation is front seat belt pretensioners, devices that automatically tighten the safety belt in the event of a crash.

Front airbags are less a threat than they once were since the automakers started de-powering them in 1997. They still pose a serious safety hazard if you're sitting in the driver or front passenger's seat and you're a small adult or teenager, a senior, or if you've recently had upper torso surgery. Side airbags can also cause serious injuries if you're improperly seated. Plus, airbags are notorious for deploying for no reason and not deploying when they should. Nevertheless, airbag benefits now outweigh their shortcomings and shouldn't be disabled unless you are in one of the vulnerable groups mentioned above.

A vehicle with a high crash-protection rating is a lifesaver. Since some vehicles are more crashworthy than others, and size doesn't always guarantee crash safety, it's important to buy one that gives you the best protection from a frontal collision. For example, the Chrysler Caravan and Ford Windstar minivans are similarly designed, but your chances of surviving a high-speed collision with the Windstar are far greater than they are with a Caravan or any other Chrysler minivan. Furthermore, the latest redesigns don't translate into the *safest* redesigns, judging by the average crash ratings given to the 2000–02 Nissan Maxima and Infiniti I30/135 by the Insurance Institute for Highway Safety (IIHS) and the 2002 Toyota Camry's poor front-seat side-impact scores released by the National Highway Traffic Safety Administration (NHTSA).

However, be wary of all of the five-star crash-rating hoopla touted by Ford and Toyota; there isn't any one vehicle that can claim a prize for being safest. Vehicles that do well in NHTSA side- and front-crash tests may not do very well in IIHS offset crash tests, or may have poorly designed head restraints that would increase the severity of neck injuries. Or a vehicle may have a high number of airbag failures, such as the bags deploying when they shouldn't or not deploying when they should.

Before making a final decision on the vehicle you want, look up its crashworthiness profile in Part Three and compare that rating with the scores given to similar vehicles.

Can I get the best for less?

Sure. Sometimes a cheaper twin or hybrid model will fill the bill. Twins are those nameplates that are virtually identical in body design and mechanical components, like the Chevrolet Cavalier and Pontiac Sunfire.

American manufacturers are a wily bunch. While beating their chests over the need to buy "American," they have joined Asian automakers in co-ventures where they know they can't compete on their own. This has resulted in manufacturing partnerships whose parentage isn't always easy to nail down but which incorporate a higher degree of quality control than the American company's efforts alone. For example, Toyota and Pontiac churn out identical Matrix and Vibe compacts in the United States and in Ontario. This year, we'll also see Daewoo's small cars sold by GM through Suzuki dealers.

Twins and hybrids are only two of the alternatives; there are plenty of others. Minivans, for example, come in two versions: a base commercial, or cargo, version and a more luxurious model for private use. The commercial version doesn't have as many bells and whistles, but it's more likely to be in stock and will probably cost much less. And if you're planning to convert it, there's a wide choice of independents that may do a better job than the dealer, and for less. Of course, you will want a written guarantee from the converter that none of the changes will invalidate the manufacturer's warranty.

Representing about one-fourth of the new-vehicle market, leasing is an alternative used to make vehicles seem more affordable, but it's really more expensive than buying, and for most people the pitfalls will far outweigh any advantages. If you have to lease, keep your losses to a minimum by leasing for the shortest time possible and by making sure that the lease is close-ended (meaning that you walk away from the vehicle when the lease period ends). The CAA estimates that 75 percent of lessees return their vehicles when the lease expires.

If long-term leasing looks to be too expensive, consider purchasing a 3- to 5-year-old vehicle with 60,000–100,000 km on the clock and some of the original warranty left. Such a vehicle is likely to be just as reliable for less than half the cost of one bought new. Parts will be easier to find, independent servicing should be a breeze, insurance premiums will be much lower, and your financial risk is lessened considerably if you get a lemon.

When new isn't "new"

Nothing will cause you to lose money more quickly than buying a car that's older than advertised, has previously been sold and taken back, has accident damage or has had the odometer disconnected or turned back.

Even if the vehicle hasn't been used, it may have been left outdoors for a considerable length of time, causing the deterioration of rubber components, premature body and chassis rusting, or severe rusting of internal mechanical parts (which leads to brake malfunction, fuel line contamination, hard starting, and stalling).

You can check a vehicle's age by looking at the date-of-manufacture plate found on the driver-side door pillar. If the date of manufacture is 7/03, your vehicle was probably one of the last 2003 models made before the September changeover to the 2004s. Redesigned vehicles or those new to the market are exceptions to this rule. They may arrive at dealerships in early spring or mid-summer, and are considered to be next year's models. They also depreciate more quickly, owing to their earlier launching, but this difference narrows over time. Also, some vehicles, like the Kia Sportage, may skip a model year.

During the 1970s, Ford and Nissan conspired with their dealers to sell previous years' cars and trucks as the latest model. Check the doorplate for the true model year and, if cheated, use jurisprudence to get your money back (see Part Two).

Sometimes, a vehicle can be too new, and cost you more in maintenance because redesign glitches haven't yet been worked out. As Honda's North American manufacturing chief, Koki Hirashima, so ably put it, carryover models generally have fewer problems than vehicles that have been significantly reworked or just introduced to the market. Newly redesigned vehicles get quality scores that are, on average, 2 percent worse than vehicles that have been around for a while, says J.D. Power. Some surprising poor performers: the 2002 Chevrolet TrailBlazer, 2002 Jaguar X-Type, 2002 Nissan Altima, 2002 Toyota Avalon and Camry, and 2001 Honda Civic.

Because they were the first off the assembly line for that model year, most vehicles assembled between September and February are called "first series" cars. "Second series" vehicles, made between March and August, incorporate more assembly-line fixes and are better built than the earlier models, which may depend on ineffective "field fixes" to mask problems until the warranty expires. Both vehicles will sell for the same price, but the post-February ones will be a far better buy, since they benefit from assembly-line upgrades and rebates.

There's also the very real possibility that the new vehicle you've just purchased was damaged while being shipped to the dealer and was fixed by the dealer during the pre-delivery inspection. It's estimated that this happens to about 10 percent of all new vehicles. There's no specific Canadian legislation allowing buyers of vehicles damaged in transit to cancel their contracts. In a more general sense, however, Canadian common-law jurisprudence *does* allow for cancellation or compensation whenever the delivered product differs markedly from what the buyer expected to receive.

Fuel economy fantasies

The hidden expense of poor gas mileage is one of the top complaints among owners of 2003 model cars and trucks. Drivers say gas mileage is seldom as high as it's hyped to be, in fact, it'll likely be 10–20 percent *less* than advertised. Environment Canada's independent research also confirms that 2003–04 models are much less fuel efficient than vehicles built 15 years ago. At that time, the fleet averaged 10.5L/100km; now it's estimated by independent analysis to be 11.2L/100km. Even more astounding, a Ford Model T got 25 mpg (11.5L/100 km) nearly a century ago. Today, the average 2003 Ford gets 23 mpg (12.5L/100 km).

Why such a contradiction between promise and performance?

It's simple: automakers cheat on their tests. They submit their own test results to the government after testing under optimum conditions. And we know Ford or GM wouldn't lie.

Sure.

Transport Canada then publishes these self-serving, "cooked" figures as its own research. One Ford service bulletin is remarkably frank in discounting the validity of these tests:

> Very few people will drive in a way that is identical to the EPA [sanctioned] tests.... These [fuel economy] numbers are the result of test procedures that were originally developed to test emissions, not fuel economy.

In theory this kind of misrepresentation is actionable, but no successful consumer lawsuit has been reported in Canada, because most people simply live with the fact that they were fooled.

And that's understandable. Although good fuel economy is important, it's hardly worth a harsh ride, excessive highway noise, side-wind buffeting, anemic acceleration, and a cramped interior. You may end up with much worse gas mileage than advertised and a vehicle that's underpowered for your needs.

If you never quite got the hang of metric fuel economy measurements, use the fuel conversion table that follows to establish how many miles to a gallon of gas your vehicle provides.

Fuel Economy Conversion Table

L/100 km	mpg	L/100 km	mpg	L/100 km	mpg
5.0	56	7.4	38	12.5	23
5.2	54	7.6	37	13.0	22
5.4	52	7.8	36	13.5	21
5.6	50	8.0	35	14.0	20
5.8	48	8.5	33	15.0	19
6.0	47	9.0	31	15.0	18
6.2	46	9.5	30	17.0	17
6.4	44	10.0	28	18.0	16
6.7	43	10.5	27	19.0	15
6.8	42	11.0	26	20.0	14
7.0	40	11.5	25	21.0	13
7.2	39	12.0	24	23.0	12

Fuel economy figures are published by Natural Resources Canada (a free copy of its Fuel Consumption Guide can be obtained by calling 1-800-387-2000 or accessing *oee.nrcan.gc.ca/autosmart/*).

Finally, let's not leave the subject of fuel economy fantasies without mentioning the Pogue Carburetor, Canada's own fuel economy urban legend.

In 1935, Canadian inventor Charles Nelson Pogue tried to patent a "miracle" carburetor that he said produced fuel economy in the range of 200-plus miles to the gallon. This claim was supported in the May 1936 issue of *Canadian Automotive Trade* magazine, which reported that a 1,879-mile trip used only 14.5 gallons of fuel. Pogue later denied the story.

In *Snopes.com* the Pogue carburetor's performance is further debunked:

> No one reputable was allowed to see the mechanical miracle in action, let alone have a chance to measure its results. After the initial excitement over Pogue's 1936 announcement had faded, more serious types began to openly doubt that the carburetor would work as described.
>
> In the December 1936 issue of *Automotive Industries* magazine, its engineering editor, P.M. Heldt, said of a sketch of the Pogue carburetor: "The sketch fails to show any features hitherto unknown in carburetor practice, and absolutely gives no warrant for crediting the remarkable results claimed." Other journalists were beginning to voice similar opinions.
>
> In response to calls to put up or shut up, Pogue's miracle carburetor was heard of no more. Faced with the choice of believing someone had made claims his invention couldn't later live up to or that a monied bad guy had bought up a technology to forever keep it off the market, at least some chose to believe the suppression theory. That the carburetor never made it to the public, they said, was proof enough of its existence.

Abusive maintenance fees

Another important hidden cost, maintenance inspections and parts costs are usually exaggerated by dealers and automakers to make up for vehicles that rarely require fixing or that are sold in insufficient numbers to support a service bay. Both Mazda and Honda owners suspect this to be the case, as the following Honda owner reports and a CBC *Marketplace* Mazda dealer survey confirm:

> The worst thing so far is the new Honda Canada maintenance schedule, which they are also applying to my 1999 Civic where it will be even more costly because it is based on a 6,000 km cycle rather than the 8,000 km cycle for my 2001 Civic.
>
> I figured it out today that it will cost me about $4,500 to get my car serviced at my Honda dealer over the 100,000 km warranty. And they get right snotty if I suggest I don't need all of the "endless inspections" they specify. For example, even though my manual says I need new coolant at 72,000 km they specify it at 60,000 km and it costs $120 with their special additive. A fuel filter installed costs $80. A timing belt is almost $400.

In an investigative report shown on February 18, 2003, CBC TV's *Marketplace* surveyed 12 Mazda dealers across the country to determine how much they charged for a "regularly scheduled maintenance inspection" as listed in the owners' manual of a 2001 Mazda MPV with 48,000 km. (*www.cbc.ca/consumers/market/files/cars/mazda_warranty/*). Even with different labour rates, Mazda says the check-up should not cost more than about $280 anywhere in Canada. Here are the prices they were quoted:

- $400 (Montreal)
- $546 (Burlington)
- $225 (Toronto)
- $300 (Toronto)
- $271 (Winnipeg)
- $700 (Winnipeg)
- $253 (Calgary)
- $340 (Calgary)
- $450 (Calgary)
- $350 (Vancouver)
- $500 (Vancouver)
- $525 (Vancouver)

The first quote for service in Winnipeg was $700. A service technician told *Marketplace*'s producer that the price varied between $400 and $500 and warned that the work is needed to maintain the warranty. However, a check of the owner's manual does not show any connection between the service and keeping the warranty in place. Gregory Young, director of Corporate Public Relations for Mazda Canada, had this to say:

> There's nothing in the owner's manual that says if you don't have this work done in its entirety at this prescribed time that automatically your warranty is void.

If this situation arises, owners are free to get their warranty service done at cheaper independent facilities as long as the automaker's schedule is followed rigorously and no warranty claim can be tied to botched servicing.

Who Can You Trust?

Car columnists for rent

Most investigative stories done on the auto industry in Canada (such as secret car warranties, dangerous airbags, and Chrysler minivan defects) have been written by women business journalists and "action line" troubleshooters, not auto beat reporters. Why?

Because car journalists are mostly on the take. Most are arrogant, self-important promoters whose writing skills aren't good enough for their paper's "Real Estate" or "Travel" sections and their sense of morality takes a back seat as they take a front seat on freebie trips to Paris and other exotic locales hosted by the auto industry.

The system chews up young, crusading auto scribes. They have to jump through hoops to get their stories out, simply because their editors or station managers have bought into many of the fraudulent practices so common to the auto industry. Haranguing staff for more "balance" is the pretext *du jour* for squelching hard-hitting stories implicating dealers and automakers.

News editors don't want truth, they want comfort.

They'll spend weeks sifting through some politico's trash cans looking for evidence of shady dealings, while ignoring the auto industry scams threaded throughout their own classified ads.

Want proof? Go ahead and try to decipher the fine print in *The Globe and Mail* or *Toronto Star* leasing ads or, better yet, tell me what the fine print scrolled at breakneck speed on television new-vehicle commercials really says. Probably something to the effect that "everything said or shown before in this ad may or may not be true." Where is the investigative reporter who will submit these ads to an optometrists' group that will confirm that the message is unreadable?

Think about this: Dealers selling used cars from residences, posing as private parties ("curbsiders"), are periodically exposed by dealer associations and "crusading" auto journalists. Yet these scam artists place dozens of ads, or more, weekly in the classified section of local newspapers, where the same phone numbers and billing addresses constantly reappear. Why isn't there an exposé by reporters working for these papers? Why don't they publish the fact that it's mostly hypocritical, whining new-vehicle dealers who supply curbsiders with their cars? That's what I'd call balanced reporting, but it'll never be done. Classified ad sellers only care about selling ads, not whether the seller is a dishonest used-car salesman or an escort agency that's a front for prostitution.

Robert Farago, a British freelance car columnist, knows exactly how the game is played between automakers and car journalists. In an article titled "Gentle Persuasion" found on his The Truth About Cars website (*www.pistonheads.com/truth/default.asp?storyId=4262*), Farago describes it this way:

> All publications want access to the hot new cars. They want a test drive or spy shot, computer mock-up, something, anything before their rivals. Commercial necessity gives manufacturers power over both the car mags and their jobbing journos. Again, I've never heard of a manufacturer saying: "I'll give you our latest car as long as you play nice." The relationship between carmakers and the press is as complicated as that of an alcoholic and his wife. But co-dependency doesn't change the basic implication: screw us this time, and we'll screw you next.

Over the years, I've talked with a dozen or so auto company PR executives who shepherd around these egocentric, freeloading, pseudo-journalists, and they tell me it's the hardest part of their job.

Nevertheless, there are some Canadian auto writers who have distinguished themselves by writing balanced investigative reports on the auto industry. *Toronto Star* business columnist Ellen Roseman writes a column that is particularly thorough in unearthing auto stories often missed by others. There's also Maryanna Lewyckyj, a *Toronto Sun* business and consumer columnist who has taken advanced auto repair classes and is more car-savvy than most automaker PR staffers. They liken being called by her to a visit to the dentist. And there's Krystyna Lagowski, an independent auto journalist (*klcreative@sympatico.ca*) who often writes for the *Toronto Star* and doesn't shy away from deflating auto executive stuffed shirts.

The "Car of the Year" scam

Once you've established a budget and selected some vehicles that interest you, the next step is to ascertain which ones have high safety and reliability ratings. Be wary of the ratings found in some enthusiast magazines and car enthusiast websites; their supposedly independent tests are a lot of baloney.

You want proof of how misleading these ratings can be? Take *Car and Driver* as an example. It rated the Ford Focus as a "Best Buy" during its first three model years, while government and consumer groups decried the car's dozen or so recall campaigns and huge number of owner safety and reliability complaints.

These so-called independent tests are compromised because they use vehicles that are supplied by the manufacturer and tuned to just the right specifications. Servicing is impeccable and the manufacturer has probably loaded the car with an assortment of expensive options to compensate for any design faults. These tests also don't account for a vehicle's vulnerability to climate, crash forces, poor servicing, or inadequate parts distribution.

Automakers can't lose with these rigged tests. And if the tester wants more free courtesy cars or current prices, specs, and photos, the published report had better gloss over the vehicle's defects and hype its mediocre features. As well, if the magazine or newspaper receives advertising from the manufacturer, any criticism that gets through the reporter's own self-censorship will be muted by the editor.

Beware of car ratings

There are dozens of organizations and magazines that rate cars for everything from their overall reliability and frequency of repairs (J.D. Power and *Consumer Reports*) to their crashworthiness and appeal to owners (NHTSA, IIHS, and TQS). For instance, BMW's Mini ranked 25th of 28 brands in J.D. Power's Initial Quality Survey (IQS). But the popular British import ranks second of 30 brands included in Strategic Vision's Total Quality Survey (TQS). As for overall front and side crash protection, the Mini scored a respectable four out of five stars rating and a "Best Pick" offset-crash rating.

Sure, the BMW Mini Cooper is cute, but is it reliable? Surveys don't agree.

Getting reliable info

Funny, as soon as they hear that you're shopping for a new car, everybody wants to tell you what to buy—relatives, co-workers, friends, and everyone else who thinks they know what's best for you. And, be especially wary of testimonials:

Have you "driven a Ford lately," Clyde?

> 10 April 1934
> Mr. Henry Ford
> Detroit, Michigan
>
> Dear Sir:
> While I still have got breath in my lungs I will tell you what a dandy car you make. I have drove Fords exclusively when I could get away with one. For sustained speed and freedom from trouble the Ford has got every other car skinned, and even if my business hasent been strickly legal it don't hurt enything to tell you what a fine car you got in the V8.
>
> Yours truly,
>
> Clyde Champion Barrow

Barrow (with his partner in crime, Bonnie Parker) was killed in a police ambush on a Louisiana highway just six weeks after his letter to Ford. He died at the wheel of his stolen Ford V8. If he drove a Ford Focus or Windstar, his crime spree would have undoubtedly been much shorter.

After a while you'll get so many conflicting opinions that it'll seem as if any choice you make will be the wrong one. Before making your decision, remember that you should invest a month of research in your $30,000-plus new-car buying project. This includes two weeks for basic research (see the following) and another two weeks to actually bargain with dealers to get the right price and equipment. The following sources provide a variety of useful information that will help you ferret out what vehicle best suits your needs and budget.

Internet and online services

Anyone with access to a computer and a modem can now obtain useful information relating to the auto industry in a matter of minutes, and at little or no cost. This is accomplished in two ways: subscribing to an online service, like America Online, or AOL, that offers consumer forums and easy Internet access—or going directly to the Internet through a low-cost Canadian Internet service provider (ISP) and cruising the thousands of sites that interactively summarize the subject you are researching.

Web surfing

There are billions of information pages on the Internet. Most are junk, but there are plenty of worthwhile sites, too. An effective, easy-to-use search engine will find the quality items.

I recommend Google (*www.google.com*). It selects websites that are most linked to by other sites and those that are the most popular and pertinent. Other popular engines are Yahoo! (*www.yahoo.com*) and MSN (*www.msn.com*).

Web shopping

The key word here is "shopping," because *Consumer Reports* magazine has found that barely 2 percent of Internet surfers actually buy a new or used car online. Yet, over 50 percent of buyers admit to using the Net to get prices and specifications before visiting the dealership. Apparently, few buyers want to purchase a new or used vehicle without seeing what's offered and knowing all money paid will be accounted for.

New-vehicle shopping through automaker and independent websites on the Internet is a quick and easy way to compare prices and model specifications. In fact, buyers now have access to information they once were routinely denied or had trouble finding, such as manufacturers' suggested retail prices, invoice prices, promotional offers, rebates, the book value for trade-ins, and some safety data. Shoppers can access comparative data online by surfing the Kelley Blue Book and Edmunds websites. The information is geared to American buyers, but Canadian shoppers can get Canadian specs and test-drive reports from *CanadianDriver.com.*

There are three places in Canada where invoice prices can be obtained—for a price. They are the Toronto Metro Credit Union, the Automobile Protection Association, and Car Cost Canada. Car Cost Canada (*carcostcanada.com* or 1-800-805-2270) will give you the invoice price, less holdbacks, disclose the subsidized financing and leasing rates, and show what rebates and dealer incentives apply. Furthermore, you are also given a dealer invoice breakdown for all major options.

2004 Sienna LE 7 Passenger, Toyota

Option	**Retail Price**
Arctic Frost pearl paint	$0
Freight	$1,260
Air tax	$100
Base price	$34,750
Total	$36,110

The list provided above contains all available options. Keep in mind that some options require particular combinations to be selected. Please check with your dealer for the specific requirements of each option.

This is the Manufacturer's Suggested Retail Price. The invoice price would likely be several thousand dollars less.

The cost of the service for one quote by phone (you have the option of getting the info faxed back to you) is $29. If ordered through the company's website, the cost drops to $19. Three vehicles quoted online cost $29, and a "six-pack" request runs $39.

Other advantages to online shopping: Some manufacturers offer a lower price to online shoppers, and the entire transaction, including financing, can be done on the Net. Buyers don't have to haggle; they merely post their best offer electronically to a number of dealers in their area code (for more convenient servicing of the vehicle) and await counteroffers. Three caveats: You will have to go to a dealer to finalize the contract and be preyed upon by the financing and insurance (F&I) sales agents; as far as bargains are concerned, *Consumer Reports* says its test shoppers found that lower prices are more frequently obtained by first visiting the dealer showroom and concluding the sale there; and only a third of online dealers respond to customer queries.

Helpful automotive websites

The first website you should access is *www.lemonaidcars.com,* the official site for the *Lemon-Aid* guides. It carries updates and important links to other sites of particular interest to new- and used-vehicle shoppers, as well as owners seeking refunds for factory-related defects. See Appendix I for a list of other useful websites. Most website addresses are prefaced with *http://www.* and most Canadian sites have the suffix *.ca.* Also, some sites listed in this guide may have changed or been dropped. If you have difficulty locating one of the websites, return to your search engine and try again, modifying the name slightly.

Canadian Automobile Association

The Canadian Automobile Association (CAA) is an efficient, competent organization that has reluctantly stuck its big toe into the hot water of consumer advocacy, mostly by promoting auto safety and calling for lower fuel taxes. The group spends much of its time starting and towing cars, selling vacations, preparing trip maps, and decrying high gasoline taxes and poor roads. Much like Canada's Better Business Bureaus, whose noble intentions were compromised early on by business-led intimidation (the threat of lawsuits and the withdrawal of financial support), CAA has traditionally looked upon consumer advocacy with a wary eye.

That said, I must add that some provincial CAA-affiliated groups, like the Alberta Motor Association, British Columbia Automobile Association, and CAA-Quebec, take their consumer advocacy roles quite seriously and have vigorously defended their members' rights. Too bad they're the exception, not the rule.

There are two major surveyors of automobile quality: J.D. Power and Associates, a private American automobile consulting organization, and Consumers Union, an American non-profit consumer organization that publishes *Consumer Reports.* J.D. Power publishes two quality reports: an Initial Quality Survey, considered by many to be its weakest effort, that notes problems reported by owners during the first 90 days of ownership; and a Long-term Quality Survey that highlights problems reported over a number of years. Often, vehicles that do poorly in one survey come out near the top on the other.

Consumer Reports and CAA's *Autopinion* (*Carguide*)

Consumer groups and non-profit auto associations are your best bets for the most unbiased auto ratings. They're not perfect, though, so it's a good idea to consult several and look for ratings that match from publication to publication. My favourites are *Consumer Reports* and the Canadian Automobile Association's *Autopinion.* Both publications list only the manufacturer's suggested retail price (MSRP), not the invoice price.

Consumer Reports (*CR*) is an American publication that has a tenuous affiliation with the Consumers' Association of Canada. Its ratings, extrapolated from Consumers Union's annual U.S. member survey, accurately mirror the Canadian experience. There are two exceptions, however. Components that are particularly vulnerable to our harsh climate usually don't perform as well as the *CR* reliability ratings indicate; and poor servicing caused by a weak dealer body in Canada can make some service-dependent vehicles a nightmare to own in Canada, whereas the American experience may be more benign.

Based on 600,000-plus American and Canadian member responses, *CR* lists used vehicles that, according to owner reports, are significantly better or worse than the industry average. Statisticians agree that *CR*'s sampling method has some room for error, but the ratings are good, conservative, and consistent guidelines for buying a new vehicle. My only criticisms of the ratings are that many models, like Volkswagen, Toyota and Honda, can do no wrong, yet service bulletins and extended warranties show they have serious engine, transmission, and electrical problems. Plus, 1994 and older vehicles are excluded from *CR*'s ratings and many of the ratings about the frequency of repair of certain components aren't specific enough. For example, don't just tell me there are problems with the fuel or electrical system. Rather, let me know about specific components—is it the fuel pumps that are failure-prone, or the injectors that clog up, or the battery that suddenly dies?

There's also the Canadian Automobile Association's annual "Vehicle Ownership Survey," found in the February issue of *Carguide* magazine. It's available from CAA and newsstands for $5.95 and is kept on display through June.

Carguide is published six times a year. It contains reams of automaker advertising and publishes mostly general-interest articles, as well as a summary of new cars and trucks. Its most useful feature is its used-vehicle ratings, based on CAA's annual sampling of 20,000 owners—that's less than 5 percent of *Consumer Reports*' survey, but at least you know they're all Canadian drivers. The *Autopinion* supplement gives you a good general idea of those vehicles that have generated the most problems for CAA members (Chrysler Neon, Ford Windstar, and Ford F-Series trucks), but its conclusions should be compared with *Consumer Reports*' or *Lemon-Aid*'s recommendations before making a definite decision.

Lemon-Aid *vs.* Consumer Reports

CR and *Lemon-Aid* ratings are often in agreement. Where they differ is in *Lemon-Aid*'s greater reliance upon NHTSA safety complaints, service bulletin admissions of defects, and owner complaints received through the Internet (rather than from a subscriber base, which may simply attract owners singing from the same hymnal).

In looking over *CR*'s best new and used-vehicle picks for 2003, as published in its December 2002 edition, there are a number of recommended vehicles that defy all logic.

Foremost is the Ford Windstar, followed by the Saturn Vue and the Jeep Liberty. It is inconceivable that *CR* isn't aware of the multiplicity of powertrain, body, and suspension failures affecting these vehicles. Then there's the assorted Chrysler lineup. Now, you'd have to live on another planet to not know that Chryslers are afflicted by chronic automatic transmission, body, brake, and AC defects. In fact, *Consumer Reports*' "Frequency of Repair" tables in the same edition give out plenty of black marks to the aforementioned models and components. Yet sloppy research failed to pick up on these contradictions.

Leasing Losses

Why leasing costs more

You don't want to lease.

Lessees usually pay the full manufacturer's suggested retail price (MSRP) on a vehicle loaded with costly options, plus hidden fees and interest charges that wouldn't be included if the vehicle were purchased instead. DesRosiers Automotive Research Inc. found that cheaper cars often cost more to lease than some luxury models. This was confirmed by the Canadian Bankers Association (CBA) after it blasted dealers' leasing contracts for charging interest rates as high as 34 percent and called for consumer protection legislation to regulate the industry.

Rob LoPresti, a Toronto-based consumer advocate and leasing consultant, has campaigned tirelessly for more disclosure in leasing contracts. He has created CarCalculator, an easy-to-use program that runs on most computers. It takes the mystery out of leasing, giving you the annual interest rate you are paying, the total cost of the lease, and how it compares to financing—facts guaranteed to frustrate any fast-talking dealer or leasing agent. For more info, check out *www.carcalculator.com.* Another useful site that takes the mystery out of leasing is run by the United States Federal Reserve Board and can be found at *www.federalreserve.gov/pubs/leasing.* This site goes into incredible detail, comparing leasing versus buying, and has a handy dictionary of terms you're most likely to encounter.

Leasing advantages

Leasing can be worthwhile for people who frequently trade in their cars, since it's more convenient and results in less sales tax. (This advantage is wiped out, however, if you lease for longer than three years or buy back the car at the end of the lease.) In some cases you may be paying too much if you don't lease, since leasing enables you to drive a new vehicle without tying up a bundle of money that you could otherwise invest or use to pay down more costly debts (such as mortgages or charge cards). Leasing may also offer some tax advantages if your car expenses are deductible from gross income, but the savings may be minimal. According to the accounting firm of PricewaterhouseCoopers, in 99 percent of the cases it has examined there wasn't much difference in the tax liability if the vehicle was bought or leased. Your chief consideration shouldn't be the tax savings, but rather the difference between the implicit interest rate in the lease and the financing charge.

Leasing disadvantages

Leased vehicles are usually overpriced, jam-packed with non-essential options, and accompanied by a hefty upfront fee. The leasing agent has another avenue to rip you off by setting the vehicle's buy-back value at much more than it's likely to be worth—check the residual values section found in Part Three for a realistic buy-back price for the vehicle you're thinking of leasing. It's also a smart idea to buy additional "gap" insurance to cover the balance owed on the lease if the vehicle is stolen or written off in an accident, but most lessors charge too much for the protection (about $200 is fair). In some cases, the leasing company will throw in gap insurance at no extra cost.

Ask the leasing firm what it's prepared to do if the car turns out to be a lemon. Most companies accept that this happens from time to time and will simply return the car to the manufacturer and get a replacement. That's not part of the standard agreement, so be sure that such a clause is included in your lease before you sign it.

Here are two other reasons why you may not wish to lease:

- You drive the Canadian average of 20,247 km a year but the leasing agency charges you up to 20 cents/km over a limit of 18,000–20,000 km.
- You always seem to have dents or scrapes on your car and can't be bothered getting regular maintenance. Responsible leasing firms allow for reasonable wear and tear on the vehicle during the leasing contract, but rip-off companies count every scratch, sticking you with a grossly inflated repair bill.

About 7 percent of consumers who try leasing don't lease again following disputes over "excess wear and tear" charges, says Art Spinella, vice president of CNW Marketing/Research, a Brandon, Oregon, consulting firm. Two years of study by CNW concluded that one in five leased vehicles were charged an average of $1,647 (U.S.) for excess wear. Mercedes-Benz Credit Corp, the leasing arm of DaimlerChrysler in the States, uses a "credit card test" to determine if charges apply: if the interior or exterior stain, scratch, or blemish can be physically covered by a credit card, there will be no extra charges. In the States, Chase Auto Finance has eliminated large end-of-lease damage charges by waiving excess wear-and-tear costs of up to $1,500 on new leases. Canadian lessees can save money by getting an independent garage to correct any excessively worn areas or by insisting that an independent arbitrator decide which repairs must be done under the wear-and-tear contract provisions.

Unfair lease restrictions

Make sure that the lease allows you to service the vehicle yourself at an independent repair facility. A maintenance lease that ties you to the leasing firm's repair shop can lead to outrageous service charges. Use a less expensive independent repair shop for routine servicing, and keep all your receipts to prove that proper maintenance was carried out (as required by the manufacturer) should a warranty dispute arise.

Be wary of unfair restrictions, excessive penalties, and hidden damage charges. Excessive penalty charges for early cancellation of the contract are horrendous and usually require a payout of three to six months' lease payments. Sometimes lessors will put you on a leasing treadmill by waiving the penalty only if another vehicle is leased. Most impose a 20,000 km per year mileage limit and levy a whopping surcharge on the excess. You may also be restricted from driving the vehicle outside Canada or lending it to a third party.

Decoding leasing ads

Take a close look at the small print found in most leasing ads. Pay particular attention to the "weasel" words relating to the model year, kind of vehicle (demonstration or used), equipment, warranty, interest rate, buy-back amount, down payment, security payment, monthly payment, transportation and preparation charges, administration fee ("acquisition" and "disposal" fees), insurance premium, number of free kilometres, and excess kilometre charge.

When and Where to Buy

When to buy

A good time to buy a new car or minivan is in the winter, between January and March, when you get the first series of rebates and dealer incentives, and production quality begins to improve. Try not to buy when there's strike action—it will be especially tough to get a bargain because there's less product to sell and the dealer has to make as much profit as possible on each vehicle remaining in his diminishing stock. Furthermore, work stoppages increase the chances that online defects will go uncorrected, and the vehicle will be delivered, as is, to product-starved dealers.

Instead, lie low for a while and then return in force in the summer and early fall, when you can double-dip from additional automakers' dealer incentive and buyer rebate programs, which can average about a thousand dollars each. Remember, too, that vehicles made between March and August offer the most factory upgrades, based on field reports from those unfortunate fleet managers and rental car agencies who bought the vehicles when they first came out.

Allow yourself at least two weeks to finalize a deal if you're not trading in your vehicle, and longer if you sell your vehicle privately. Visit the showroom at the end of the month, just before closing, when the salesperson will want to make that one last sale to meet the month's quota. If sales have been terrible, the sales manager may be willing to do some extra negotiating.

Where to buy

Good dealers aren't always the ones with the lowest prices. Dealing with someone who gives honest and reliable service is just as important as getting a good price. Check a dealer's honesty and reliability by talking with motorists who drive vehicles purchased from that dealer (identified by the nameplate on the trunk). If these customers have been treated fairly, they'll be glad to recommend him or her. You can also ascertain the quality of new-vehicle preparation and servicing by renting one of the dealer's minivans or pickups for a weekend or by getting your trade-in serviced.

How can you tell which dealers are the most honest and competent? Well, judging from the thousands of reports I receive each year, dealerships in small suburban and rural communities are more fair than big-city dealers, because they're more vulnerable to negative word-of-mouth advertising and to poor sales—when their vehicles aren't selling, good service takes up the slack. Prices may also be more competitive, but don't count on it.

Dealers selling more than one manufacturer's product line present special problems. Overhead can be quite high, and cancellation of a dual dealership by an automaker in favour of an exclusive franchise elsewhere is an ever-present threat. Parts availability may also be a problem, because a dealer with two separate vehicle lines must split his inventory and may therefore have an inadequate supply on hand.

The quality of new-vehicle service is directly linked to the number and competence of dealerships within the network. If the network is weak, parts are likely to be unavailable, repair costs can go through the roof, and the skill level of the mechanics may be questionable. Among foreign manufacturers, Asian automakers have the best overall dealer representation across Canada, except for Mitsubishi and Kia.

Kia's dealer network is very weak, having been left by its owner, Hyundai, to fend for itself for the past three years. The small number of Mitsubishi dealers are scattered around the country, piggybacked onto existing Chrysler franchises.

European automakers are almost all crowded into Quebec, Ontario, and British Columbia, leaving owners in the Maritimes and Western Canada without

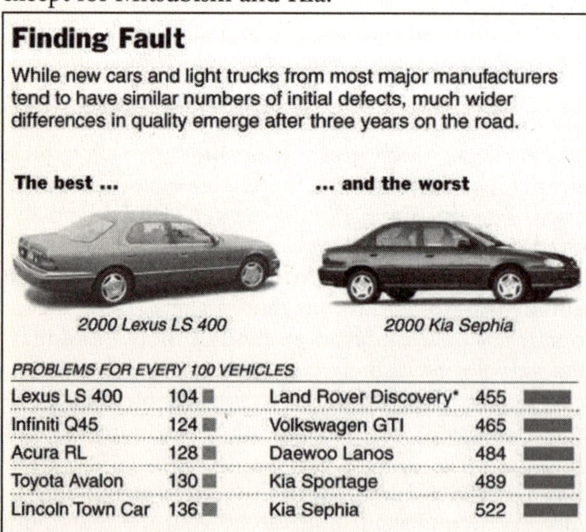

PROBLEMS FOR EVERY 100 VEHICLES

Lexus LS 400	104	Land Rover Discovery*	455
Infiniti Q45	124	Volkswagen GTI	465
Acura RL	128	Daewoo Lanos	484
Toyota Avalon	130	Kia Sportage	489
Lincoln Town Car	136	Kia Sephia	522

Source: J.D. Power

Hyundai has promised to upgrade its Kia product lineup for 2004.

much choice for servicing. This is particularly troublesome because most European imports are highly dealer dependent for parts and service. Furthermore, since parts aren't easily found outside of the dealer network, they tend to be pricier than Japanese and American components. This became quite evident with Volkswagen recently when the company had to replace faulty ignition coils on its 2001–02 lineup. Not only was servicing in Canada shown to be woefully inadequate, VW's lack of a Canadian customer assistance office made a bad situation worse.

And, talking about bad situations, Mercedes-Benz, with 318 problems per 100 vehicles, finished below the average Power rankings for the second consecutive year. The study found U.S. and European brands started out equal in terms of initial quality, but substantial quality and durability gaps soon appeared. U.S. automakers had 49 fewer problems per 100 vehicles than Europeans after three years of ownership.

No matter where your car comes from, it's always a good idea to patronize dealerships that are accredited by auto clubs such as CAA affiliates or consumer groups like the APA. Auto club accreditation is no iron-clad guarantee of honest or competent business practices, but if you're cheated or fall victim to poor servicing from one of their recommended garages (look for the accreditation symbol in their phone book ads or on their shop windows), the accreditor is one more place to take your complaint and apply additional mediation pressure. And, as you'll see under "Repairs" in Part Two, plaintiffs have won in court by pleading that the auto club is legally responsible for the consequences of the recommendations it makes.

Automobile brokers/vehicle buying services

Brokers are independent agents who try to find the new or used vehicle you want at a price below what you'd pay at a dealership (including the extra cost of the broker's services). Broker services appeal to buyers who want to save time and money while simultaneously avoiding most of the stress and hassle associated with the dealership experience, which for many people is like a swim in shark-infested waters.

Brokers get new vehicles through dealers, while used vehicles may come from dealers, auctions, private sellers, and leasing companies. Basically, brokers find an appropriate vehicle to meet the client's expressed needs, and then negotiate its purchase (or lease) on behalf of their client. The majority of brokers tend to deal exclusively in new vehicles, with a small percentage dealing in both new and used vehicles. Ancillary services vary among brokers, and may include such things as comparative vehicle analysis and price research.

The cost of hiring a broker ranges anywhere from a flat fee of a few hundred dollars to a percentage of the value of the vehicle (usually 1–2 percent). A flat fee is usually best because it encourages the broker to keep the selling price low. Reputable brokers are not beholden to any particular dealership or make, and will disclose their flat fee up front or tell the buyer the percentage amount they charge on a specific vehicle.

Finding the right broker

This is a toughie, because good brokers are hard to find, particularly in Western Canada and British Columbia. Buyers who are looking for a broker should first ask friends and acquaintances if they can recommend one. Word-of-mouth referrals are often the best, because people won't refer others to a service with which they were dissatisfied. Your local credit union or the regional CAA office is a good place to get a broker referral from people who see their work on an everyday basis.

Toronto's Metro Credit Union (contact David Lawrence or Rob LoPresti at 1-800-777-8507 or 416-252-5621) has a vehicle counselling and purchasing service (Auto Advisory Services Group, comprising CarFacts and AutoBuy) where members can hire an expert "car shopper" who will do the legwork—including the tedious and frustrating dickering with sales staff—and save members time and hassle. This program can also get that new or used vehicle at a reduced (fleet) rate, arrange top-dollar prices for trade-ins, provide independent advice on options like rustproofing and extended warranties, carry out lien searches, and even negotiate the best settlement with insurance agents. The credit union also holds regular car-buying seminars throughout the year in the Greater Toronto Area and maintains a website (see Appendix I).

Buying club smoke and mirrors

Be wary of buying clubs or referrals that aren't backed by a national organization or a non-profit financial institution. Except for some auto associations, buying clubs that promise huge savings on purchases, or claim to get new vehicles at dealer's cost, seldom survive close scrutiny. The savings they promote are often illusory at best, because the so-called wholesale or dealer's cost price is usually no different from the MSRP.

Furthermore, many referral agencies make "sweetheart" deals with dealers that send them under-the-table kickbacks or predetermined commissions. Whether dealing with the APA, a credit union, or an online referral service, find out first if the agency receives a fee for giving out referrals to particular sellers.

Some credit union or union buying clubs do a very good job in cutting prices on all vehicles. They use their buying clout in a particular region to exact substantial concessions from local dealers. They generally charge minimal membership fees, and they don't get too chummy with the local dealers because they put their members' interests first.

Safety First

Highway deaths in 2002 were at the highest level since 1990, and 82 percent of that increase from 2001 is from fatal rollover crashes. The number of people killed in SUV rollovers rose 14 percent in one year.

Unfortunately, there are no more easy safety solutions. According to NHTSA, 59 percent of the vehicle occupants who died in 2002 were not wearing seat belts, and 42 percent of the highway deaths were in alcohol-related crashes. Although these figures show a dramatic reduction in fatalities and injuries over the past three decades, safety experts feel additional safety measures will, henceforth, pay small dividends.

Active safety

Advocates of active safety generally blame the victim and believe that driving skill can prevent accidents or reduce the severity of injuries. This has been the prevailing attitude for much of this century among automakers who maintained that accidents were caused by the proverbial unskilled "nut behind the wheel" and believed that safe driving could best be taught through schools or private driving courses.

Ralph Nader illustrates this attitude in his book *Unsafe at Any Speed* by quoting a 1954 letter written by Howard Gandelot, GM's safety engineer, in response to the following Buick owner's query:

> The other day I had to step quickly on the brake to avoid hitting a little kitten, and in so doing, my son, eight, was thrown against the dash and broke off a front second tooth. If some padding can be applied it will help save faces and maybe lives. This is just a suggestion for safer motoring for all.

Gandelot replied:

> Driving with young children in an automobile always presents some problems. As soon as the youngsters get large enough to be able to see out when standing up, that's what they want to do – and I don't blame them. When this time arrived with both our boys I made it a practice to train them so that at the command "Hands!" they would immediately place their hands on the instrument panel if standing in the front compartment, or on the back of the front seat if in the rear, to protect themselves against sudden stops. This took a little effort and on a couple of occasions I purposely pumped them a trifle when they didn't respond immediately to the command so that they learned quickly. Even now when either one of them is in the front seat, at the command of "Hands!" they brace themselves. I frequently give these commands even when there is no occasion to do so, just so we all keep in practice.

Active safety components are generally those mechanical systems, such as anti-lock brake systems (ABS), high-performance tires, and traction control, that may, indeed, help avoid accidents if the driver is skillful and mature.

That's a big "if."

The theory of active safety has several drawbacks. First, there are dozens of independent studies that show driving schools have a negative impact upon highway safety because they often funnel immature or cognitive-impaired drivers on our roads. Even if a young driver learns how to master defensive-driving techniques, there's still no assurance that this training will override immature behaviour, or that a confused senior won't mistake the accelerator for the brake pedal.

Phil Berardelli's report in the September 8, 1998 edition of the *Washington Post* drew a similar conclusion:

> A 1994 report to Congress by NHTSA, concludes, "Novice driver education is not doing a very good job in motivating youngsters to drive safely." Additionally, a recent report by the Traffic Injury Research Foundation of Canada on driver education programs in several countries concluded that there is no evidence linking formal instruction programs to reduced crash rates. Moreover, an intensive study of teen drivers in the DeKalb County, Georgia, school system [1983 and 1987] concluded that formal instruction of students "was not associated with reliable or significant decreases in crash involvement."
>
> This is one reason why some organizations—the Insurance Institute included—have been pushing a concept called "graduated licensing," in which driving privileges are phased in as teens gain more experience behind the wheel.... Chief among the states' new provisions is a midnight-to-5 a.m. curfew for the first 18 months of a license, and a requirement that a young driver receive at least 40 hours of adult-supervised time behind the wheel before a license can be acquired.

Following recent events in Toronto and in Santa Monica, California, where seniors have killed or maimed pedestrians, there's a renewed clamour for graduated driver licensing of seniors throughout North America. Older drivers would face additional driving restrictions as their physical abilities deteriorate.

Second, since over 40 percent of all fatal accidents are caused by drivers who are under the influence of alcohol or drugs, all the high-performance options and specialized driving courses in the world will not provide much protection from impaired drivers who draw a bead on your vehicle. Finally, because active safety components get a lot of use—you're likely to need anti-lock brakes 99 times more often than an airbag—they have to be well-designed and well-maintained to remain effective.

They usually aren't.

Passive safety

I believe in passive safety features that assume you will be in life-threatening situations and should be warned in time to avoid a collision or automatically protected from collision forces when they occur. Daytime running lights and a third, centre-mounted brake light are two passive safety features that have paid off handsomely in reduced injuries and lives saved.

Passive safety assumes some accidents aren't avoidable and that when an accident occurs, the vehicle should provide as much protection as possible to the driver, the vehicle's other occupants, and other vehicles that may be struck—without depending on the driver's reactions. Passive safety components that have consistently proven to reduce vehicular deaths and injuries are seat belts, vehicle structures that enhance crashworthiness by absorbing or deflecting crash forces away from the vehicle's occupants, and laminated windshields.

"Unsafe" Safety Features

In the late '60s, Washington forced automakers to include worthwhile and effective safety features like collapsing steering columns and safety windshields in their cars. As the years have passed, the number of mandatory safety features has increased to include seat belts, airbags, and crashworthy construction. These improvements met with public approval until a decade ago, when reports of deaths and injuries caused by anti-lock brake systems (ABS) and airbag failures showed that defective components and poor engineering negated the potential life-saving benefits associated with these devices.

For example, one out of every five ongoing NHTSA investigations into possible defects in cars and light trucks concerns inadvertent airbag deployment, failure of the airbag to deploy, or injuries suffered when the airbag did go off. In fact, airbags are the agency's single largest cause of current investigations, exceeding even the full range of brake problems, which runs second.

Anti-lock brake system (ABS)

ABS brakes are way overrated. They are often ineffective, plus there are thousands of reports that they are failure prone and expensive to service.

The following complaint from a 2000 Dodge Caravan owner to NHTSA is typical:

> Consumer was travelling about 25 mph [40 km/h] on a ramp. Applied brakes and the ABS braking system didn't work, and pedal didn't go down to the floor. He hit vehicle in front.

ABS should prevent a vehicle's wheels from locking when the brakes are applied in an emergency situation, thus reducing skidding and the loss of directional control. When braking on wet and dry roads, your stopping distance will be about the same as with conventional braking systems. But in gravel, slush, or snow, your stopping distance will be greater.

The most important feature of ABS is that it preserves steering control. As you brake in an emergency, ABS will release the brakes if it senses wheel lock-up. Braking distances will lengthen accordingly, but at least you'll have some steering control. On the other hand, if you start sliding on glare ice, don't expect ABS to help you out very much. The laws of physics (no friction, no stopping!) still apply on ABS-equipped vehicles. You can decrease the stopping distance, however, by removing your all-season tires and installing four snow tires that are the same make and size.

Transport Canada studies also show ABS effectiveness is highly overrated. IIHS—an American insurance research group that collects and analyzes insurance claims data—says that cars with ABS brakes are actually more likely to be in crashes where no other car is involved but a passenger is killed. Other insurance claim statistics show that ABS brakes aren't producing the overall safety benefits that were predicted by the government and automakers. One IIHS study found that a passenger has a 45 percent greater chance of dying in a single-vehicle crash in a car with anti-lock brakes than in the same car with old-style brakes. On wet pavement, where ABS supposedly excels, that figure rises to a 65 percent greater chance of being killed. In multi-vehicle crashes, ABS-equipped vehicles have a passenger death rate 6 percent higher than vehicles not equipped with ABS.

The high cost of ABS maintenance is one disadvantage that few safety advocates mention, but consider the following: original parts can cost five times more than regular braking components, and many dealers prefer to replace the entire ABS unit rather than troubleshoot what is a very complex system.

Keep in mind that anti-lock brakes are notoriously unreliable on all makes and models. They often fail completely, resulting in no braking whatsoever, or they may extend stopping distance by 30 percent. This phenomenon is amply documented throughout NHTSA's complaint database found at *www.lemonaidcars.com.*

Chrysler, beset with thousands of complaints of ABS failures, struck a deal with the U.S. government (Canada tagged along) to service its early systems free of charge and guarantee the most expensive components for 10 years (check Part Two's "Secret Warranties" and the NHTSA website for ammo to show your service manager).

Airbags (unsafe at any speed?)

No airbag is safe, but some model years are safer than others. Pre-'97 airbags explode too forcefully and can seriously injure or kill occupants vulnerable to injury. Later, de-powered models are less hazardous.

Granted, in high-speed collisions airbags save lives, particularly if you're an unbelted male. However, in fender-benders where no one would have been hurt if the airbag hadn't deployed, full-power, early airbag systems can kill women, seniors, and children—or leave them horribly scarred or deaf. And you can never be sure when an airbag will suddenly explode in your face. In fact, millions of vehicles have been recalled because their airbags go off when they shouldn't and don't go off when they should.

If you are hit from the rear and thrown within 25 cm of the steering wheel as the airbag deploys, you risk severe head, neck, or chest trauma, or even death. If you are making a turn, with your arms crossing in front of the steering wheel, you are out of position and risk fractures to both arms. If you drive with your thumbs extended a bit into the steering hub area, as I often do, you risk losing both thumbs when the housing cover explodes.

Sometimes, just being female is enough to get you killed.

Two startling Transport Canada studies were uncovered in October 1999 as part of a CBC *Marketplace* investigation into airbag safety. The consumer TV show unearthed government-financed research that showed airbags reduce the risk of injury by *only* 2 percent for adults who wear seat belts. Even more incredible, the studies confirm that airbags actually *increase* the risk of injury to women by 9 percent and the risk of death for children by 21 percent.

The research was conducted in 1996 and 1998 on early fully powered airbags. Fearing the public disclosure of the findings, it took Transport Canada four months to release the studies to *Marketplace.* American government officials have refused to comment upon the Canadian research, despite the fact that it contradicts the basic premise of airbag use—they are safe at any speed, regardless of gender.

In fact, the danger to women is so great that an earlier 1996 Transport Canada and George Washington University study of 445 drivers and passengers drew these frightening conclusions:

> While the initial findings of this study confirm that belted drivers are afforded added protection against head and facial injury in moderate to severe frontal collisions, the findings also suggest that these benefits are being negated by a high incidence of bag-induced injury. The incidence of bag-induced injury was greatest among female drivers. Furthermore, the intervention of the airbag can be expected to introduce a variety of new injury mechanisms such as facial injuries from "bag slap," upper extremity fractures, either directly from the deploying airbag module or from arm flailing, and thermal burns to the face and arms.

A frightening admission. Don't look for the above study on Transport Canada's website. It's not there. Other supporting studies can be accessed from NHTSA at *www.nhtsa.dot.gov/esv/search.html*; *www.lemonaidcars.com*; *www.airbagonoff.com*; and *www.plescia.org/indexair.htm.*

Inadvertent airbag deployment

Airbags frequently go off for no apparent reason, usually due to faulty sensors. Causes of sudden deployment include passing over a bump in the road; slamming the car door; having wet carpets; or, in some Chrysler minivans, simply putting the key in the ignition.

This happens more often than you might imagine, judging by the frequent recalls and thousands of complaints recorded on NHTSA's website at *www.nhtsa.dot.gov/cars/problems/complain/complaintsearch.cfm.* Incidentally, insurers are refusing to pay for damage to the car or for airbag replacements unless there has been a collision. Meanwhile, automakers deny responsibility on the grounds that the vehicle collided with something. In the end, the driver is faced with a hefty repair bill and no means of proving the automaker's liability.

Airbag deployment for no apparent reason is bad enough if the vehicle is parked. However, if the airbag explodes while the vehicle is being driven, it will likely cause an accident and then be of no use during the ensuing impact because it has already deflated. Says Michael Leshner, a U.S. forensic engineer, "The airbag knocks them silly—*then* they have an accident."

Transport Canada allows car owners to have a mechanic disengage airbags for several reasons: having to sit fewer than 25 cm from the steering wheel, having to put children in the front seat, or having adult passengers with medical conditions that may induce airbag injuries. Motorists are required by regulation to submit the forms, but Transport Canada does not verify the accuracy of the information. People can have the work done without waiting for a response from the government. The department does not track people who disconnect their bags without going through the official route.

Side airbags—good and bad

Side airbags are designed to protect drivers and passengers in side-impact crashes, estimated to account for almost a third of vehicular deaths. They also have been shown to help keep unbelted occupants from being ejected in rollovers.

About 35 percent of all vehicles sold carry some kind of side airbag protection, says the Automotive Occupants Restraints Council, but many protect only the body, not the head. Only about 10 percent of vehicles sold will have side head protection.

IIHS's side crash tests of 2003 compact SUVs, has found that side-impact bags that provide head protection can make the difference between a "Poor" and a "Good" crash rating.

Because side airbags aren't required by federal regulation in the States or in Canada, neither government has developed any tests to measure how safe they are for children and small adults. The IIHS hopes its test results will goad government regulators and automakers into standardizing side airbag design and increasing their effectiveness and safety.

In the meantime, keep in mind that preliminary safety studies show side airbags may be killers to children or any occupant not sitting in the correct position. Research carried out in 1998 by safety researchers (Anil Khadikar, Biodynamics Engineering Inc., and Lonney Pauls, Springwater Micro Data Systems, "Assessment of Injury Protection Performance of Side Impact Airbags") shows there are four hazards that have yet to be addressed by automakers:

1. Inadvertent airbag firing (short circuit, faulty hardware or software)
2. Unnecessary firing (sometimes opposite side airbag will fire; airbag may deploy when a low-speed side-swipe wouldn't have endangered occupant safety)
3. A three-year-old restrained in a booster seat could be seriously injured
4. Out-of-position restrained occupant could be seriously injured

The researchers conclude with the following observation: "Even properly restrained vehicle occupants can have their upper or lower extremities in harm's way in the path of an exploding [side] airbag."

The above study and dozens of other scientific papers confirm that small and tall restrained drivers face death or severe injury from frontal and side airbag deployments for the simple reason that they are outside of the norm of the 5'8", 180 lb., male test dummy.

These studies also debunk the safety merits of ABS brakes, so it's no surprise they go unheralded by Transport Canada and other government and private safety groups. With a bit of patience, though, you can find this research at *www.nhtsa.dot.gov/esv/*.

In the meantime, don't forget NHTSA's side airbag warning issued on October 14, 1999:

> Side impact airbags can provide significant supplemental safety benefits to adults in side impact crashes. However, children who are seated in close proximity to a side airbag may be at risk of serious or fatal injury, especially if the child's head, neck, or chest is in close proximity to the airbag at the time of deployment.

Protect yourself

Because not all airbags function, or malfunction, the same way, *Lemon-Aid* has done an exhaustive analysis of U.S. and Canadian recalls, crash data, and owner complaints to determine which vehicles and which model years use airbags that may seriously injure occupants or deploy inadvertently. That data can be found in Part Three's model ratings.

Additionally, you should take the following steps to reduce the danger from airbag deployment:

- Don't buy a vehicle with side airbags unless they offer head protection and you are confident all occupants will remain properly positioned out of harm's way.
- Make sure that seat belts are buckled and all head restraints are properly adjusted (about ear level).
- Make sure the head restraints are rated "good" by IIHS in Part Three.
- Insist that passengers who are frail, short, or have recently had surgery sit in the back.

- Make sure that the driver's seat can be adjusted for height and has tracks with sufficient rearward travel to allow short drivers to remain a safe distance away from the bag's deployment (over 25 cm) and still reach the accelerator and brake.
- If you are short-statured, consider buying aftermarket pedal extensions from auto parts retailers or optional adjustable accelerator and brake pedals to keep you a safe distance away from a deploying airbag.
- Buy a vehicle that comes with manual passenger-side airbag disablers.
- Buy a vehicle that uses sensors to detect the presence of an electronically tagged child safety seat in the passenger seat and disables the airbag for that seat.
- Pop the airbag housing covers to ensure the airbags haven't been removed.
- Try to have a retrofitted cut-off switch installed by a dealer, but don't be surprised to find Canadian dealers reluctant to install the devices due to their unfounded fear of lawsuits (see Appendix I for *Marketplace*'s website, which lists which automakers will install retrofitted devices).
- Have a retrofitted cut-off switch installed by an independent garage.

Bob Edwards, owner of Airbag Service (Burlington) Ontario Inc., de-activates two to three airbags a week. Transport Canada gives out his name (sssh, don't tell anyone) to drivers whose dealers won't accommodate their requests. Simply disabling the airbag costs from $150, and he will travel to the customer's home or office to do the job. Edwards says he can buy aftermarket on-off switches in the States that fit most car models for about $250 and install them for about $200 more, while some automakers charge $700–$1,600.

For information, call Bob at 905-331-9601 or 1-800-519-3531 (from some selected Ontario regions). You can email him at *airbagservice@cogeco.ca.* If you live in other parts of Canada, contact *www.airbagservice.com*, the national franchiser.

Protect your children

Small-statured children, or those 12 years of age or younger, are best protected from collision death or injury by being buckled up in the middle or right rear seat in a crashworthy vehicle. The U.S. NHTSA rates child safety seats on ease of use at *www.nhtsa.dot.gov/CPS/CSSRating/.* To date, 68 seats have been rated by NHTSA, representing about 95 percent of the seats available to consumers. Convertible seats were rated in both the rear-facing and forward-facing mode and combination seats were rated in both the forward-facing and booster modes.

What's the Best Protection?

<table>
<tr><th>AGE/ WEIGHT</th><th>SEAT TYPE/ SEAT POSITION</th><th>USAGE TIPS</th></tr>
<tr><td colspan="3">INFANTS</td></tr>
<tr><td>Birth to 1 year/
Less than 20 lbs</td><td>Infant only seat/rear facing or convertible seat used rear facing</td><td>Never use in front seat where air bag is present
Tightly install child seat in rear seat, facing the rear, child seat should recline at approximately 45°</td></tr>
<tr><td>Less than 1 year
20-35 lbs</td><td>Convertible seat used rear facing (select one recommended for heavier infants.)</td><td rowspan="2">Harness straps/slots at or below shoulder level (use lower set of slots for most convertible seats)
Harness straps snug on child, harness clip at armpit level</td></tr>
<tr><td colspan="2">Seats should be secured to the vehicle by safety belts or the LATCH system</td></tr>
<tr><td colspan="3">PRESCHOOLERS/TODDLERS</td></tr>
<tr><td>1 to 4 years
at least 20 lbs to approx. 40 lbs</td><td>Convertible Seat/ forward facing or
Forward Facing Only or
High back Booster with harness.</td><td rowspan="2">Tightly install child seat in rear seat, facing forward
Harness straps/slots at or above child's shoulders (use upper set of slots for most convertible seats)
Harness straps snug on child, harness clip at armpit level</td></tr>
<tr><td colspan="2">Seats should be secured to the vehicle by safety belts or the LATCH system</td></tr>
<tr><td colspan="3">Young Children</td></tr>
<tr><td>4 to at least 8 yrs/ unless they are at least 4'9" (57") tall</td><td>Belt-Positioning Booster (no back, base only) or High Back Belt-Positioning Booster</td><td rowspan="2">Booster base used with adult lap and shoulder belt in rear seat
High back booster offers head/neck support for vehicles with rear seats having low or no head support
Shoulder belt should rest snugly across chest, and should NEVER be placed under the arm or behind the back
Lap belt should rest low, across lap/ upper thigh area – not across the stomach</td></tr>
<tr><td colspan="2">Never use with lap-only belts—Belt positioning boosters are always used with both LAP and SHOULDER belts.</td></tr>
</table>

Other factors to consider

- **Two-door vehicles**—It can be difficult to install a child safety seat correctly because you have to get into the back seat yourself to install it securely. It can also be difficult to get your child in and out of the child safety seat.
- **Small back seats**—The back seats of small cars and many pickup trucks are too small to properly accommodate some child safety seats, especially those in the rear-facing reclined position. The base of the child safety seat must not be wider than the space between the seat buckle and safety belt attachments. Wide bases may also block access to buckles for outboard lap and shoulder belts, making child safety seat installation difficult. In this case, try a child safety seat with a narrow base.

- **Deep bucket seats**—Many child safety seats will not fit in vehicle seats with deep buckets. Try a child safety seat with a narrow base or top tether strap.
- **Slope of back seat**—Rear-facing child safety seats should be reclined at about a 45-degree angle. The slope of the vehicle seat may raise the back of the child safety seat too much, putting the infant in an upright rather than reclined position. To remedy this situation, place a tightly rolled towel or firm, solid-core foam between the vehicle's seat cushion and the seat back to help achieve the correct angle. Always check the child safety seat instructions and vehicle owner's manual for correct installation.
- **Contour of back seat**—While the centre of the rear seat may seem the safest place for a child, many back seats have a hump in the centre, making it difficult to install a child safety seat correctly. The safest position in the back seat is where the child safety seat fits securely.
- **Splits in bench seats**—Splits in wide bench seats can make it difficult to install a child safety seat correctly. Move it to a different rear seating position.
- **Forward-anchored belts**—If the safety belt extends from the seat forward of where the back and seat cushions meet, the child safety seat may be too loose. Move it to a different rear seating position, or try a different style child safety seat with a top tether strap, provided the vehicle has a top tether anchorage.
- **Pickup truck jump seats/extended cabs**—Child safety seats will not fit properly in many pickup truck rear seats. There is not enough space between the rear of the front seat and the child to allow forward motion in the event of a crash or even a sudden stop. Side-facing jump seats are not safe for a child safety seat under any circumstances.
- **Vehicles manufactured before September 1, 1995**—Some vehicles may have safety belt systems—such as automatic safety belts—that require additional hardware to install child safety seats correctly. Be sure to read both the vehicle and child safety seat manuals and any safety belt labels.
- **Vehicles manufactured after September 1, 1995**—Vehicles are now equipped with safety belt locking features—such as locking or switchable retractors—that make installation of a child safety seat easier. To be sure you are installing the seat correctly, read the vehicle owner's manual and any safety belt labels.

Crashworthiness (bigger *is* better)

Crashworthiness is the one safety improvement over the past 30 years that everyone agrees has paid off handsomely without presenting any additional risks to drivers or passengers. Plus, there's no extra charge for the additional protection.

By surrounding occupants in a protective cocoon and deflecting crash forces away from the interior, auto engineers have successfully created safer vehicles without increasing size or cost. And purchasing a vehicle with the idea that you'll be involved in an accident some day is not unreasonable. According to IIHS, the average car will likely have two accidents before ending up as scrap and is twice as likely to be in a severe front-impact crash as a side-impact crash.

Two Washington-based agencies monitor how vehicle design affects crash safety: NHTSA and IIHS. Crash information from these two groups doesn't always correspond because, while IIHS's results incorporate all kinds of accidents, including offset crashes and bumper damage sustained from low-speed collisions, NHTSA's figures mostly relate to 57 km/h (35 mph) frontal and some side collisions. Offset testing has just begun. The frontal tests are equivalent to two vehicles of equal weight hitting each other head-on while travelling at 57 km/h (35 mph), or to a car slamming into a parked car at 114 km/h (70 mph). Bear in mind that a vehicle providing good injury protection may also cost more to repair because its structure, not the occupants, absorbs most of the collision forces. That's why safer vehicles don't always have lower insurance rates.

Cars vs. trucks

Occupants of large vehicles have fewer severe injury claims than do occupants of small vehicles. This was proven conclusively in a 1996 NHTSA study that showed collisions between light trucks or vans and small cars resulted in an 81 percent higher fatality rate for the occupants of the small cars than occupants of the light trucks or vans.

Vehicle weight protects you principally in two-vehicle crashes. In a head-on crash, for example, the heavier vehicle drives the lighter one backward, which decreases forces inside the heavy vehicle and increases forces in the lighter one. All heavy vehicles, even poorly designed ones, offer this advantage in two-vehicle collisions. However, they may not offer good protection in single-vehicle crashes.

Crash test figures show that SUVs, vans, and trucks also offer more protection to adult occupants than passenger cars in most crashes, because their higher set-up allows them to ride over other vehicles (Ford's 2002 4X4 Explorer lowered its bumper height to prevent this hazard). Conversely, due to their high centre of gravity, easily overloaded tires, and unforgiving suspensions, these vehicles have a disproportionate number of single-vehicle rollovers, which are far deadlier than frontal or side collisions (think Ford Explorer/Firestone). In the case of the Ford Explorer, Bridgestone/Firestone CEO John Lampe testified on August 2001 that 42 of 43 rollovers involving Ford Explorers in Venezuela were on competitors' tires—shifting the rollover blame to the Explorer's design and crashworthiness.

IIHS figures show that for every 450 kg (1,000 lb.) added to a vehicle's mass, the risk of injury to an unrestrained driver is lowered by 34 percent, and the risk of injury to a restrained driver is lowered by 25 percent. GM's series of two-car crash tests carried out a decade ago dramatically confirm this fact. Engineers concluded that if two cars collide, and one weighs half as much as the other, the driver in the lighter car is 10 times more likely to be killed than the driver in the heavier one. This held true no matter how many stars the smaller car was awarded in government crash tests.

GM is now using this data to fight federal proposals to make vehicles more fuel efficient. As their argument goes, fuel consumption can only be reduced by cutting vehicle weight, which means more collision fatalities. On the other hand, engineers say reducing bumper height; developing "smart" seat belts, airbags, and head restraints; and preventing seatbacks from collapsing in low-speed collisions will save lives and allow for considerable weight reduction.

Interestingly, a vehicle's size or past crashworthiness rating doesn't always guarantee that you won't be injured. U.S. government crash tests show, for example, that a small 1991 Ford Escort gives better full-frontal collision protection than a 1994 Mercury Villager or Nissan Quest, two large minivans that could have been easily engineered to crash safely at moderate speeds of 57 km/h (35 mph). Also, different years of the same model or size variation can produce startlingly different scores. For example, Chrysler's 1995 Caravan and Voyager minivans earned better driver crash protection scores than did the 1996 and 1997 Grand Caravan and Grand Voyager. And the 2002 GM Cavalier and Sunfire two-door compact's crash scores aren't as good as the '97 versions.

Unsafe designs

Although it sounds hard to believe, automakers *will* deliberately manufacture a vehicle that will kill or maim simply because, in the long run, it costs less to stonewall complaints and pay off victims than to make a safer vehicle. I learned this lesson after reading the court transcripts of *Grimshaw v. Ford* and listening to court testimony of GM engineers who deliberately placed fire-prone "sidesaddle" gas tanks in millions of pickups to save $3 per vehicle (see *consumerlawpage.com/article/gm-exploding-tank.shtml*). In *Grimshaw*, the court found that in designing the Pinto, Ford "engaged in cost-benefit analysis which balanced life and limb against corporate savings and profits." Consequently, the company rejected a $4-per-car repair that would have made the Pinto less prone to explode from a rear-ender.

More recent examples of corporate greed triumphing over public safety: pre-1997 airbag designs that maim or kill women, children, and seniors; anti-lock brake systems that don't brake (a major problem with GM minivans, trucks, and sport-utilities, and Chrysler sport-utilities and minivans); flimsy front seats and seatbacks; the absence of rear head restraints; and fire-prone GM pickup and Ford Crown Victoria fuel tanks.

Faulty head restraints

Offered as a standard feature on all front and some rear seats, the main function of head restraints is to prevent neck whiplash injuries during a rear collision when the impact forces snap your head back. According to a 1999 joint study carried out by IIHS and State Farm Insurance, over a quarter of accident claims involve neck injury. Whiplash injuries are not only a literal pain in the neck, they're also affecting more female drivers and younger drivers than ever before. The study also revealed that, in mid-sized vehicle accidents, women have more neck injuries than men (30 percent versus 23 percent) and, surprisingly, drivers 65 years and older had lower neck-injury rates than younger drivers.

Whiplash injuries are also closely related to head restraint design. Your vehicle's head restraints may be either fixed or adjustable. Fixed restraints are preferred, but adjustable ones are acceptable as long as they're adjusted high enough (at ear level). Make sure that rear visibility isn't obstructed by the front or rear restraints. The State Farm study showed that drivers struck in rear collisions in cars with head restraints rated "Good" by IIHS are 24 percent less likely to sustain neck trauma than drivers using head restraints rated "Poor." The problem is, 70 percent of head restraints are rated "Poor" by IIHS; only 3 percent are rated "Good." Current ratings can be found at the IIHS website: *www.highwaysafety.org/ vehicle_ratings/head_restraints/head.htm.*

Flying seats and seatbacks

Seat anchorages have to conform to government load regulations, but the regulations are so minimal that seats can easily collapse or tear loose from their anchorages, leaving drivers and passengers vulnerable in accidents. NHTSA's complaint database is replete with instances where occupants report severe injuries caused by collapsing seatbacks that have injured front seat occupants or crushed children in safety seats placed in the rear seat.

Driver and passenger seatbacks frequently collapse in rear-end collisions. In a recent Illinois Appeal Court decision (*Carillo v. Ford*, Cook County Circuit Court), a $14.5 million jury verdict was sustained against Ford after a driver's 1991 Explorer seatback failed, leaving the driver paralyzed from the chest down.

Experts testified that the seatback design was unreasonably dangerous in high-speed, rear-impact collisions. Ford's experts countered that the "yielding seats" were reasonably safe and met the NHTSA safety standard for seatback strength established in 1971, based on a 1963 recommendation by the Society of Automotive Engineers. This standard requires a seat to withstand 200–300 lb. of force. However, in a typical rear-impact crash, seats are subjected to forces four or five times that, according to Kenneth Saczalski, an engineering consultant who specializes in seat design.

The Appeal Court rejected Ford's plea.

Top 20 safety defects

The U.S. federal government's online safety complaints database contains well over 100,000 entries, going back to vehicles made in the late '70s. Although originally intended to record incidents of component failures that only relate to safety, you will find every problem imaginable dutifully recorded by clerks working for NHTSA.

A perusal of the listed complaints shows that some safety-related failures occur more frequently than others and often affect one manufacturer more than another. Here is a summary of some of the more commonly reported failures, in order of frequency:

1. Airbags not deploying when they should or deploying when they shouldn't
2. ABS total brake failure; wheel lock-up
3. Tire tread separation
4. Electrical or fuel system fires
5. Sudden acceleration
6. Sudden stalling
7. Sudden electrical failure
8. Transmission failing to engage or suddenly disengaging
9. Transmission jumping from Park to Reverse or Neutral; vehicle rolling away when parked
10. Steering or suspension failure
11. Seat belt failures
12. Collapsing seatbacks
13. Defective sliding door, door locks, and latches
14. Poor headlight illumination
15. Dash reflecting onto windshield
16. Hood flying up
17. Wheel falling away
18. Steering wheel lifting off
19. Transmission lever pulling out
20. Exploding windshields

Choosing an Inexpensive Vehicle

As far as reliability and dependability are concerned, sport-utilities, vans, and pickups are more problem-prone than cars simply because there are more things that can go wrong, like malfunctioning minivan power sliding doors, ABS brake failures, and noisy, failure-prone powertrains.

One saving grace is that rear-drive cars and minivans often cost less to repair because key components are easily accessible, parts have been around for years and can be purchased from independent jobbers, and knowledgeable do-it-yourselfers abound.

Watch the warranty

On average, 20 percent of a dealer's parts and service revenue is generated by warranty work, but there can be a big difference between warranty promise and warranty performance. Most automakers offer bumper-to-bumper warranties good for at least the first 3 years/60,000 km and some problem-prone models and luxury makes get additional base coverage up to 5 years/100,000 km. DaimlerChrysler, and South Korean automakers are also using more comprehensive 5- to 7-year warranties as important marketing tools to give their cars luxury cachet, or allay buyers' fears of poor quality/reliability. It's also becoming an industry standard for car companies to pay for roadside assistance, a loaner car, or hotel accommodations if your vehicle breaks down while you're away from home and it's still under warranty. *Lemon-Aid* readers report few problems with these ancillary warranty benefits.

However, just like the performance ads you see on TV, what you see isn't always what you get. For example, bumper-to-bumper coverage usually excludes stereo components, brake pads, clutch plates, and many other expensive components.

Insurance deals

Insurance costs can average between $700 and $2,000 per year, depending on the type of vehicle you own, your personal statistics and driving habits, and whether you can obtain coverage under your family policy. Even though it seems unfair, small SUVs and pickups are usually classed in the high-risk category (accident and theft) and therefore usually cost more than passenger cars to insure.

Researchers for the Ontario-based Consumer's Guide to Insurance website have found that it pays to shop around for cheap auto insurance rates. In February 2003, the group discovered that the same insurance policy could vary in cost by a whopping 400 percent.

For example, a 41-year-old married female driving a 2002 Honda Accord and a 41-year-old married male driving a 1998 Dodge Caravan, both with unblemished driving records, should pay no more than $1,880, but some companies surveyed asked as much as $7,515.

And that's what they would have paid, if they hadn't shopped around.

The Consumer's Guide to Insurance is a free service that tells consumers which companies have the lowest car insurance rates. The service can be accessed at *www.insurancehotline.com* or by telephone at 416-686-0531.

Use "secret" warranties

Automobile manufacturers are reluctant to publicize their secret warranty programs because they feel that such publicity would weaken consumer confidence in their products and increase their legal liability. The closest they come to an admission is to send out a "goodwill policy," "special policy," or "product update" service bulletin for dealers' eyes only. These bulletins admit liability and propose free repairs for defects that include faulty paint, air conditioning, and engine and transmission failures on Chrysler, Ford, and GM vehicles (see Part Two).

If you're refused compensation, keep in mind that secret warranty extensions are, first and foremost, an admission of manufacturing negligence. You can usually find them in technical service bulletins (TSBs) that are sent daily to dealers by automakers. Your bottom-line position should be to accept a pro rata adjustment from the manufacturer, whereby you share a third of the repair costs with the dealer and automaker. If polite negotiations fail, challenge the refusal in court on the grounds that you should not be penalized for failing to make a reimbursement claim under a secret warranty that you never knew existed!

Service bulletins are great guides for warranty inspections (especially the final one), and they're useful in helping you decide when it's best to trade in your car. They're written by automakers in "mechanic-speak" because service managers relate better to them that way, and manufacturers can't weasel out of their obligations by claiming that they never wrote such a bulletin.

If your vehicle is out of warranty, show these bulletins to less expensive independent garage mechanics, so they can quickly find the trouble and order the most recent *upgraded* part, ensuring that you don't replace one defective component with another.

Because these bulletins are sent out by U.S. automakers, Canadian service managers and automakers may deny, at first, that they even exist (are you listening, Chrysler and Ford?). However, when they're shown a copy they usually find the appropriate Canadian part number or bulletin in their files. The problem and its solution don't change from one side of the border to another. Imagine American and Canadian tourists being towed across the border because each country's technical service bulletins were different. Mechanical fixes do differ in cases where a bulletin is for California only, or relates to a safety or emissions component used only in the States. But these instances are rare indeed.

The best way to get bulletin-related repairs carried out is to visit the dealer's service bay and attach the specific ALLDATA-supplied service bulletin covering your vehicle's problems to a work order.

Getting your vehicle's service bulletins

Free summaries of automotive recalls and technical service bulletins listed by year, make, model, and engine can be found at the ALLDATA (*www.alldata.com/TSB/*) and NHTSA websites. Like the NHTSA summaries, ALLDATA's summaries are so short and cryptic that they're of limited use . You can download the complete contents of all the bulletins applicable to your vehicle from ALLDATA at *www.alldatadiy.com*, however, if you pay the $24.95 (U.S.) annual subscription fee.

Buy Asian

Many well-meaning Canadian buyers want to encourage domestic auto production by purchasing a product made by the Detroit Big Three, yet they know Asian vehicles have a better reputation for quality. Actually, this should be less of a dilemma now that Chrysler's owned by German automaker Daimler and so many vehicles and components are manufactured in the United States, Brazil, China, and Mexico. Although Chrysler's Daimler connection hasn't resulted in better quality scores (Daimler has its own quality problems), automakers have found that offshore factories can actually raise quality to higher levels than what's turned out in their own countries. For example, American auto manufacturers have known for decades that Canadian plants build higher-quality vehicles than factories in the United States. That's one of the reasons why I recommend the Cambridge, Ontario-built Matrix over GM's identical American-built Pontiac Vibe. Surprisingly, BMW's highest-quality factory is found in Rosslyn, South Africa.

Nevertheless, most studies show that, in spite of attempted improvements over the past two decades, vehicles made by DaimlerChrysler, Ford, and to a lesser extent, GM still don't measure up to Japanese and some South Korean (Hyundai) products in terms of quality and technology. This is particularly evident in SUVs and minivans, where Honda, Nissan, and Toyota have long retained the highest reliability and dependability ratings.

Whether you buy domestic or imported, overall vehicle quality has improved a great deal during the past three decades. Premature rusting is less of a problem and reliability is on the rise. On the other hand, repairs are outrageously expensive and complicated to do. Cars and minivans made by GM Ford, and Chrysler still report serious powertrain deficiencies, often during the first year in service. These defects include electrical system failures caused by faulty computer modules; malfunctioning ABS systems, brake rotor warpage, and early pad wearout; failure-prone air conditioning and automatic transmissions; and defective engine head gaskets, intake manifolds, fuel systems, suspensions, steering, and paint.

Don't buy the myth that parts for imports are overpriced or hard to find. It's actually easier to find parts for Japanese vehicles than for domestic vehicles due to the large number of units produced, the presence of hundreds of independent suppliers, the ease with which relatively simple parts can be interchanged from one model to another, and the large reservoir of used parts stocked by junkyards. Incidentally, when a part is hard to find, the *Mitchell Manual* is a useful guide to substitute parts that can be used for many different models. It's available in some libraries, most auto parts stores, and practically all junkyards.

Customer relations is the Japanese automakers' Achilles heel. Dealers are spoiled rotten by several decades of easy sales and have developed a "take it or leave it" showroom attitude often accompanied by a woeful ignorance of their own model lineup. This was once a frequent complaint of Honda and Toyota shoppers, though recent APA undercover surveys show a big improvement among Toyota dealers.

Where it has gotten worse is in getting straight answers in the service bay and in getting after-warranty assistance for well-known factory-related defects.

South Korean automakers

There's no problem with discourteous or ill-informed South Korean dealers—instead, poor quality has been their bugaboo. Yet, like Honda and Toyota's recovery following their own first missteps, Hyundai has made considerable progress in bringing quality up to an acceptable level. Kia has been less successful. Nevertheless, the Kia Sportage SUV has had fewer factory-induced defects recorded than some popular Detroit models like the Ford Explorer.

Up to the mid-'90s, South Korean vehicles were merely cheap, poor-quality knock-offs of their Japanese counterparts. They would start to fall apart after their third year due to subpar body construction, unreliable automatic transmission and electrical components, and parts suppliers who put low prices ahead of reliability and durability. This has been particularly evident with Hyundai's Excels and early Sonatas. During the past several years, though, Hyundai's product lineup has been extended and refined, and quality is no longer a worry. Plus, Hyundai's comprehensive base warranty protects owners from most of the more expensive breakdowns that may occur.

Hyundais are easily repaired by independent garages, and their rapid depreciation doesn't mean much because they cost so little initially. Entry-level buyers are known to keep their cars longer than most, thereby easily amortizing the higher depreciation rate.

Daewoo lives!

Not really. Its models are returning this year as Suzukis and Chevrolets. GM's Daewoo Automotive & Technology Co. is resuming shipments halted in 2001 when the former Daewoo Motor Co. teetered on the brink of bankruptcy and was bought by General Motors. GM is using its Suzuki partnership to relaunch Daewoo models like the mid-sized Magnus and Verona sedan, sold in Canada as the Chevrolet Epica.

Daewoo is the weakest of the South Korean brands and it's doubtful whether GM's takeover will benefit either company. In the meantime, owners of previously bought Daewoos will have additional service facilities with the new jury-rigged Suzuki-Chevrolet outlets.

European automakers

Heck, even the Germans have abandoned their own products. For example, a 2002 J.D. Power survey of 15,000 German car owners found German drivers are happiest at the wheel of a Japanese car. This conclusion included compacts cars, luxury cars, and off-roaders. Toyota won first place on quality, reliability, and owner satisfaction, while Nissan's Maxima headed the luxury class standings. BMW was the first choice among European offerings.

Money-losing Land Rover, now owned by Ford, is still struggling since it moved into the North American market with its overpriced, quality-challenged, and under-serviced SUVs. But, Land Rover of North America shrugs off its quality problems, and Ford isn't likely to improve things much while it struggles with its own mismanagement and poor-quality woes.

For many years, Land Rovers got a free ride from fawning car columnists on this side of the Atlantic (British press criticism had been merciless) due to the vehicle's African safari cachet and marginal sales figures. Nevertheless, when Matt Joseph, a Madison, Wisconsin newspaper and radio car reviewer, published a mildly critical review of the 2000 Discovery, he paid for his audacity—with his job.

Still, Matt got the last laugh after all: A 2003 study of 34,000 car owners with vehicles up to eight years old published by Britain's Consumers' Association found that less than half of British owners would recommend a British-made Rover or Vauxhall to a friend.

The most highly rated cars in the study were the Japanese Subaru, Isuzu, and Lexus. Over 85 percent of drivers would recommend them.

Here's another surprise: Mercedes-Benz's quality control isn't equal to that company's pretensions. After stumbling badly when it first launched its rushed-to-production, American-made ML320 for the 1998 model year, the automaker has sent out many urgent service bulletins that seek to correct a surprisingly large number of production deficiencies. Judged by its own confidential service bulletins, Mercedes' quality control is the worst of the German automakers.

Volkswagen isn't a whole lot better than Mercedes. True, it has always been early on the scene with great concepts, but they have always been accompanied by poor execution and a weak servicing network. With its failure-prone and under-serviced Eurovan and Camper, the company hasn't been a serious minivan player since the late '60s, and VW's few sport-utility and pickup variants have been resounding duds. Almost two decades ago, VW imported a mini sport-utility called "The Thing." Ugly, underpowered, and unreliable, it lasted only a couple of model years (wonder if its designers moved over to Pontiac's Aztek SUV project).

With European models, your service options are limited and customer relations staff can be particularly insensitive and arrogant. Plus, you can count on lots of aggravation and expense due to the unacceptably slow distribution of parts and their high markup. Because these companies have a quasi-monopoly on replacement parts, there are few independent suppliers you can turn to for help. And auto wreckers, the last-chance repository for inexpensive car parts, are unlikely to carry European parts for vehicles that are more than three years old or were manufactured in small numbers.

These vehicles also age badly. The weakest areas remain the drivetrains, electronic control modules, electrical and fuel systems, brakes, accessories (sound system, AC, etc.), and body components.

Other Buying Considerations

Depreciation

Depreciation is the biggest—and most often ignored—expense you encounter when you trade in your vehicle or when an accident forces you to buy another vehicle before the depreciated loss can be amortized. Most new cars depreciate a whopping 30–40 percent during the first two years of ownership. Fortunately, minivans, vans, trucks, and sport-utilities don't lose as much of their value before their third year in use. The best way to use depreciation rates to your advantage is to choose a vehicle listed as being both reliable and economical to own and keep it for five to 10 years. Alternatively, you may buy insurance to protect you from depreciation's bite.

Performance

Take the phrase "carlike handling" with a large grain of salt when buying a minivan. Rear-drives are built on a modified truck chassis and use truck steering and suspension components, making them handle more like trucks than cars, in spite of automakers' claims to the contrary. Also, what you see is not necessarily what you get—additional options are usually a prerequisite to make them safe and comfortable to drive.

Front-drives handle better than rear-drives, but the additional size and weight requires a whole new set of driving skills when cornering at moderate speeds, parking, or turning. And, they are more likely to roll over than passenger cars.

Front-drive

Front-drives direct engine power to the front wheels, which pull the vehicle forward while the rear wheels simply support the rear. The biggest benefit of front-drives is foul-weather traction. With the engine and transmission up front there's lots of extra weight pressing down on the front-drive wheels, increasing tire grip in snow and on wet pavement. But when you drive up a steep hill or tow a boat or trailer the weight shifts, and you lose the traction advantage.

Although I recommend a number of front-drive vehicles in this guide, I don't like them as much as rear-drives. Granted, front-drives provide a bit more interior room (no transmission hump), more carlike handling, and better fuel economy than do rear-drives, but damage from potholes and fender-benders is usually more extensive, and maintenance costs (especially premature front tire and brake wear) are much higher than with rear-drives.

Servicing front-wheel drives can be a real nightmare—and a wallet buster. Entire steering, suspension, and drivetrain assemblies have to be replaced when only one component is defective. Downtime is considerable, the cost of parts is far too high, and the drivetrain and its components aren't designed for the do-it-yourself mechanic. A new front-drive transmission assembly (called a transaxle) can cost about $3,000 to repair, compared to $700 for a rear-drive transmission. And having to make such a repair isn't that remote a possibility, particularly if you own a 1989–2001 Chrysler minivan or a Ford Taurus, Sable, or Windstar.

Rear-drive

Rear-drives direct engine power to the rear wheels, which push the vehicle forward. The front wheels steer and also support the front of the vehicle. With the engine up front, the transmission in the middle, and the drive axle in the rear, there's plenty of room for larger and more durable drivetrain components. This makes for less crash damage, lower maintenance costs, and higher towing capacities than front-drives.

On the other hand, rear-drives don't have as much weight over the rear wheels as do the front-drives (no, putting concrete blocks in the bed or trunk will only void your transmission warranty), and therefore they can't provide as much traction on wet and icy roads unless they're equipped with an expensive traction-control system.

Four-wheel drive (4X4)

Four-wheel drive (4X4) directs engine power through a transfer case to all four wheels, which *pull* and *push* the vehicle forward, giving you twice as much traction. On most models, when the 4X4 drive isn't engaged, the vehicle reverts to rear-drive. The large transfer case housing makes the vehicle sit higher, giving you additional ground clearance.

Keep in mind that extended driving over dry pavement with 4X4 engaged will cause the driveline to bind and result in serious damage. Some buyers are turning instead to rear-drive pickups equipped with winches and large, deep-lugged rear tires.

Many 4X4 customers have been turned off by the typically rough and noisy driveline; a tendency for the vehicle to tip over when cornering at moderate speeds (a Ford Bronco and Isuzu Rodeo specialty); vague, trucklike handling; high repair costs; and poor fuel economy. No wonder car-based SUVs like the Toyota RAV4 and the Honda CRV are so popular: buyers want versatility without sacrificing fuel economy, comfort, or handling.

All-wheel drive (AWD)

Essentially, this is 4X4 drive *all* the time. Used mostly in sedans and minivans, all-wheel drive (AWD) never needs to be de-activated when running over dry pavement and doesn't require the heavy transfer case (although some sport-utilities and pickups do use a special transfer case) that raises ground clearance and cuts fuel economy. AWD-equipped vehicles aren't recommended for off-roading because of their lower ground clearance and fragile driveline parts, which aren't as rugged as 4X4 components.

Preventing rust

First off, remember that the best rustproofing protection is to keep your vehicle in a dry, unheated garage or outside. Never bring your car in and out of a heated garage during the winter months. The months when temperatures are just a bit above freezing are the worst for rust promotion; keep your vehicle especially clean and dry during that time.

If you live in an area where roads are heavily salted in winter, or in a coastal region, make sure you have undercoating sprayed annually and include the rocker panels (inside the door bottoms), the rear hatch's bottom edge, the tail-gate, and the wheelwells. This treatment, costing less than $100, will protect vital suspension and chassis components, make the vehicle ride more quietly, and allow you to ask a higher price at trade-in time. The only downside, which can be checked by asking for references, is that the undercoating may give off an unpleasant odour for months and it may drip, soiling your driveway.

Hybrid cars

Automakers are offering hybrid vehicles like the Toyota Prius and Honda Hybrid that use an engine/electric motor for maximum fuel economy and low emissions while providing the driving range of a comparable small car. Yet this latest iteration of the electric car still has serious drawbacks, which may drive away even the most "green"-minded buyers.

The hybrid cars' battery components, such as lead and cadmium, can be toxic to the environment, safety in crashes is not assured, price isn't competitive, reliability and repairs may be problematic, and interior room is cramped. Hybrid cars are usually lighter than their non-hybrid predecessors—the Honda Insight is 746 kg less—though some hybrids, like the Toyota Prius, may be heavier (137 kg more). This could put them at a safety disadvantage if hit by a larger vehicle.

If you find the limitations of an electric hybrid too daunting, why not simply buy a more fuel-efficient small car? Here are the top eight environmentally friendly cars recommended by Toronto-based Environmental Defence Canada (asterisks indicate models that are recommended by *Lemon-Aid)*:

1. *Toyota Echo
2. *Honda Civic
3. *Toyota Corolla
4. Saturn SC
5. Saturn SL
6. *Nissan Sentra
7. Ford Focus
8. *Hyundai Accent

Surviving the Options Jungle

Dealers make more than three times as much profit selling options as they do selling most cars (50 percent vs. 15 percent). No wonder their eyes light up when you start perusing their options list. If you must have some options, compare prices with independent retailers and buy where the price is lowest and the warranty is the most comprehensive. Buy as few options as possible from the dealer, since you'll get faster service, more comprehensive guarantees, and lower prices from independent suppliers. Remember, extravagantly equipped vehicles hurt your pocketbook in three ways: they cost more to begin with, they cost more to maintain, and they often consume extra fuel.

A heavy-duty battery and suspension and perhaps an upgraded sound system will generally suffice for American-made vehicles; most imports already come well-equipped. An engine block heater with a timer—favoured by 39 percent of new car shoppers—isn't a bad idea, either. It's an inexpensive investment that ensures winter starting and reduces fuel consumption by allowing you to start out with a semi-warm engine. Factory-installed in-line heaters are usually more efficient and durable.

It's hard to buy and even harder to lease a new vehicle that isn't loaded with unnecessary options. This is particularly evident with SUVs and trucks (commercial, bare-bones vans are available). If you're unsure as to what optional equipment like sound systems and anti-theft devices should cost, shop around and compare prices with independent suppliers.

When ordering parts, remember that purchases from American outlets can be slapped with a small customs duty, in spite of the free trade agreement, if the part isn't made in the United States. Then you'll pay the inevitable GST levied on the part's cost and customs duty. Finally, your freight carrier may charge a $15–$20 brokerage fee for representing you at the border.

Smart Options

Adjustable pedals and extensions

A safety feature first offered by Ford, but now available from most automakers, this device moves the brake and accelerator pedals forward or backward about 10 cm (4 in.) to accommodate and protect short-statured drivers from airbag-induced injuries.

If the manufacturer of your vehicle doesn't offer optional power-adjustable pedals, there are several companies selling inexpensive pedal extensions by mail order through the Internet; go to *stores.yahoo.com/hdsmn/pedex.html.* If you live in Toronto or London, Ontario, check out *www.kinomobility.com.*

Adjustable steering wheel

This option facilitates access to the driver's seat and permits a more comfortable driving position. It's particularly useful if a vehicle will be driven by more than one person.

Air conditioning

Whether you choose air conditioning should depend more on where you live and your comfort requirements than on saving money. In the same way that buying an underpowered van to save on fuel costs can leave you regretting its poor performance, your savings will be the last thing on your mind when you're sweltering in your vehicle some summer day.

At highway speeds, air conditioning increases fuel consumption by 3–4 percent. It provides extra comfort, reduces wind noise (from not having to roll down the windows), and improves window defogging. But air conditioning units are just as failure-prone (expect repairs of about $1,000 around the 5-year mark) as power window motors, easy prey for premature corrosion, and an excellent incubator for airborne bacteria and allergens. If you must have air conditioning, opt for a factory-installed unit. You'll get a longer warranty and reduce the chance that other mechanical components will be damaged during installation.

Anti-theft systems

Unrecovered stolen vehicles and stolen components cost Canadian consumers nearly $600 million per year in insurance premiums. There's a one-in-130 chance that your car will be stolen and only a 60 percent chance that you'll ever get it back. It's no wonder that over one-third of new vehicles are equipped with standard or optional alarms, with a clear preference for fuel cut-off devices and electronic alarms among the four main types of security systems: active and passive units either armed by the driver or self-arming, ignition disablers, parts identification, and security keys.

The most effective theft-deterrent systems aren't always the most expensive ones. Don't waste your money on costly anti-theft devices that depend solely upon your car's horn or lights to scare thieves away. They often go off at the wrong time and can be easily de-activated. Furthermore, they frighten the thief away only after the vehicle has been damaged. When it comes to stealing your car's contents, no alarm system can resist a brick through a side window. It takes just 12 minutes for thieves to strip a car seat, radio, and body parts, and few citizens are brave enough to personally stop a theft or testify in court.

Since most cars are stolen by amateurs, the best theft deterrent is a visible device that complicates the job while immobilizing the vehicle and sounding an alarm. For less than $150, you can install both a steering wheel lock and a hidden remote-controlled ignition disabler. These clublike devices cost between $50 and $75, and deter thieves by forcing them to carry a hacksaw and adding about a half-minute to the time it takes to steal the average car or make off with the airbag (they have to cut through the steering wheel or bust the lock). Ignition disablers are also inexpensive and very effective. They sell for $30–$90, depending on their sophistication.

Battery (heavy-duty)

The best battery for northern climates is the optional heavy-duty type offered by many manufacturers for about $80. It's a worthwhile purchase, especially for vehicles equipped with lots of electric options. Most standard batteries last only two winters; heavy-duty batteries give you an extra year or two for about 20 percent more than the price of a standard battery.

Block heater

A block heater is an inexpensive investment at about $75. Factory-installed in-line models are effective and durable. Cheaper ($40 installed) dipstick or freeze plug models are not recommended.

Central locking control

With a price tag of around $200, this option is most useful for families with small children, car-poolers, or drivers of pickups, minivans, and vans who can't easily slide across the seat to lock the other doors. Look for a remote feature that provides automatic locking and then unlocking when the inside door handle is pulled.

Child safety seat (integrated)

Integrated safety seats are designed to accommodate any child more than one year old or weighing over 9 kg (20 lb.). Since it's permanently integrated into the seatback, it takes the fuss out of installing and removing the safety seat and finding some place to store it. When not in use, it quickly folds away out of sight, becoming part of the seatback. Two other safety benefits: you know the seat has been properly installed, and your child gets used to having his or her "special" seat in back where it's usually safest to sit.

Engines

Choose the most powerful 6- or 8-cylinder engine available if you're going to be doing a lot of highway driving, plan to carry a full passenger load and luggage on a regular basis, or intend to load up the vehicle with convenience features like air conditioning. Keep in mind that multipurpose vehicles with larger engines are easier to resell and retain their value the longest. For example, Honda's '96 Odyssey minivan was a sales dud in spite of its bulletproof reliability, mainly because buyers didn't want a minivan with an underpowered 4-cylinder power plant. Some people buy underpowered vehicles in the mistaken belief that increased fuel economy is a good trade-off for decreased engine performance. It isn't.

Engine and transmission cooling system (heavy-duty)

This relatively inexpensive option provides extra cooling for the transmission and engine. It can extend the life of these components by preventing overheating when heavy towing is required.

Extended warranties

A waste of $1,000–$1,500 for vehicles recommended in *Lemon-Aid* or vehicles sold by automakers that have written goodwill warranties covering engine and transmission failures (Ford and GM). If you can get a great price for a vehicle rated just "Average" or "Above Average," but want to protect yourself from costly repair bills, frequent garages that offer lifetime warranties on parts listed here as failure-prone such as powertrains, exhaust systems, and brakes. Only buy an extended warranty as a last resort and make sure you know what it covers and for how long. Check Auto Warranty Reviews Online (*www.autowarrantyreviews.org*) for information regarding the track record and solvency of American and Canadian companies selling extended warranties.

Keyless entry (remote)

A safety and convenience option. You don't need to fiddle with the key in a dark parking lot or take off a glove in cold weather to unlock or lock the vehicle. Try to get a keyless entry system combined with anti-theft capability, such as an ignition kill switch or some other disabler.

Paint colour

Choosing a popular colour can make your vehicle easier to sell at a good price. DesRosiers Automotive Consultants says blue is the preferred colour overall, but green and silver are also popular with Canadians. Remember that certain colours require particular care. For example:

Black (and other dark colours): These are most susceptible to sun damage because of their heavy absorption of ultraviolet rays.

Pearl-toned colours: These are the most difficult to work with. If the paint needs to be retouched, it must be matched to look right from both the front- and side-angle views.

Red: This also shows sun damage, so keep your car in a garage or shady spot whenever possible.

White: Although grime looks terrible on a white car, it's the easiest colour to care for.

Power-assisted sliding doors, mirrors, windows, and seats

Merely a convenience feature with cars, power-assisted windows, doors, and seats are a necessity with minivans—crawling across the front seat a few times to roll up the passenger-side window or lock the doors will quickly convince you of their value. Power seats with memory are particularly useful if a vehicle is driven by more than one person. Automatic window and seat controls currently have few reliability problems, and they're fairly inexpensive to install, troubleshoot, and repair. Power-sliding doors are especially failure-prone on all makes and shouldn't be purchased by families with children.

Side airbags

A worthwhile feature if you are the right size and properly seated, side airbags are presently way overpriced and aren't very effective unless the head and upper torso are protected. Side airbags are often featured as a $700 add-on to the sticker price, but you would be wise to bargain aggressively, since a supplier source told *Automotive News* that automakers pay about $100 for a pair of side curtain airbags.

Suspension (heavy-duty)

Always a good idea, this inexpensive option pays for itself by providing better handling, allowing additional ride comfort (though a bit on the firm side), and extending shock life an extra year or two.

Tires

NHTSA studies show that 10 percent of the vehicles on the road will likely have at least one bald tire and 50 percent will probably have at least one tire with half-worn tread. Tires with half-worn tread have almost 15 percent less snow traction, longer wet-weather stopping distance, and hydroplane at slower speeds. Interestingly, worn tires perform better on dry pavement than full-tread tires.

There are three rules to remember when purchasing tires. First, neither brand nor price is a reliable gauge of performance, quality, or durability. Second, choosing a tire recommended by the automaker may not be in your best interest, since traction and long tread life are often sacrificed for a softer ride and maximum EPA mileage ratings. And third, don't buy any new tire that's older than two years, since the rubber compound may have deteriorated due to poor handling and improper storage (near electrical motors). You can check the date of manufacture on the side wall.

There are two types of tires: all-season and performance. Touring is just a fancier name for all-season tires. All-season radial tires cost between $90 and $150 per tire. They're a compromise, since according to Transport Canada they won't get you through winter with the same margin of safety as snow tires will, and they don't provide the same durability on dry surfaces as do regular summer tires. In areas with low to moderate snowfall, however, these tires are adequate as long as they're not pushed beyond their limits.

Mud or snow tires provide the best traction on snowy surfaces, but traction on wet roads is actually decreased. Treadwear is also accelerated by the use of softer rubber compounds. Beware of using wide tires for winter driving; 70-series or wider give poor traction and tend to "float" over snow.

Spare tires

Be wary of space-saver spare tires. They often can't match the promised mileage and seriously degrade steering control. Furthermore, they are usually stored in spaces inside the trunk that won't hold a normal-sized tire. Where the spare is stored can also have safety implications. Watch out for spares stowed under the chassis or mounted on the rear hatch on some sport-utilities. Frequently, the attaching cables and bolts rust out or freeze, permitting the spare to fall off or making it next to impossible to access when needed.

Which tires are best?

There is no independent Canadian agency that evaluates tire performance and durability. However, U.S.-based NHTSA rates treadwear, traction, and resistance to sustained high temperatures and etches them onto all tires sold in the States and Canada, and also regularly posts its findings on the Internet (*www.nhtsa.dot.gov*). The treadwear grade is fixed at a base 100 points and the tire's wear rate is measured after the tire is driven through a course that approximates most driving conditions. A tire rated 300 will last three times as long as one rated 100.

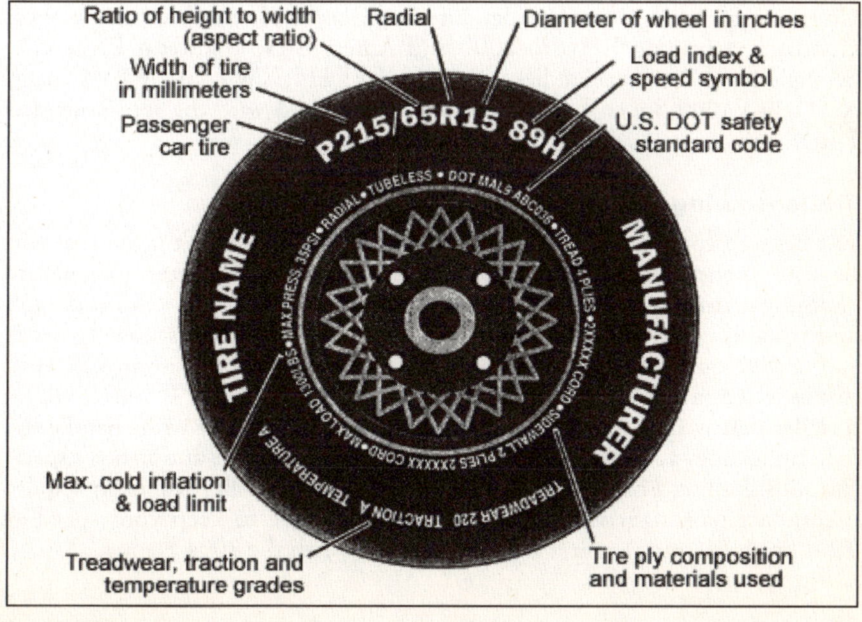

I've come up with the following tire ratings after researching government tests, consumer comments, and industry insiders. Remember, some of the brand names may be changed.

Dunlop D65 Touring and SP20 AS: Treadwear rated 520, these tires are the bargain of the group. They provide excellent wet and dry cornering and good steering response. Cost: $146 list; sells for $100.
Goodrich Control TA M65: Treadwear rated 360, this tire excels at snowbelt performance and is bargain priced at $120.
Goodyear Aquatred #3: Treadwear rated 340, this tire is a bit noisy and its higher-rolling resistance cuts fuel economy. Still, it's an exceptional performer on wet roads and works especially well on front-drive cars where the weight is over the front tires. Average performance on dry pavement. Costs about $130; no discounts.
Goodyear Regatta #2: Treadwear rated 460, this $100 tire does everything well, including keeping tire noise to a minimum.
Michelin MX4: Treadwear rated 320, this tire gives a smoother ride and a sharper steering response than the Aquatred. Cost: $144 list; often discounted to $100. MX1 is also an excellent winter performer.
Pirelli P300 and P400: Treadwear rated 460 and 420, these are two of the best all-around all-season tires. Cost: $173 list; sells for about $120.
Yokohama TC320: Treadwear rated 300, these are two other Aquatred knock-offs that perform almost as well for half the price.

Other good tire choices: Dayton Timberline A/T, General XP, Goodrich Touring T/A HR4, Goodyear Eagle LS, Kelly Navigator Platinum TE, Pirelli P3000, and Scorpion A/T. Winter tires: Futura Euro-Metric, Goodyear Ultra Grip Ice, Michelin Arctic Alpin, and Pirelli Winter Ice Assimmetrico.

The following tires aren't recommended: Bridgestone Potenza RE 92; Continental truck tires; Cooper Lifeline Classic II; all Firestone makes; Goodrich Advantage; Hydro 2000 and Ameri G4S; Goodyear Eagle GA, Wrangler, and WeatherHandler; Michelin XGT H4, XW4, and MXV4 Green X; Pirelli P4000 Super Touring (not to be confused with the recommended P400); and Toyo 800 Plus.

Trailer-towing equipment

Just because you need a vehicle with towing capability doesn't mean that you have to spend big bucks. The first things you should determine before choosing a towing option are whether a pickup or small van will do the job and whether your tires will handle the extra burden. For most towing needs (up to 900 kg/2,000 lb.), a passenger car, small pickup, or minivan will work just as well as a full-sized pickup or van (and cost much less). If you're pulling a trailer that weighs more than 900 kg, most passenger cars won't handle the load unless they've been specially outfitted according to the automaker's specifications. Pulling a heavier trailer (up to 1,800 kg/4,000 lb.) will likely require a compact passenger van. You may, however, have to keep your speed at 72.4 km/h (45 mph) or less as Toyota suggests with the 2004 Sienna.

Automakers reserve the right to change limits whenever they feel like it, so make any sales promise an integral part of your contract (see "False Advertising" in Part Two). A good rule of thumb is to reduce the promised tow rating by 20 percent. In assessing towing weight, factor in the cargo, passengers, and equipment of both the trailer and the tow vehicle. Keep in mind that five people and luggage add 450 kg (1,000 lb.) to the load, and that a full 227-litre (50-gallon) water tank adds another 225 kg (500 lb.). The manufacturer's gross vehicle weight rating (GVWR) takes into account the anticipated average cargo and supplies that your vehicle is likely to carry.

Automatic transmissions are fine for trailering, although there's a slight fuel penalty to pay. Manual transmissions tend to have greater clutch wear due to towing than do automatic transmissions. Both transmission choices are equally acceptable. Remember, the best compromise is to shift the automatic manually, for maximum performance going uphill and to maintain control while not overheating the brakes when descending mountains.

Unibody vehicles (those without a separate frame) can handle most trailering jobs as long as their limits aren't exceeded. Front-drives aren't the best choice for pulling heavy loads in excess of 900 kg (2,000 lb.), since they lose some steering control and traction with all the weight concentrated in the rear.

Whatever vehicle you choose, keep in mind that the trailer hitch is crucial. It must have a tongue capacity of at least 10 percent of the trailer's weight; otherwise it may be unsafe to use. Hitches are chosen according to the type of tow vehicle and, to a lesser extent, the weight of the load.

Most hitches are factory-installed, even though independents can install them more cheaply. Expect to pay about $200 for a simple boat hitch and a minimum of $600 for a fifth-wheel version.

Equalizer bars and extra cooling systems for the radiator, transmission, engine oil, and steering are a prerequisite for towing anything heavier than 900 kg (2,000 lb.). Heavy-duty springs and brakes are a big help, too. Separate brakes for the trailer may be necessary to increase your vehicle's maximum towing capacity.

Transmission: automatic, manual, 5-speed, and overdrive automatic

A transmission with four or more forward speeds is usually more fuel efficient than one with three forward speeds, and manual transmissions are usually more efficient than automatics, although this isn't always the case. Nevertheless, most motorists prefer to pay extra to have a transmission that shifts by itself, even though this convenience saps the performance of small engines, requires expensive repairs, and makes you less alert to driving conditions. This last point is particularly important, because a manual transmission makes you aware of the traffic flow and requires that you shift gears in anticipation of changes.

Despite its many advantages, the manual transmission is an endangered species in North America, where manuals equip only 12.2 percent of all new vehicles (mostly econocars, sports cars, and budget trucks), whereas European buyers opt for a manual transmission almost 90 percent of the time.

Unnecessary Options

Most luxury cars like the Mercedes E-Class and GM's 2004 DeVille (above) offer automatic levelling suspensions.

Automatic level control

A useless option, unless you're planning to carry heavy loads or pull a trailer. It's expensive to repair and not easily adjusted.

Cruise control

Mainly a convenience feature; auto-makers provide this $250–$300 option to motorists who use their vehicles for long periods of high-speed driving. Some fuel is saved, owing to the constant rate of speed, and driver fatigue is lessened during long trips. Still, the system is particularly failure-prone and expensive to repair, can lead to driver inattention, and can make the vehicle hard to control on icy roadways. Malfunctioning cruise control units are also one of the major causes of sudden acceleration incidents. At other times, cruise control can be very distracting, especially to inexperienced drivers unaccustomed to sudden speed fluctuations.

Electronic instrument readout

If you've ever had trouble reading a digital watch face or resetting your VCR, you'll feel right at home with this electronic gizmo. Gauges are presented in a series of moving digital patterns that are confusing, distracting, and unreadable in direct sunlight. This system is often accompanied by a trip computer and a vehicle monitor that computes fuel use and kilometres to empty, indicates average speed, and signals component failures. Figures are frequently in error or slow to catch up.

Foglights

A pain in the eyes for other drivers, foglights aren't necessary for most drivers with well-aimed original-equipment headlights.

Gas-saving gadgets/fuel additives

Because full-sized sport-utilities, vans, and trucks are notorious fuel burners, the accessory market has been flooded with hundreds of gas-saving gadgets and fuel additives that purport to make these vehicles less fuel-thirsty. There isn't one on the Canadian market that works, according to Transport Canada, and the use of any of these products is a quick way to render your warranty invalid.

ID etching

This \$150–\$200 option is a scam. The government doesn't require it and thieves and joyriders aren't deterred by the etchings. If you want to etch your windows for your own peace of mind, several private companies will sell you a \$15–\$30 kit that does an excellent job (*www.autoetch.net*), or you can wait for your municipality or local police agency to conduct one of their periodic free VIN ID etching sessions in your locality.

GPS navigation systems

This navigation aid links a GPS satellite unit to the vehicle's cellular phone and electronics. For a monthly fee, the unit connects drivers to live operators who will help them with driving directions, give repair or emergency assistance, or relay messages. If the airbag deploys or the car is stolen, satellite-transmitted signals are automatically sent from the vehicle to operators who will notify the proper authorities of the vehicle's location.

Many of the systems are obtrusive, distracting, washed out in sunlight, and hard to calibrate. A portable Garmin GPS unit is one of the more user-friendly models that's also reasonably priced.

Night Vision

Cadillac's DeVille Night Vision uses an infrared camera to project the heat images of animals, pedestrians, and other cars on your windshield that may be just beyond the reach of your headlights. In theory, the \$2,000 option is a significant safety improvement. In practice, however, it's distracting, works poorly on twisting, mountainous roads, and presents a grainy, ill-defined image.

Paint and fabric protectors

Factory paint jobs have become less and less durable over the years, as automakers cheapen their product. Once a problem mainly with Chrysler, Ford, and GM vehicles, now many of Honda's American-built Accords and Mazda's new models are showing premature paint delamination and peeling within the first three years of ownership.

Don't kid yourself, though. Paint protection products sold by auto dealers won't make much of a difference. They aren't just overpriced—they also don't work. Selling for $200–$300, these "sealants" add nothing to a vehicle's resale value. Although paint lustre may be temporarily heightened, this treatment is less effective and more costly than regular waxing, and it may also invalidate the manufacturer's guarantee at a time when the automaker will look for any pretext to deny your paint claim.

Consumer Reports and the Canadian Automobile Protection Association believe that auto fabric protection products are nothing more than Scotchguard variations, which can be bought in aerosol cans for a few dollars, instead of the $50–$75 charged by dealers.

Power-assisted minivan sliding doors

Not a good idea if you carry children. These doors have a high failure rate: opening or closing for no apparent reason and injuring children caught between the door and post.

Rooftop carrier

Although this inexpensive option provides additional baggage space and may allow you to meet all your driving needs with a smaller vehicle, a loaded roof rack can increase fuel consumption by as much as 5 percent. An empty rack cuts fuel economy by about 1 percent.

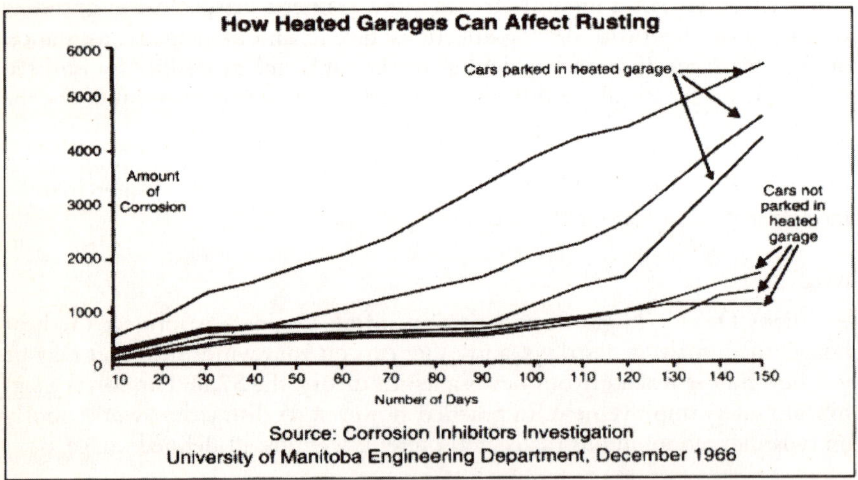

In 1966, University of Manitoba researchers proved heated garages accelerate rusting.

Rustproofing

Rustproofing is no longer necessary now that the automakers have extended their own rust warranties. In fact, you have a greater chance of seeing your rustproofer go belly up than having your untreated vehicle ravaged by premature rusting. Even if the rustproofer stays in business, you're likely to get a song and dance about why the warranty won't cover so-called "internal" rusting, or why repairs will be delayed until the sheet metal is actually rusted through.

If you live in an area where roads are heavily salted in winter, or in a coastal region, have your vehicle washed every few weeks and undercoated annually, paying particular attention to rocker panels (door bottoms) and wheelwells. Also, don't use a heated garage in winter; Canadian studies show that a heated garage will accelerate the damage caused by corrosion.

Sunroof

Unless you live in a temperate region, the advantages of a sunroof are far outweighed by its disadvantages. You're not going to get better ventilation than a good air-conditioning (AC) system would provide, and a sunroof may grace your environment with painful wind noises, rattles, water leaks, and road dust accumulation. Gas consumption is increased, night vision is reduced by overhead highway lights shining through the roof opening, and several inches of headroom can be lost. Flip-up styles are particularly prone to leaks, while electronically controlled sliding sunroofs often fall prey to short circuits. Without a manual override, your car will be vulnerable to theft and the weather.

Factory installation of a sunroof is far more costly than having it done by an independent ($1,000 vs. $250), with little difference in the quality of the job or the warranty.

Tinted glass

Tinting jeopardizes your safety by reducing night vision. On the other hand, it does keep the interior cool in hot weather, reduces glare, and hides the car's contents from prying eyes. Factory applications are worth the extra cost, since cheaper aftermarket products (costing about $100) distort visibility and peel away after a few years. Some tinting done in the United States can run afoul of provincial highway codes that require more transparency.

Cutting the Price

Buy by Internet or fax—avoid the "showroom shakedown"

Of the 15.1 million new vehicles sold last year in North America, 2 million were sold over the Internet. This figure isn't all that surprising when you consider that hundreds of Internet-based new-car shopping services (like Autobytel) have sprung up during the past several years. These services are for consumers who want a fair price, loathe haggling, or are just too busy to visit a number of showrooms. For a service fee, these firms encourage dealers to bid against each other for your business through electronic mail. In other cases, you can go directly to a number of dealer websites and initiate the bidding process on your own. Shoppers report modest savings, particularly in regions where there aren't that many dealers and the local dealer has a "take it or leave it" attitude.

J.D. Power's 2003 Sales Satisfaction Index study says that a quarter of the customers who left a dealership without buying did so because they didn't like the sales staff's attitude. Even worse, half of those customers who were unhappy with the way they were treated bought a vehicle from another automaker. Isuzu, Kia, Subaru, Suzuki, and Toyota dealers had the worst sales satisfaction scores.

Fax bidding

Huge price cuts are only a fax machine away. Simply fax an invitation for bids (a cover letter with your company logo would help) to area dealerships, asking them to give their bottom-line price for a specific make and model and clearly stating that all final bids must be faxed within a week. Because no salesperson is acting as a commission-paid intermediary, the dealers' first bids are likely to start off a few hundred dollars less than advertised. When all the bids are received, the lowest bid is faxed to the other dealers to give them a chance to beat that price. After a week of bidding, the lowest price gets your business.

This simple idea really works.

Dozens of *Lemon-Aid* readers have told me how buying by fax kept the price down and dispensed with the showroom song-and-dance routine between the sales agent and the sales manager.

This *Lemon-Aid* reader, from Alberta (a tough province for dealer-haggling), paid a thousand dollars less than the list price for a 2003 Mazda6.

> I took an approach highlighted in your *Lemon-Aid* guide to buying new cars and minivans. After narrowing my selection down to a new Mazda6, I faxed out a request to every Mazda dealer in Alberta. The dealer I had been talking to in Calgary flatly stated that there was no negotiating room on the price, because the car was very new to the market, and was very popular. After receiving quotes from several dealers, I selected the winning dealer, which was out of Edmonton, with a winning bid almost $1,100.00 lower than the Calgary dealer. The sales process went smoothly, and the car is great. I also noticed a couple of other things as a result of going the fax approach:
>
> - The Calgary area Mazda dealers generally ignored my fax. The one salesman I had been talking to emailed me a quote, for sticker price, plus the dealer extras. I later heard that many of the dealerships are owned by one family, with various brothers owning various dealerships. So much for competition.
> - Every Edmonton and area dealer responded to my fax. They varied from $300 in savings to almost $1,300 in savings. (The lowest bid did not get selected due to a mistake on my part, as I had assumed they had not included a couple of extras.) Most of the responses were on the same day my fax went out.
> - The Lethbridge dealer responded to my fax a week late. There must have been some mixup, or they have the slowest fax machine in the world, so they missed the opportunity for my business.

- Several dealers inquired why I was not dealing with the Calgary dealers, but after I mentioned that I was simply looking for a better price, they responded with a quote. (Some grudgingly responded.)
- Dealers expect to get jerked around by people doing the fax best bid process, so if you treat them with fairness, they will be pleasantly surprised.

P.J.

Fax Bid Request

Date

Attention: New Car Sales Manager:

I am planning on buying a new____________and am requesting a price quote for the above vehicle with the following equipment:

* __________________________
* __________________________
* __________________________
* __________________________
* __________________________

I am interested in the following colours, in order of preference:

* __________________________
* __________________________

Please include the price of all items, and the extras, such as tire tax, air conditioning tax, delivery, and inspection, etc.

I am sending this fax to several dealers, and the best quote will get my business. I am willing to drive out of town if the savings warrant.

If you are interested in selling me a car, please send me a quote by fax or email by ______________. I shall give you a second opportunity to beat the lowest bid within the week and will contact the winning dealer to confirm the order.

I am flexible to purchase the car off your lot, or on order, whatever works best for you.

Thank you in advance for your assistance.

Jane Consumer
Fax: 555-555-5555

The myth of "no haggle" pricing

It's all smoke and mirrors. In effect, all dealers bargain. They hang out the "No dickering, one price only" sign simply as a means to discourage customers from asking for a better deal. Like parking lots and restaurants that claim they won't be responsible for lost or stolen property, they're bluffing. Still, you'd be surprised by how many people believe that if it's posted, it's non-negotiable.

There are several price guidelines, however, and dealers use the one that will make the most profit on each transaction. Two of the more common prices quoted are the MSRP (what the automaker advertises as a fair price) and the dealer's invoice cost, which is supposed to indicate how much the dealer paid for the vehicle. Both price indicators leave considerable room for the dealer's profit margin, along with some extra padding in the form of inflated transportation and preparation charges. If presented with both figures, go with the MSRP, since it can be verified by calling the manufacturer. Any dealer can print up an invoice and swear to its veracity. If you want an invoice price from an independent source, contact Carcostcanada.com (at 1-800-805-2270).

Buyers who live in rural areas and Western Canada are often faced with grossly inflated auto prices compared to those charged in major metropolitan areas. A good way to get a more competitive price without buying out of province is to buy a couple of out-of-town newspapers (the Saturday *Toronto Star* "Wheels" section is especially helpful) and demand that your dealer bring his selling price, preparation, and transportation fees into line with the prices advertised.

Another tactic is to take a copy of a local competitor's car ad to a competing dealer selling the same brand and ask for a better price. Chances are, they've already lost a few sales due to the ad, and will work a little harder to match the deal; if not, they're almost certain to reveal the tricks in the competitor's promotion to make the sale.

Getting a Fair Price

What's the dealer's cut?

Most new car salespeople are reluctant to give out information on the amount of profit figured into the cost of each new car, but a few years back *Automotive News* gave American dealer markups based on MSRP. I've reduced the percentages a bit to reflect what Canadian dealers now receive and I've included negotiable freight, PDI, and administrative fees, which you should bargain down.

Dealer Markup (American Vehicles)

small cars: 10–12%
mid-sized cars: 15–18%
large cars: 20%
sports cars: 17–20%
high-end sports cars: 20+%

luxury cars: 22+%
high-end luxury cars: 24+%
minivans: 16+%
high-end minivans: 17+%
base pickups: 14+%
high-end pickups: 16+%
vans and sport-utility vehicles: 16+%
fully equipped, top-of-the-line vans and SUVs: 20+%

Dealer Markup (Japanese Vehicles)

small cars: 8–12%
mid-sized cars: 10–15%
large cars: 15–17%
sports cars: 15–17%
high-end sports cars: 18+%
luxury cars: 18%
high-end luxury cars: 20+%
minivans: 11+%
high-end minivans: 13+%
base pickups: 12+%
high-end pickups: 14+%
vans and sport-utility vehicles: 13+%
fully equipped, top-of-the-line vans and SUVs: 15+%

2004 Cavalier Z24 Sedan, Chevrolet

Option	Retail Price
OnStar - requires US8 - includes one year safe and sound pland (UE1)	$995
side impact airbags - seated mounted driver and fron passenger (AJ7)	$515
transmission - 4 speed automatic with overdrive (MXO)	$1,180
wheel equipment - 16 x 6 chromed cast aluminum (PFC)	$845
Freight	$850
Air tax	$0
Base price	$21,730
Total	$26,115

The 2004 Cavalier will be virtually unchanged except for a name change to the Cobalt.

European vehicle markups are slightly lower than those charged by American dealers. In addition to the dealer's markup, some Detroit Big Three vehicles from previous years may also have a 3 percent carryover allowance paid out in a dealer incentive program.

Holdback

Ever wonder how dealers who advertise vehicles for "a hundred dollars over invoice" can make a profit? They are counting mostly upon the manufacturer's holdback.

In addition to the MSRP, the invoice price, dealer incentives, and customer rebates (available to Canadians from the *www.apa.ca* website), another key element in every dealer's profit margin is the manufacturer's holdback—quarterly payouts dealers depend upon when calculating gross profit.

The holdback was set up almost 40 years ago by General Motors as a guaranteed profit for dealers tempted to bargain away their entire profit to make a sale. It usually represents 1–3 percent of the sticker price (MSRP) and is seldom given out by Asian or European automakers, who use dealer incentive programs instead. There are several free Internet sources for holdback information: the most recent and comprehensive are *www.edmunds.com* and *www.kbb.com*, two websites geared toward American buyers. Although there may be a difference in the holdback percentage between American automakers and their Canadian subsidiaries, it's usually not significant.

Some GM dealers maintain that they no longer get a holdback allowance. They are being disingenuous—the holdback may have been added to special sales "incentive" programs, which won't show up on the dealer's invoice. Options are the icing on the cake, with their average 35–65 percent markup.

Can you get a fair price?

Yep, but you'll have to keep your wits about you and time your purchase well into the model year—usually in late winter or fall.

New-car negotiations aren't wrestling matches where you have to pin the sales agent's shoulders to the mat to win. If you feel that the overall price is fair, don't jeopardize the deal by giving no quarter. For example, if you've brought the contract price 10 percent or more below the MSRP and the dealer sticks you with a $200 "administrative fee" at the last moment, let it pass. You've saved money and the sales agent has saved face.

Of course someone will always be around to tell you how he or she could have bought the vehicle for much less. Let that pass, too.

To come up with a fair price, subtract one-half the dealer markup from the MSRP and trade the carryover and holdback allowance for a reduced delivery and transportation fee. Compute the options separately and sell your trade-in privately. Buyers can more easily knock $1,000–$2,000 off a $20,000 base price if they wait until January or February when sales are stagnant, choose a vehicle in stock, and resist unnecessary options.

Once you and the dealer have settled on the vehicle's price, you aren't out of the woods yet. Like a tag-team wrestling match, you'll then be handed over to an F&I (financing and insurance) specialist, whose main goal is to convince you to buy additional financing, loan insurance, paint and seatcover protectors, rustproofing, and extended warranties. These items will be presented on a computer screen as costing only "a little bit more each month."

Compare the dealer's insurance and financing charges with an independent agency that may offer better rates and better service. Often the dealer gets a kickback for selling insurance and financing, and guess who pays for it? Additionally, remember that if the financing rate looks too good to be true, you're probably paying too much for the vehicle. The F&I closer's hard-sell approach will take all your willpower and patience to resist, but when he gives up, your trials are over.

Add-on charges are the dealer's last chance to stick it to you before the contract is signed. Dealer pre-delivery inspection (PDI) and transportation charges, "documentation" fees, and extra handling costs are ways that the dealer gets extra profits for nothing. Dealer preparation is often a once-over-lightly affair, with a car seldom getting more than a wash job and a couple of dollars of gas in the tank. It's paid for by the factory in most cases and, when it's not, should cost no more than 2 percent of the car's selling price. Reasonable transportation charges are acceptable, although dealers who claim that the manufacturer requires the payment often inflate them.

Dealer incentives and customer rebates

When vehicles are first introduced in the fall they're generally overpriced. Later on, near the end of summer, automakers offer customer cash rebates of $750–$3,000, plus increased dealer cash incentives for almost as much. Incentives are first offered during the winter months and boosted in late summer or early fall when dealer showroom traffic has fallen off. Smart shoppers who buy during these months can shave an additional 10 percent off a vehicle's list price (about $3,000 on a $30,000 MSRP) by double-dipping from both of these automaker rebate programs or by using zero percent financing.

In most cases, the manufacturer's rebate is straightforward and mailed directly to the buyer from the automaker. There are other rebate programs that require a financial investment on the dealer's part, however, and these shared programs tempt dealers to offset losses by inflating the selling price or pocketing the manufacturer's rebate. Therefore, when the dealer participates in the rebate program, demand that the rebate be deducted from the MSRP, and not from some inflated invoice price concocted by the dealer.

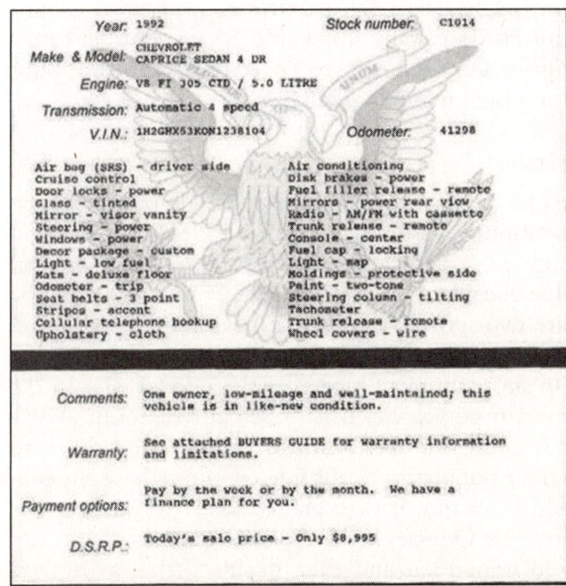

Year: 1992 Stock number: C1014

Make & Model: CHEVROLET CAPRICE SEDAN 4 DR

Engine: V8 FI 305 CID / 5.0 LITRE

Transmission: Automatic 4 speed

V.I.N.: 1H2GHX53KON1238104 Odometer: 41298

Air bag (SRS) - driver side	Air conditioning
Cruise control	Disk brakes - power
Door locks - power	Fuel filler release - remote
Glass - tinted	Mirrors - power rear view
Mirror - visor vanity	Radio - AM/FM with cassette
Steering - power	Trunk release - remote
Windows - power	Console - center
Decor package - custom	Fuel cap - locking
Light - low fuel	Light - map
Mats - deluxe floor	Moldings - protective side
Odometer - trip	Paint - two-tone
Seat belts - 3 point	Steering column - tilting
Stripes - accent	Tachometer
Cellular telephone hookup	Trunk release - remote
Upholstery - cloth	Wheel covers - wire

Comments: One owner, low-mileage and well-maintained; this vehicle is in like-new condition.

Warranty: See attached BUYERS GUIDE for warranty information and limitations.

Payment options: Pay by the week or by the month. We have a finance plan for you.

D.S.R.P.: Today's sale price - Only $8,995

Phony price stickers like these are sold on the Internet.

Some rebate ads will include the phrase "from dealer inventory only." If your dealer doesn't have the vehicle in stock, you won't get the rebate. Keep in mind that the manufacturer's rebate is considered to be part of the fair value and is not deductible from the purchase price prior to determining the payable retail sales tax.

Sometimes automakers will suddenly decide that a rebate no longer applies to a specific model, even though their ads continue to include it. When this happens, take all brochures and advertisements showing your eligibility for the rebate plan to provincial consumer protection officials. They can use false advertising statutes to force automakers to give rebates to every purchaser who was unjustly denied one.

If you are buying a heavily discounted vehicle, be wary of "option packaging" by dealers who push unwanted protection packages (rustproofing, paint sealants, and upholstery finishes), or who levy excessive charges for preparation, filing fees, loan guarantee insurance, and credit life insurance.

Rebate "false bargains"

To come out ahead, you have to know how to play the rebate game. Customer and dealer incentives are frequently given out to stimulate sales of year-old models that are unpopular, scheduled to be redesigned, or headed for the axe. By carefully choosing which rebated models you buy, it's easy to realize important savings with little downside risk. For example, the 2004 Toyota Sienna has been substantially improved over the 2003 model. If you plan on keeping the car five years or longer, get the upgraded 2004 version and amortize the extra cost. If you expect to trade in the minivan after a few years, go for the rebated 2003 model and pocket the upfront savings.

But some rebate deals can be false bargains, like the steep discounts and other incentives offered with Chrysler's failure-prone, mid-sized sedans (Intrepid and Concorde) and Ford's equally unreliable Focus, Taurus, Sable, and Freestar. The cheaper price doesn't begin to cover these vehicles' poor quality, rapid depreciation, and high maintenance costs.

Inflated or deflated prices

Generally, vehicles are priced according to what the market will bear and then are discounted as the competition heats up (Chrysler and GM minivans and Ford trucks are prime examples). A vehicle's stylishness, scarcity, or general popularity can inflate its value considerably. For example, Pontiac's 2004 Vibe and Toyota's 2004 Matrix are two sporty wagons that have prevailing market values higher than their suggested selling price, mainly because they're leaders in a hot market niche and in short supply. Once sales slow down later in the new year, popular though overpriced vehicles usually sell at a discount (VW's New Beetle, for example). A good rule of thumb for vehicles that have an inflated value is to wait for their popularity to subside or to purchase the previous year's version. On the other hand, if your choice has a deflated market value (like underpowered first-year Odysseys or overly maligned South Korean models), find out why it's so unpopular and then decide if the savings are worth it. Vehicles that don't sell because of their weird styling are no problem, but poor quality control (think Chrysler Neon) can cost big bucks.

Leftovers

In the fall, at the beginning of each new model year, most dealers still have a few of last year's vehicles left. Some are new and some are demonstrators (that is, registering a few thousand kilometres, but never titled to a third party). The factory gives the dealer a 3–5 percent rebate on late-season vehicles, and dealers will often pass on some of these savings to clients. But are these leftovers really bargains?

They might be, if you can amortize the first year's depreciation by keeping the vehicle for six to 10 years. But if you're the kind of driver who trades every two or three years, you're likely to come out a loser by buying an end-of-the-season van or sport-utility vehicle. The simple reason is that as far as trade-ins are concerned, a leftover is a used vehicle that has depreciated at least 20 percent. The savings the dealer gives you may not equal that first year's depreciation (a cost you'll incur without getting any of the first year's driving benefits). If the dealer's discounted price matches or exceeds the 20 percent depreciation, then you're getting a pretty good deal. But if the next year's model is only a bit more expensive, has been substantially improved, or is covered by a more extensive, comprehensive warranty, it could represent a better buy than a cheaper leftover.

Ask the dealer for all work orders relating to the vehicle and make sure that the odometer readings follow in sequential order. Remember as well that most demonstrators should have less than 5,000 km on the ticker, and the original warranty has been reduced from the day the vehicle was first put on the road. Have the dealer extend the warranty or lower the price accordingly—about $100 for each month of warranty that has expired. If the vehicle's file shows that it was registered to a leasing agency or any other third party, you're definitely buying a used vehicle disguised as a demo. You should walk away from the sale because you're dealing with a crook.

Mid-year models

Entirely new models often make their debut in midsummer, and getting any kind of a discount from the dealer for mid-year models is like pulling teeth. Dealers know that a seller's market exists during the first few months of a model's launch, and they're not likely to sell their few sample vehicles for anything less than full price. Furthermore, you're likely to lose an extra year of depreciation, since mid-year vehicles are depreciated back to the beginning of the model year in which they were introduced.

Cash vs. Financing

Let's clear up one myth right away: dealers won't treat you better if you pay cash. They want you to buy a fully loaded vehicle and finance the whole deal. Paying cash is not advantageous to the dealer, since kickbacks on finance contracts represent an important part of the F&I division's profits. Actually, barely 8 percent of new-car buyers pay cash. They may be making a big mistake. Financial planners say it may be smarter to borrow the money to purchase a new vehicle even if you can afford to pay cash, because if you use the vehicle for business a portion of the interest may be tax deductible. The cash that you free up can then be used to repay debts that aren't tax deductible (mortgages or credit card debt, for example).

Rebates vs. low or zero percent financing

If you are buying an expensive vehicle, like a luxury car or an SUV, and going for longer financing, the low-rate financing will be a better deal than the rebate. A zero percent loan will save you $80 per $1,000 financed over 24 months, or $120 per $1,000 financed over 36 months, compared with 7.5 percent financing. If you were financing a $30,000 car for two years, you'd multiply $80 x 30 and save about $2,400.

Low-financing programs have the following disadvantages:

- Buyer must have exceptionally good credit.
- Shorter financing period means higher payments.
- Cash rebates are excluded.
- Only fully equipped or slow-selling models are eligible.
- Buyer pays full retail price.

Remember, to get the best price, whether you're paying cash or financing the purchase, first negotiate the price of the vehicle without disclosing whether you are paying cash or financing the purchase (simply say you haven't yet decided). Once you have a fair price, then take advantage of the financing.

Getting a loan

Borrowers must be at least 18 years old (age of majority), have a steady income, prove that they have discretionary income sufficient to make the loan payments, and be willing to guarantee the loan with additional collateral, or their parent or spouse as a co-signer.

Before applying for a loan, you should have established a good credit rating via a paid-off credit card and have a small savings account with your local bank, credit union, or trust company. Then, prepare a budget listing your assets and obligations. This will quickly show whether or not you can afford a car. Next, pre-arrange your loan with a simple phone call. This will protect you from much of the smoke and mirrors showroom shenanigans.

Incidentally, if you do get in over your head and require credit counselling, contact Credit Counselling Service (CCS), a non-profit organization located in many of Canada's major cities (*www.creditcanada.com*).

Hidden loan costs

Don't trust anyone. The Montreal-based Automobile Protection Association's undercover shoppers have found the most deceptive deals often involve major banking institutions, rather than automaker-owned companies.

In your quest for an auto loan, remember that the Internet offers help for people who need an auto loan and want quick approval, but don't like to face a banker. The Bank of Montreal (*www.bmo.com*) was the first Canadian bank to allow vehicle buyers to post a loan application on its website, and it promises to send a loan response within 20 seconds. Other banks, such as the Royal Bank, are offering a similar service. Loans are available to any web surfer, including those who aren't current Montreal or Royal customers.

Be sure to call various financial institutions to find out the following:

- The annual percentage rate on the amount you want to borrow and for the duration of your repayment period
- The minimum down payment that the institution requires
- Whether taxes and licence fees are considered part of the overall cost and, thus, are covered by part of the loan
- Whether lower rates are available for different loan periods or for a larger down payment
- Whether discounts are available to depositors and, if so, how long you must be a depositor before qualifying

When comparing loans, consider the annual rate and calculate the total cost of the loan offer; that is, how much you'll pay above and beyond the total price of the vehicle.

Dealers can finance your purchase at interest rates that are competitive with the banks' because of the rebates they get from the manufacturers and some lending institutions. Some dealers, though, mislead their customers into thinking they can borrow money at as much as five percentage points below the prime rate. Actually, they're jacking up the retail price to more than make up for the lower interest charges. Sometimes, instead of boosting the price, dealers reduce the amount they pay for the trade-in. In either case, the savings are illusory.

When dealing with banks, keep in mind that the traditional 36-month loan has now been stretched to 48 or 60 months. Longer payment terms make each month's payment more affordable, but over the long run they increase the cost of the loan considerably. Therefore, take as short a term as possible.

Be wary of lending institutions that charge a "processing" or "document" fee ranging from $25 to $100. Sometimes consumers will be charged an extra 1–2 percent of the loan up front in order to cover servicing. This is similar to lending institutions adding "points" to mortgages, except that with auto loans it's totally unjustified.

Some banks will cut the interest rate if you're a member of an automobile owners association or if loan payments are automatically deducted from your chequing account. This latter proposal may be costly, though, if the chequing-account charges exceed the interest-rate savings.

Finance companies affiliated with GM, Ford, and Chrysler have been offering low-interest loans many points below the prime rate. In many cases, this low rate is applicable only to hard-to-sell models or vehicles equipped with expensive options. The low rate frequently doesn't cover the entire loan period. If vehicles recommended in this book are covered by low-interest loans, however, then the automaker-affiliated finance companies become a useful alternative to regular banking institutions.

Loan protection

Credit insurance guarantees that the vehicle loan will be paid if the borrower becomes disabled or dies. There are three basic types of insurance that can be written into an installment contract: credit life, accident and health, and comprehensive. Most bank and credit union loans are already covered by some kind of loan insurance, but dealers sell the protection separately at an extra cost to the borrower. For this service, the dealer gets a hefty 20 percent commission. The additional cost to the purchaser can be significant. The federal 7 percent GST is applied to loan insurance, but PST may be exempted in some provinces.

Collecting on these types of policies isn't easy. There's no payment if your illness is due to some condition that existed prior to your taking out the insurance. Nor will the policy cover strikes, layoffs, being fired, etc. Generally, credit insurance is unnecessary if you're in good health, you have no dependents, and your job is secure.

The Royal Bank has two interesting loan programs. One protects motorists from depreciation losses arising from an accident. For example, if a vehicle is scrapped after an accident, the Royal Bank will reimburse its depreciated value. The other program keeps monthly payments low, except for a final balloon payment.

There are plenty of advantageous programs available elsewhere. Personal loans from financial institutions now offer lots of flexibility. Most offer financing (with a small down payment), fixed or variable interest rates, a choice of loan terms, and no penalties for prepayment. Precise conditions depend on your personal credit rating. Finally, credit unions can also underwrite new vehicle loans that combine a flexible payment schedule with low rates.

Leasing contracts are less flexible. There's a penalty for any prepayment, and rates aren't necessarily competitive.

Financing was turned down

This scam usually begins after you have purchased the car and left your trade-in with the dealer. A few days later you are told that your loan was rejected and that you now must put down a larger down payment and accept a higher monthly payment. Of course, your trade-in has already been sold.

Protect yourself from this rip-off by getting a signed agreement that stipulates the financing has been approved and that monthly payments can't be readjusted. Don't give up your trade-in until that agreement has been reached.

Dealer offers to pay off your existing lease or loan

He will. And then he'll add what he paid to your new loan at a much higher interest rate. Plus, early termination of your lease will likely expose you to substantial penalty costs.

Negotiating the Contract

How likely are you to be cheated when buying a new car? Automobile Protection Association (APA) staffers posing as buyers visited 42 dealerships in four Canadian cities in early 2002. The APA says that almost half the dealers (45 percent) they visited flunked their test, and (hold onto your cowboy hats) auto buyers in Western Canada are especially vulnerable to dishonest dealers.

In Vancouver and Edmonton, dealer ads left out important information, vehicles in the ads weren't available, or they were selling at higher prices. Fees for paperwork and vehicle preparation were frequently excessive, with Chrysler dealerships in Vancouver and Toronto charging the most ($299–$632). In some cases, the dealers may have double-billed the buyer. In Toronto, pre-delivery charges of $343 and $89 were levied on top of Chrysler's $955 PDI/transport fee.

Chrysler and Ford dealerships performed the worst overall, Toyota and General Motors performed best, and Mazda and Hyundai dealers were mediocre (they charged extra for items other automakers include in the base price). The APA found Toyota dealers demonstrated a superior level of product knowledge, covered all the bases more consistently, and applied the least pressure to make a sale. But Toyota dealers in the regions with ACCESS fixed selling prices appeared to charge substantially more.

Watch what you sign, since any document that requires your signature is a contract. Don't sign anything unless all the details are clear to you and all the blanks have been filled in. Don't accept any verbal promises that you're merely putting the vehicle on hold. And when you are presented with a contract, remember, it doesn't have to include all the clauses found in the dealer's preprinted form. You and the sales representative can agree to strike some clauses and add others.

When the sales agent asks for a deposit, make sure that it's listed on the contract as a deposit and try to keep it as small as possible (a couple hundred dollars at the most). If you decide to back out of the deal on a vehicle taken from stock, let the seller have the deposit as an incentive to cancel the contract (believe me, it's cheaper than a lawyer and probably equal to the dealer's commission).

Scrutinize all references to the exact model (there is a heck of an upgrade from base to LX or Limited), prices, and delivery dates. Delivery can sometimes be delayed three to five months, and you'll have to pay all price increases announced during the interim unless you specify a delivery date in the contract, which protects the price.

Make sure that the contract indicates that your new vehicle will be delivered to you with a full tank of gas. Once this was the buyer's responsibility, but now, with drivers spending almost $30,000 for the average new vehicle, dealers usually throw in the tank of gas.

Essential clauses

You can put things on a more equal footing by negotiating the inclusion of as many clauses as possible from the sample additional contract clauses found below. To do this, write in a "Remarks" section on your contract, and add "See attached clauses, which form part of this agreement." Then attach a photocopy of the "Additional Contract Clauses" page and persuade the sales agent to initial as many of the clauses as possible. Although some clauses may be rejected, the inclusion of just a couple of them can have important legal ramifications later on if you want a full or partial refund.

Additional Contract Clauses

1. **Original contract:** This is the ONLY contract; i.e., it cannot be changed, retyped, or rewritten, without the specific agreement of both parties.
2. **Financing:** This agreement is subject to the purchaser obtaining financing at ______% or less within _______ days of the date below.
3. **"In-service" date and mileage:** To be based on the closing day, not the day the contract was executed and will be submitted to GM for warranty and all other purposes. The General dealership will have this date corrected by GM if it should become necessary.
4. **Delivery:** The vehicle is to be delivered by ______, failing which the contract is cancelled and the deposit will be refunded.
5. **Cancellation:**
   - **(a)** The purchaser retains the right to cancel this agreement without penalty at any time before delivery of the vehicle by sending a notice in writing to the vendor.
   - **(b)** Following delivery of the vehicle, the purchaser shall have two days to return the vehicle and cancel the agreement in writing, without penalty. After two days and before thirty-one days, the purchaser shall pay the dealer $25 a day as compensation for depreciation on the returned vehicle.
   - **(c)** Cancellation of contract can be refused where the vehicle has been subjected to abuse, negligence or unauthorized modifications after delivery.
   - **(d)** The purchaser is responsible for accident damage and traffic violations while in possession of the said vehicle.
6. **Protected Price:** The vendor agrees not to alter the price of the new vehicle, the cost of preparation or the cost of shipping.
7. **Trade-in:** The vendor agrees that the value attributed to the vehicle offered in trade shall not be reduced, unless it has been significantly modified or has suffered from unreasonable and accelerated deterioration since the signing of the agreement.
8. **Courtesy Car:**
   - **(a)** In the event the new vehicle is not delivered on the agreed-upon date, the vendor agrees to supply the purchaser with a courtesy car at no cost. If no courtesy vehicle is available, the vendor agrees to reimburse the purchaser the cost of renting a vehicle.

(b) If the vehicle is off the road for more than five days for warranty repairs, the purchaser is entitled to a free courtesy vehicle for the duration of the repair period. If no courtesy vehicle is available, the vendor agrees to reimburse the purchaser the cost of renting a vehicle of equivalent or lesser value.

9. **Work Orders:** The purchaser will receive duly completed copies of all work orders pertaining to the vehicle, including warranty repairs and the pre-delivery inspection (PDI).
10. **Dealer Stickers:** The vendor will not affix any dealer advertising, in any form, on the vehicle.
11. **Fuel:** Vehicle will be delivered with a free full tank of gas.
12. **Excess Mileage:** New vehicle will not be acceptable and the contract will be void if the odometer has more than 50 km at delivery/closing.
13. **Tires:** Original equipment Firestone, Bridgestone, or Goodyear tires are not acceptable.

________	________	________
Date	Vendor's Signature	Buyer's Signature

"We can't do that"

Don't believe it when you're told, "we're not allowed to do that"—heard most often in reference to your adding the above contract clauses, cancelling the contract, or reducing the pre-delivery inspection (PDI), transportation, or administrative fees.

Sales are cancelled all the time. In fact, Saturn made a big deal of its money-back warranty where only a few dozen purchasers sought refunds over the several years during which the program was in effect. If Saturn dealers can do it, other dealers can too. As far as PDI/transportation fees are concerned, some dealers have been telling *Lemon-Aid* readers that they are "obligated" by the automaker to charge a set fee and could lose their franchise if they charge less. This is pure hogwash. No dealer has ever had their franchise licence revoked for cutting prices. Furthermore, the automakers clearly state that they don't set a bottom price, since doing so would violate Canada's Competition Act—that's why you always see them putting disclaimers in their ads saying the dealer can charge less.

The pre-delivery inspection

Many new vehicle orders are screwed up by the factory, so the pre-delivery inspection is critical to making sure that corrections are made before taking delivery. Because about 10 percent of new vehicles are damaged in transit from the factory, the PDI can also spot the damage and determine the extent of repairs needed. If the repair costs are substantial, the dealer can be forced to exchange the vehicle or give the buyer a rebate. Every auto manufacturer expects the dealer to carry out a PDI on every vehicle sold.

The PDI allowance is figured into the suggested retail price, so whatever you pay the dealer is profit. Dealers who don't get top dollar for a vehicle are tempted to skip the PDI. According to testimony before the U.S. Federal Trade Commission and the U.S. Senate Subcommittee on Antitrust and Monopoly, many dealers deliver vehicles straight from the factory to their customers with only a cursory inspection. This practice wouldn't have such serious consequences if all vehicles were delivered from the factory in reasonably good shape. Unfortunately, they're not. The PDI serves as a last chance to catch the five or so major and minor defects that *Consumer Reports* estimates afflict most new vehicles. If they aren't corrected before delivery, there's a good chance the dealer will charge to fix them later, and if these minor defects aren't caught in time, they can quickly become major failures.

The best way to ensure that the PDI will be done is to write in the sales contract that you'll be given a copy of the completed PDI sheet when the vehicle is delivered to you. Then, with the PDI sheet in hand, verify some of the items that were to be checked. If any items appear to have been missed, refuse delivery of the vehicle. Once you get home, check out the vehicle more thoroughly and send a registered letter to the dealer if you discover multiple major defects.

Selling Your Trade-In

To sell or not to sell?

It doesn't take a genius to figure out that the longer one keeps a vehicle, the less it costs to own—up to a point. The Hertz Corporation has estimated that a small car equipped with standard options, driven 16,000 km (10,000 miles), and traded each year, costs approximately 6 cents/km more (9.6 cents/mile more) to run than a comparable compact traded after five years. A small car kept for 10 years and driven 16,000 km a year would cost 6.8 cents/km less (10.8 cents/mile less) than a similar vehicle kept for five years, and a whopping 12.8 cents/km less (20.4 cents/mile less) than a comparable vehicle traded in each year. That would amount to savings of $20,380 over a 10-year period.

If you're happy with your vehicle's styling and convenience features, and it's safe and dependable, there is no reason to get rid of it. But when the cost of repairs becomes equal to or greater than the cost of payments for a new car, then you need to consider trading it in. Shortly after your vehicle's fifth birthday (or whenever you start to think about trading it in), ask a mechanic to look at it to give you some idea of what repairs, replacement parts, and maintenance work it will need in the coming year. Find out if dealer service bulletins show that it will need extensive repairs in the near future. (See Appendix I for how to order bulletins from ALLDATA.) If it's going to require expensive repairs, you should trade the vehicle right away; if expensive work isn't necessary, you may want to keep it. Auto owner associations provide a good yardstick. They estimate that the annual cost of repairs and preventive maintenance for the average vehicle is between $700 and $800. If your vehicle is five years old and you haven't spent anywhere near $3,500 in maintenance, it would pay to invest in your old vehicle and continue using it for another few years.

Consider whether your vehicle can still be serviced easily. If it's no longer on the market, the parts supply is likely to dry up and independent mechanics will be reluctant to repair it.

Don't trade for fuel economy alone. Most fuel-efficient vehicles, such as front-drives, offset the savings through higher repair costs. Also, the more fuel-efficient vehicles may not be as comfortable to drive, due to excessive engine noise, lightweight construction, stiff suspension, and torque steer.

Reassess your needs. Does your family growth require a different vehicle? Are you driving less? Are you taking fewer long trips? Let your car or minivan rust in peace and pocket the savings if its deteriorating condition doesn't pose a safety hazard and isn't too embarrassing. On the other hand, if you're in sales and are constantly on the road, it makes sense to trade every few years—in that case, the vehicle's appearance and reliability become a prime consideration, particularly since the increased depreciation costs are mostly tax deductible.

Getting the most for your trade-in

Customers who are on guard against paying too much for a new vehicle often sell their trade-ins for too little. Before agreeing to any trade-in amount, read Part Three of *Lemon-Aid Used Cars and Minivans.* The guide will give your vehicle's dealer and private selling price and offer a formula to figure out regional price fluctuations.

Now that you've nailed down your trade-in's approximate value, here are some tips on selling it with a minimum of stress:

- Never sign a new vehicle sales contract unless your trade-in has been sold—you could end up with two vehicles.
- Negotiate the price from *retail* (dealer price) down to *wholesale* (private sales).
- If you haven't sold your trade-in after two weekends, you might be trying to sell it at the wrong time of year or have it priced too high.

Private sales

If you must sell your vehicle and want to make the most out of the deal, consider selling it yourself and putting the profits toward your next purchase. You'll likely come out hundreds of dollars ahead—buyers will pay more for your vehicle because they won't have to pay the 7 percent GST on a private sale. The most important thing to remember is that there's a large market for used vehicles in good condition in the $4,000–$5,000 range. Although most people prefer buying from individuals rather than from used car lots, they may still be afraid that the vehicle is a lemon. By using the following suggestions, you should be able to sell your vehicle quite easily.

1. Know its value. Study dealers' newspaper ads and compare them with the prices listed in this book. Undercut the dealer price by $300–$800 and be ready to bargain down another 10 percent for a serious buyer. Remember,

prices can fluctuate wildly depending on which models are trendy, so watch the want ads carefully.

2. Enlist the aid of the salesperson who's selling you your new car. Offer her a few hundred dollars if she finds you a buyer. The fact that one sale hinges on the other, and the prospect of making two commissions, may work wonders.
3. Post notices on bulletin boards at your office or local supermarkets and place a "For Sale" sign in the window of the vehicle itself. Place a newspaper ad only as a last resort.
4. Don't give your address right away to a potential buyer responding to your ad. Instead, ask for the telephone number where you may call that person back.
5. Be wary of selling to friends or family members. Anything short of perfection and you eat Christmas dinner alone.
6. Don't touch the odometer. You may get a few hundred dollars more—and a criminal record.
7. Paint the vehicle. Some specialty shops charge only $300 and give a guarantee that's transferable to subsequent owners.
8. Make minor repairs. This includes a minor tune-up and patching up the exhaust. Again, if any repair warranty is transferable, use it as a selling point.
9. Clean the vehicle. Go to a reconditioning firm or spend the weekend scrubbing the interior and exterior. First impressions are important. Clean the chrome, polish the body, and peel off old bumper stickers. Remove butts from the ashtrays and clean out the glove compartment. Make sure all tools and spare parts have been taken out of the trunk. Don't remove the radio or speakers—the gaping holes will lower the vehicle's worth much more than the cost of radio or speakers. Replace missing or broken dash knobs and window cranks.
10. Change the tires. Recaps are good buys.
11. Let the buyer examine the vehicle. Insist that the vehicle be inspected by an independent garage, and accompany the prospective buyer to the garage.
12. Keep important documents handy. Show prospective buyers the sales contract, repair orders, owner's manual, and all other documents that show how the vehicle has been maintained. Authenticate your claims about fuel consumption.
13. Don't mislead the buyer. If the vehicle was in an accident, or some financing is still to be paid, admit it. Any misleading statements may be used later against you in court. It's also advisable to have someone witness the actual transaction in case of a future dispute.
14. Sell to a dealer who sells the same make. He'll give you more because he can easily sell your trade-in to customers who are interested only in that make of vehicle.
15. Write an effective ad.

Using the want ads

The most effective ads give all the basic information, including make, model, year, mileage, optional features, asking price, and whether the price is firm or negotiable. The ad should also give a number where the seller can be reached.

Don't use hyped-up or come-on advertising. An ad that screams in big, bold type "Sharp! Must see to believe!" does not inspire confidence. Stick with a more sober ad that conveys all the needed information without the hoopla. Don't say anything that can be interpreted as a guarantee, and be cautious of such catch-all phrases as "mechanic's special," which is to a vehicle buyer what a "handyman's special" is to a home buyer, and can mean anything. Another common error found in used car ads is the innocent misrepresentation of what the vehicle will do from a performance standpoint. Refrain from making fuel economy claims, even if you believe them to be true.

How to write a good want ad

- Begin your advertisement with the make of the vehicle you're selling. State the manufacturer, model name, and year.
- Give information clearly, including price. Readers react more quickly and favourably when they're given complete and definite information.
- Make it easy for the prospect to reach you. Always include your telephone number and state a preferred time for prospects to contact you.
- Use consecutive insertions for better exposure. A six-day order is best and costs less per insertion.
- Place yourself in the reader's position. Ask yourself what you would want to know if you were buying a used vehicle.

Want ads that fail usually do so because they're carelessly worded and don't contain enough information to get prompt action. Even an interested reader may pass up an ad that doesn't include a telephone number where the seller can be reached.

Selling to dealers

Selling to a dealer means that you're likely to get 20 percent less than if you sold your vehicle privately, unless the dealer agrees to participate in an accommodation sale. Most sellers will gladly pay some penalty to the dealer, however, for the peace of mind that comes with knowing that the eventual buyer won't lay a claim against them. This assumes that the dealer hasn't been cheated by the seller—if the vehicle is stolen, isn't paid for, has had its odometer spun back (or forward to a lower setting), or is seriously defective, the buyer or dealer can sue the original owner for fraud.

Bill of Sale
Used Vehicles

1. The seller agrees to sell, and the buyer agrees to buy a:
   a) ________________, b) ________________.
   Year Serial Number

2. The seller is selling the motor vehicle:
   ° without a warranty
   ° with the following warranty ____________________.

3. The buyer:
   ° has test driven the motor vehicle
   ° has not test driven the motor vehicle.

4. The purchase price in full is $________.

5. The seller acknowledges receiving from the buyer a deposit in the amount of $_______.

6. The seller warrants and guarantees that there are no liens, chattel mortgages, or security agreements outstanding with respect to the motor vehicle or any equipment and/or accessories, and that the motor vehicle and any equipment and/or accessories has/have not been given as collateral on any loan.

7. The seller and the buyer agree that the buyer was allowed to take the motor vehicle for an inspection by a mechanic before the signing of this agreement.

8. The seller warrants to the buyer that to the best of his knowledge:
   a) the odometer reading on the motor vehicle is accurate,
   b) the motor vehicle has not been damaged in a collision, and
   c) there are no outstanding traffic violations with respect to the motor vehicle.

____________________ ____________________
Date City

____________________ ____________________
Buyer Seller

Drawing up the contract

Photocopy the preceding sample bill of sale and date it. Identify the vehicle (including the serial number), its price, whether a warranty applies, and the nature of the examination made by the buyer. The buyer may ask you to put in a lower price than what was actually paid in order to reduce the sales tax. If you agree to this, don't be surprised when a Revenue Canada agent comes to your door. Although the purchaser is ultimately the responsible party, you're an accomplice in defrauding the government. Furthermore, if you turn to the courts for redress, your own conduct may be put on trial.

Summary

Purchasing a used vehicle saves you the most money. Paying cash (or with the biggest down payment you can afford), using zero percent financing programs, and piling up as many kilometres and years as possible on your trade-in are the best ways to save money with new vehicles. Remember that safety is another consideration that depends largely on the type of vehicle you choose. Focus on the following objectives.

Buy safe

Safety features to look for:

1. A high NHTSA and IIHS crashworthiness rating and low rollover potential.
2. Good-quality tires; be wary of "all-season" tires and Firestone makes.
3. Three-point belts with belt pretensioners and adjustable shoulder belt anchorages.
4. Integrated child safety seats and seat anchors, safety door locks, and override window controls.
5. De-powered dual airbags with a cut-off switch, side airbags with head protection, unobtrusive, effective head restraints, and pedal extenders for short drivers.
6. Front driver's seat with plenty of rearward travel and a height adjustment.
7. Good all-around visibility; dash doesn't reflect onto the windshield.
8. An ergonomic interior with an efficient heating and ventilation system.
9. Adequate headlights that aren't too bright for oncoming drivers.
10. Dash gauges that don't wash out in sunlight or produce windshield glare.
11. Head restraints for all seating positions.
12. Delaminated side window glass.
13. Easily accessed sound system and climate controls.
14. Navigation systems that don't require an MIT degree to calibrate.
15. Manual sliding doors in vans (if children are transported).

Buy smart

1. Buy the vehicle you need and can afford, not the one someone else wants you to buy, or one loaded with options that you'll probably never use. Take your time. Price comparisons and test-drives may take a month, but you'll get a better vehicle and price in the long run.
2. Buy in winter or later in the new year to double-dip from dealer incentives and customer rebate or low-cost financing programs.

3. Sell your trade-in privately.
4. Arrange financing before buying your vehicle.
5. Test-drive your choice by renting it overnight or for several days.
6. Buy through the Internet or by fax, or use an auto broker if you're not confident in your own bargaining skills, lack the time to haggle, or want to avoid the "showroom shakedown."
7. Ask for at least a 5 percent discount off the MSRP, and cut PDI and freight charges by at least 50 percent. Insist on a specific delivery date written in the contract, as well as a protected price in case there's a price increase between the time the contract is signed and when the vehicle is delivered. Also ask for a free tank of gas.
8. Order a minimum of options and seek a 30 percent discount on the entire option list. Try not to let the total option cost exceed 15 percent of the vehicle's MSRP.
9. Avoid leasing. If you must lease, choose the shortest time possible, drive down the MSRP, and refuse to pay an "acquisition" or "disposal" fee.
10. Japanese vehicles made in North America, co-ventures with American automakers, and re-badged imports often cost less than imports and are just as reliable. However, some Asian and European imports may not be as reliable as you might imagine—Kia's Sportage and Mercedes' M-Class sport-utilities, for example. Get extra warranty protection from the automaker if you're buying a model that has a poorer-than-average repair history. Use auto club references to get honest, competent repairs at a reasonable price.

Now that you know how to get the best for less, Part Two will show you how to get your money back when that "dream car" turns into a nightmare.

PART TWO

LEGAL RIGHTS AND WRONGS 2

A Comforting Thought

The first thing we do, let's kill all the lawyers.

William Shakespeare

Second Part of King Henry VI, Act IV, Scene II

Many 2001–03 VW and Audi owners are angry that their cars' defective ignition coils can cause their vehicles to suddenly stop and not restart. Over half a million vehicles are afflicted. VW has asked them to be patient.

One chance in 10, it's a lemon

Runzheimer Consultants says that one out of every 10 American vehicles produced by the Detroit Big Three is a "lemon." I would guess that owners of Ford and GM vehicles with faulty plastic engine intake manifold gaskets would double that figure.

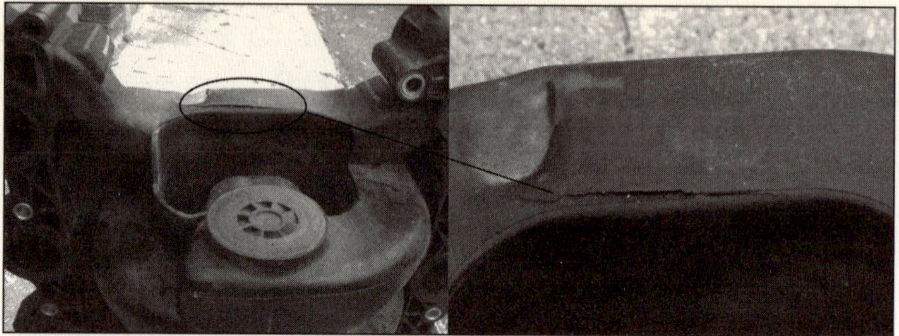

This $600–$1,000 intake manifold repair is required after only 3–5 years of use. Ford and GM are still using these plastic parts in their engines.

If you've bought an unsafe vehicle, or if you've been forced to pay for repairs to correct factory-induced defects, this section's for you. Its intention is to help you get your money back—without going to court or getting frazzled by the dealer's broken promises or "benign neglect." But if going to court is your only recourse, you'll find the jurisprudence you need to cite in your complaint to get an out-of-court settlement or to win your case without spending a fortune on lawyers and research.

Four Ways to Get Your Money Back

Remember the "money-back" guarantee? Well, automakers are reluctant to offer any warranty that requires them to take back a defective truck or van. Fortunately, our provincial consumer protection laws have filled the gap so that now, any sales contract for a new or used vehicle can be cancelled—or free repairs can be ordered—if the vehicle:

- is unfit for the purpose for which it was purchased
- is misrepresented
- is covered by a secret warranty or a "goodwill" warranty extension; or
- hasn't been reasonably durable, considering how well it was maintained, the mileage driven, and the type of driving done (particularly applicable to engine, transmission, and paint defects)

These four legal concepts enumerated above can lead to the contract being cancelled, the purchase price partially refunded, and damages awarded.

For example, if the seller says a minivan can pull a 900 kg (2,000 lb.) trailer and you discover that it can barely tow half that weight, or won't reach a reasonable speed, you can cancel the contract for misrepresentation. The same principle applies to a seller's exaggerated claims concerning a vehicle's fuel economy or reliability, as well as to "demonstrators" that are in fact used cars with false (rolled-back) odometer readings. GM's secret paint warranties and Ford's Windstar engine head gasket "goodwill" programs have all been successfully challenged in small claims court, as the jurisprudence in this section attests. And reasonable durability is an especially powerful legal argument that allows a judge to determine what the dealer and auto manufacturer will pay to correct a premature failure long after the original warranty has expired.

Here's how a 1996 Nissan Sentra XE owner from Nova Scotia prepared his small claims court petition alleging premature structural rust-out of his car's A-pillar:

> Phil,
>
> Thanks for the continuing advice.... I have to get a few independent estimates done and give Nissan Canada something in writing to which they can respond.
>
> I've started some research.... Nissan Canada is listed in the Nova Scotia Registry of Joint Stock Companies as an extra-provincial corporation, headquartered in Ontario. Within the list of directors and officers, there is a "recognized agent" based in Nova Scotia. I've checked with the small claims court clerk's office, and the clerk said I could serve the claim to the agent in Halifax, but have the claim proceed in Sydney where I live as this is where the cause of action (the rusting) took place.
>
> **Small Claims Court Act. R.S., c. 430, s. 1.**
> **Commencement of claim**
>
> 19 (1) A claim before the Court shall be commenced in the county in which
> (a) the cause of action arose; or
> (b) the defendant or one of several defendants resides or carries on business,
> by filing a claim in the form prescribed by the regulations, accompanied by the prescribed fee, with the prothonotary of the Supreme Court in the proper county.
>
> My claim will be based on the:
>
> **Consumer Protection Act. R.S., c. 92, s. 1.**
> **Implied conditions or warranties**
>
> 26 (3) Notwithstanding any agreement to the contrary, the following conditions or warranties on the part of the seller are implied in every consumer sale:
> (j) a condition that the goods shall be durable for a reasonable period of time having regard to the use to which they would normally be put and to all the surrounding circumstances of the sale.
>
> I will also cite the legal precedents that you provide in the *Lemon-Aid* guide and will likely subpoena the service manager that did the initial assessment and told me he sees lots of these, and that Nissan

Canada wanted more repairs rather than replacements. I may try to subpoena minutes and materials from the April meeting in which Nissan Canada told dealers to repair rather than replace A-pillar rusting.

Here is a copy of the completed complaint letter, which can serve as an example for anyone who owns a defective new car or minivan:

Senior Customer Service Representative—Eastern Canada
Nissan Canada
5290 Orbitor Drive
Mississauga, Ontario L4W 4Z5
Fax: (905) 629-6553

Dear Mr. Henriques:
Re: 1996 Nissan Sentra, VIN————————.

I am dissatisfied with the response you have provided me with respect to the structural corrosion on my Sentra. The car is barely out of warranty and the condition clearly began within the warranty period. Further, the rust is a result of a factory defect that renders the car not reasonably durable for its intended use.

In compliance with the Nova Scotia Consumer Protection Act ("...that the goods shall be durable for a reasonable period of time having regard to the use to which they would normally be put...") and the "implied warranty" set down by the Supreme Court of Canada in *Donoghue v. Stevenson* and *Longpre v. St.-Jacques Automobile,* I hereby request that the defects be repaired without charge.

Should you fail to repair these defects in a satisfactory manner and within a reasonable period of time, I reserve the right to have the repairs done at a facility of my choosing and to claim reimbursement for the repairs and consequential damages in court, without further delay.

To assist you in making the right decision, I have attached the estimate from Village Auto Body, which is the body shop of Lloyd MacDonald Nissan in Sydney, N.S.

I have personally owned two Nissans, and my family has owned a total of five. The current situation is an exception to the type of performance we have come to expect of Nissan vehicles, Nissan Canada, and its dealers. While I cannot guarantee that my next vehicle will be a Nissan even if you repair the vehicle, I can 100 percent guarantee that there will never again be a Nissan in our family if you do not. I'm sure you understand and appreciate the value of customer loyalty, as I'm sure you appreciate that future profits on any new vehicles sold to my family will far outweigh the relatively modest costs of repairing my Sentra.

A response by September 6 would be appreciated.

Unfair contracts

New and used sales contracts aren't fair, nor are they meant to be. Dealer and automakers' lawyers spend countless hours making sure their clients are well protected with ironclad standard-form contracts.

Called "contracts of adhesion," judges look upon these agreements with a great deal of skepticism. They know these are contracts in which you have little or no bargaining power, such as loan documents, insurance contracts, and automobile leases. So when a dispute arises over terms or language, provincial consumer protection statutes require that judges interpret these contracts in the way most favourable to the consumer.

Don't say "hearsay"

It's essential that printed evidence and/or witnesses (relatives are not excluded) are available to confirm that a false representation actually occurred, or that a part is failure-prone, or that its replacement is covered by a secret warranty. Stung by an increasing number of small claims court defeats, automakers are now asking small claims court judges to disallow evidence from *Lemon-Aid*, service bulletins, or memos on the pretext that it's hearsay (not proven), unless confirmed by an independent mechanic, or unless the document is recognized by the automaker or dealer's representative at trial ("Is this a common problem? Do you recognize this service bulletin? Is there a case-by-case 'goodwill' plan covering this repair?"). This is why you should bring in an independent garage mechanic or body expert to buttress your allegations. Sometimes, though, the service manager or company representative will make key admissions if questioned closely by you, a court mediator, or the trial judge. That questioning can be particularly effective, if you call for the exclusion of witnesses until they're called (let them mill outside the courtroom wondering what their colleagues have said).

Automakers often blame owners for having pushed their vehicle beyond its limits. Therefore, when you seek to set aside the contract or get a repair reimbursed, it's essential that you get the testimony of an independent mechanic and co-workers in order to prove that the vehicle's poor performance isn't caused by negligent maintenance or abusive driving.

It should have lasted longer!

The reasonable durability claim is your ace in the hole. It's probably the easiest allegation to prove, since all automakers have benchmarks as to how long body components, trim and finish, and mechanical and electronic parts should last (see the durability chart on pages 104–105). Vehicles are expected to be reasonably durable and merchantable. What is reasonably durable depends on the price paid, kilometres driven, the purchaser's driving habits, and how well the vehicle was maintained by the owner. Judges carefully weigh all these factors in awarding compensation or cancelling a sale.

Whatever the reason you use to get your money back, don't forget to conform to the "reasonable diligence" rule that requires you to file suit within a reasonable time after purchase or after you've discovered the defect. If there have been no negotiations with the dealer or automaker, this period cannot exceed a few months. If either the dealer or the automaker has been promising to correct the defects for some time, or has carried out repeated unsuccessful repairs, the delay for filing the lawsuit can be extended.

Refunds for other expenses

Unless you can quote a court decision for support, it's hard to obtain compensation for hotel and travel costs, or general inconvenience. Fortunately, this year we have the judgment you'll need. Automakers and dealers hate to pay these kinds of consequential expenses because they can't control the amount of the refund. (Towing expenses, however, are usually accepted.) Courts are more generous, having ruled that all expenses (damages) flowing from a problem covered by a warranty or service bulletin are the manufacturer's/dealer's responsibility under both common law (all provinces except Quebec) and Quebec civil law. Fortunately, when legal action is threatened—usually through small claims court—automakers quickly back down from their refusal to pay consequential damage claims.

Okay, here's that important Canadian judgment that'll help you. It gives consequential damages to a motorist fed up with his "lemon" Cadillac: *Wharton v. Tom Harris Chevrolet Oldsmobile Cadillac Ltd. and General Motors of Canada Limited*, B.C. Supreme Court, Date: 19991202, Docket: C982104, Registry: Vancouver.

In his decision, Judge Leggatt threw the book at GM and the dealer:

(a) Hotel accommodations: $217.17

(b) Travel to effect repairs at 30 cents per kilometre: The plaintiff claims some 26 visits from his home in Ucluelet to Nanaimo. Some credit should be granted to the defendants since routine trips would have been required in any event. Therefore, the plaintiff is entitled to be compensated for mileage for 17 trips (approximately 400 km from Ucluelet to Nanaimo return) at 30 cents per kilometre.

$2,040.00

TOTAL: $2,257.17

[20] The plaintiff is entitled to non-pecuniary damages for loss of enjoyment of their luxury vehicle and for inconvenience in the sum of $5,000.

Warranties

The manufacturer's or dealer's warranty is a written legal promise that a vehicle will be reasonably reliable, subject to certain conditions. Regardless of the number of subsequent owners, this promise remains in force as long as the warranty's original time/kilometre limits haven't expired. Unfortunately, these warranties are full of so many loopholes ("you abused the car; it was poorly maintained; it's normal wear and tear") that they may be useless when a vehicle breaks down.

Thankfully, car owners get another kick at the can. As clearly stated in *Frank v. GM*, every vehicle sold new or used in Canada is also covered by an *implied* warranty—a collection of federal and provincial laws and regulations that protect you from hidden defects, misrepresentation, and a host of other scams. Furthermore, Canadian law presumes that car dealers, unlike private sellers, are aware of the defects present in the vehicles they sell. That way, they can't just pass the ball to the automakers and walk away from the dispute.

Treacherous tires

Tires aren't usually covered by car manufacturers' warranties (except for GM and Ford), and are warranted instead by the tiremaker on a pro-rated basis. This isn't such a good deal, because the manufacturer is making a profit by charging you the full list price. If you were to buy the same replacement tire from a discount store you'd likely pay less, without the pro-rated rebate.

But consumers have gained additional rights following Bridgestone/Firestone's massive tire recall in 2001. Due to the confusion and chaos surrounding Firestone's handling of the recall, Ford's 575 Canadian dealers stepped into the breach and replaced the tires with any equivalent tires dealers had in stock. No questions asked.

This is an important precedent that tears down the traditional liability wall separating tire manufacturers from automakers in product liability claims. In essence, whoever sells the product can now be held liable for damages. In the future, Canadian consumers will have an easier time holding the dealer, automaker, and tiremaker liable, not just for recalled products, but for any defect that affects the safety or reasonable durability of that product.

This is particularly true now that the Supreme Court of Canada (*Winnipeg Condominium v. Bird Construction* [1995] 1S.C.R.85) has ruled that defendants are liable in negligence for any designs that resulted in a risk to the public for safety or health. The Supreme Court reversed a long-standing policy and provided the public with a new cause of action that had not existed before in Canada. Prior to this Supreme Court ruling, companies dodged liability for falling bridges and crashing planes by warranty exclusion and "entire-agreement" contract clauses. In the Winnipeg Condominium case, the Supreme Court held that repairs made to prevent serious damage or accidents could be claimed from the designer/builder for the cost of repair in tort, from any subsequent purchaser. Consumers with tire or other safety-related claims relating to the safety of their vehicles would be wise to insert the above court decision (with explanation) in their claim letter and mail or fax it to the automaker's "legal affairs" or "product liability" department. A copy should be also deposited with the clerk of the small claims court if you have to use that recourse.

Other warranty repairs

Safety restraints such as airbags and safety belts have warranty coverage extended for the lifetime of the vehicle, following an agreement made between U.S. automakers and importers. In Canada, though, many automakers try to dodge this responsibility, alleging that they are separate entities, distinct from their American counterparts. That distinction is both disingenuous and dishonest, and wouldn't likely hold up in small claims court—probably the reason why most automakers relent when threatened with legal action.

Aftermarket products and services—such as gas-saving gadgets, rustproofing, paint protectors, air conditioning, and van conversions—can render the manufacturer's warranty invalid, so be sure to check with your dealer before purchasing any optional equipment, products or services from an independent supplier.

How fairly a warranty is applied is more important than how long it remains in effect. Once you know the normal wear rate for a mechanical component or body part, you can demand proportional compensation when you get less than normal durability—no matter what the original warranty said.

Some dealers tell customers that they need to have original equipment parts installed in order to maintain their warranty. A variation on this theme requires that routine servicing—including tune-ups and oil changes (with a certain brand of oil)—be done by the selling dealer, or the warranty is invalidated.

Nothing could be further from the truth.

Canadian law stipulates that whoever issues a warranty cannot make that warranty conditional on the use of any specific brand of motor oil, oil filter, or any other component, unless it's provided to the customer free of charge.

Sometimes dealers will do all sorts of minor repairs that don't correct the problem, and then after the warranty runs out they'll tell you that major repairs are needed. You can avoid this nasty surprise by repeatedly bringing your vehicle in to the dealership before the warranty ends. During each visit, insist that a written work order include the specific nature of the problem as *you* see it and that the work order carry the notation that this is the second, third, or fourth time the same problem has been brought to the dealer's attention. Write it down yourself, if need be. This allows you to show a pattern of non-performance by the dealer during the warranty period and establishes that it's a serious and chronic problem. When the warranty expires, you have the legal right to demand that it be extended on those items consistently reappearing on your handful of work orders. *Lowe v. Fairview Chrysler* (see pages 128–129) is an excellent judgment that reinforces this important principle. In another lawsuit, *François Chong v. Marine Drive Imported Cars Ltd. and Honda Canada Inc.* (see page 135), a Honda owner used the B.C. small claims courts to force Honda to fix his engine seven times—until they got it right.

A retired GM service manager gave me another effective tactic to use when you're not sure a dealer's warranty "repairs" will actually correct the problem for a reasonable period of time after the warranty expires. Here's what he says you should do:

When you pick up the vehicle after the warranty repair has been done, hand the service manager a note to be put in your file that says you appreciate the warranty repair, however, you intend to return and ask for further warranty coverage if the problem reappears before a reasonable amount of time has elapsed—even if the original warranty has expired. A copy of the same note should be sent to the automaker.... Keep your copy of the note in the glove compartment as cheap insurance against paying for a repair that wasn't fixed correctly the first time.

Extended (supplementary) warranties

Supplementary warranties providing extended coverage may be sold by the manufacturer, dealer, or an independent third party, and are automatically transferred when the vehicle is sold. They cost between $1,000 and $1,500, and should be purchased only if the vehicle you're buying is off its original warranty, if it has a reputation for being unreliable or expensive to service (see Part Three), or if you're reluctant to use the small claims courts when factory-related trouble arises. Don't let the dealer pressure you into deciding right away.

Generally, you can purchase an extended warranty anytime during the period in which the manufacturer's warranty is in effect, or, in some cases, shortly after buying the vehicle from a used-car dealer. An automaker's supplementary warranty is the best choice but will likely cost about a third more than warranties sold by independents. And in some parts of the country, notably British Columbia, dealers have a quasi-monopoly on selling warranties with little competition from the independents.

Because up to 60 percent of the warranty's cost represents dealer markup, dealers love to sell extended warranties, whether you need them or not. Out of the remaining 40 percent comes the sponsor's administration costs and profit margin, calculated at another 15 percent. What's left to pay for repairs is a minuscule 25 percent of the original amount. The only reason why automakers and independent warranty companies haven't been busted for operating this warranty Ponzi scheme is because only half of the car buyers who purchase an extended service contract actually use it.

It's often difficult to collect on supplementary warranties because independent companies frequently go out of business or limit the warranty's coverage through subsequent mailings. Provincial laws cover both situations. If the bankrupt warranty company's insurance policy won't cover your claim, take the dealer to small claims court and ask for the repair cost and the refund of the original warranty payment. Your argument for holding the dealer responsible is a simple one: by accepting a commission for acting as an agent of the defunct company, the dealer took on the obligations of the company as well. As for limiting the coverage after you have bought the warranty policy, this is illegal, and allows you to sue both the dealer and the warranty company for a refund of both the warranty and repair costs.

Emissions control warranties

These little-publicized warranties can save you big bucks if major engine or exhaust components fail prematurely. They come with all new vehicles and cover major components of the emissions control system for up to 8 years/130,000 km. Unfortunately, although owner's manuals vaguely mention the emissions warranty, most don't specify which parts are covered. Fortunately, the U.S. Environmental Protection Agency has intervened on several occasions with hefty fines against Chrysler and Ford after ruling that all major motor and fuel-system components are covered. These include fuel metering, ignition spark advance, restart, evaporative emissions, positive crankcase ventilation, engine electronics (computer modules), and catalytic converters, as well as hoses, clamps, brackets, pipes, gaskets, belts, seals, and connectors. Canada, however, has no government definition, and it's up to each manufacturer and the small claims courts to decide which components are covered. Use the EPA guidelines at *www.epa.gov/otaq/consumer/warr95fs.txt* to make your case.

Many of the confidential technical service bulletins listed in Part Three show which parts failures are covered under the emissions warranty, even though motorists are routinely charged for their replacement. The following example, applicable to Ford's 1996–98 cars, minivans, vans, and trucks, shows the automaker will pay for major repairs to correct the loss of engine coolant, or engine oil contaminated with coolant. Unfortunately, few owners will ever see these bulletins, and will end up paying for repairs that are really Ford's responsibility.

Coolant Loss/Engine Oil Contamination, 3.8L, 4.2L

Article No. 99 - 20 - 7
10/04/99
^ COOLING SYSTEM - 3.8L - UNDETERMINED LOSS OF COOLANT
^ COOLING SYSTEM - 4.2L - UNDETERMINED LOSS OF COOLANT
^ ENGINE - 3.8L - ENGINE OIL CONTAMINATED WITH COOLANT
^ ENGINE - 4.2L - ENGINE OIL CONTAMINATED WITH COOLANT

FORD:
1996–97 THUNDERBIRD
1996–98 MUSTANG, WINDSTAR
1997–98 E-150, E-250, F-150

MERCURY:
1996–97 COUGAR

This TSB article is being republished in its entirety to correct the front cover/water pump bolt torque values and the front cover gasket part number.

ISSUE:
Engine coolant may be leaking into the engine oil on some vehicles. The internal coolant leak may be difficult to identify. This may be caused by the lower intake manifold side gaskets and/or front cover gaskets allowing coolant to pass into the cylinders and/or the crankcase.

ACTION:
Revised lower intake manifold side and front cover gaskets have been released for service. Refer to the following text and Application Chart for details.

WARRANTY STATUS:
Eligible under the provisions of bumper-to-bumper warranty coverage and emissions warranty coverage.

Make sure you get your emissions system checked out thoroughly by a dealer or independent garage before the emissions warranty expires and before having the vehicle inspected by provincial emissions inspectors. In addition to ensuring you pass provincial tests, this precaution could save you up to $1,000 if both your catalytic converter and other emissions components are faulty.

Tracking secret warranties

1st Stage—Service advisories are posted on an automaker's internal computer network. They offer troubleshooting tips and allow the dealer to bill the manufacturer for the repair. This info is seldom shared with the customer.

2nd Stage—If the defect grows in scope and a more involved solution is needed (for example, one requiring upgraded parts), automakers then draw up a technical service bulletin (TSB) and distribute it to dealers and U.S. and Canadian government agencies. The TSB is only issued after the manufacturer thinks it has a solution for the defect. TSBs issued by Chrysler, Ford, and GM will usually spell out clearly which base warranty will cover the repair (emissions warranty, bumper-to-bumper, etc.). Interestingly, Asian and European automakers are vague in describing their warranty obligations. Honda, for example, uses the term "goodwill" as a euphemism to describe its warranty extensions.

3rd Stage—More and more customers hear through *Lemon-Aid*, ALLDATA, friends, and relatives that some TSBs recognize that a common factory-related defect exists, and find that the base warranty is clearly inadequate to deal with the scope of the problem. Dealers and customers exert pressure for additional after-warranty assistance. This, in turn, results in a second TSB, sent only to dealers, extending the warranty coverage to correct the defect and leaving the amount of the customer's refund to the dealer's discretion.

Now, customer dissatisfaction builds to a crescendo, since the dealers and automakers keep the extended guidelines to themselves and customers get widely divergent refunds. This only angers the owners more, brings in the media, and leads to a proliferation of Internet gripe sites and lawsuits (small claims and class actions).

4th Stage—Finally, the aggravation is too great and the automaker decides to make a press release followed by an owner notification letter (sent to first owners only, at their last known address) that clearly spells out what all owners will get and which vehicles are involved. A special bulletin or letter is also sent out to dealers to ensure they follow the guidelines 100 percent. Ford calls these Owner Notification Programs, GM calls them Special Policies, and Chrysler calls them Owner Satisfaction Notifications. No matter the euphemism, they are all an extension of the original warranty and apply to vehicles purchased new or used.

Remember, second owners and repairs done by independent garages are included in these secret warranty programs. Large, costly repairs, such as blown engines, burned transmissions, and peeling paint, are often covered. Even mundane little repairs, which can still cost you a hundred bucks or more, are frequently included in these programs. Take, for example, the elimination of foul, musty, or mildew odours emitted by your air conditioning unit. Despite what the dealer may say, it's covered by the base warranty.

If you have a TSB but you're still refused compensation, keep in mind that secret warranties are an admission of manufacturing negligence. Try to compromise with a pro rata adjustment from the manufacturer. If polite negotiations fail, challenge the refusal in court on the grounds that you should not be penalized for failing to make a reimbursement claim under a secret warranty you never knew existed!

Here are a few examples of the latest, most comprehensive secret warranties that have come across my desk in the last several years.

Acura/Honda

1994–97 Acura CL and Honda Accord, Prelude, and Odyssey models equipped with 4-cylinder engines

Problem: Defective engine oil seals can slip, causing the engine to drain of oil and eventually seize. **Warranty coverage:** Honda will inspect and replace the seal and install a clip to ensure the seal cannot move.

1998–2003 Accord, Odyssey, and Pilot models equipped with 6-cylinder engines

Problem: Defective aluminum engine block. **Warranty coverage:** Repair or replace engine under a "goodwill" program.

Honda: V6 Engine Oil Leaks 01-009

September 10, 2002

Applies To:
1998–2002 Accord V6 - ALL
1999–2003 Odyssey - ALL
2003 Pilot - ALL

SYMPTOM: An oil leak from the front, middle, or rear of the engine.

PROBABLE CAUSE: The cast aluminum engine block may be porous at the front, middle, or rear.

CORRECTIVE ACTION: Depending on the location of the leak, seal it with JB Weld or with 3-Bond-coated sealing bolts.

In warranty: The normal warranty applies.

Out of warranty: Any repair performed after warranty expiration may be eligible for goodwill consideration by the District Parts and Service Manager or your Zone Office. You must request consideration, and get a decision, before starting work.

1999–2003 Acura CL, TL; Honda Odyssey, and Prelude (1999 models are covered on a case-by-case basis)

Problem: Defective automatic transmission and torque converter. **Warranty coverage:** 7 years/100,000 miles (160,000 km). A "goodwill" program that was formally confirmed in the August 4, 2003 edition of *Automotive News*. Defective transmissions will be repaired or replaced free of charge, whether the vehicle was bought new or used. Signs of trouble are:

Audi/VW

2001–03 cars equipped with 1.8L engines, which includes the Audi TT and A4; and the VW Golf/GTI, Jetta, New Beetle, and Passat. The companies also included the Passat W8 engine, all VWs equipped with the 2.8L VR6; as well as the Audi 3.0L V6 engine.

Problem: Defective ignition coils. When these coils fail, the vehicle suddenly stalls and won't start. **Warranty coverage:** VW will fix every single car by replacing the coils whether they are broken or not. There is no mileage or time limit on this warranty extension. Owners complain that a backlog of parts has resulted in long waits and repeat dealer visits.

1998–2002 New Beetles and 1999–2002 Golfs, GTIs, and Jettas. 850,000 vehicles are affected

Problem: If window clamp malfunctions, it prevents the window from being raised or lowered. **Warranty coverage:** The new clamp and the work to install it will be covered free of charge under this special warranty.

Chrysler

1991–2001 models with A604, 41TE, 42LE, and later automatic transmissions

Problem: It's very likely this problem will carry over to the 2002–04 models.

Report Date : July 21, 2003 at 03:57 PM

TYPE: Vehicle

YEAR: 2003

MAKE: Dodge

MODEL: Caravan

Service Bulletin Num: 13

Date of Bulletin: Apr 01, 2003

Component: Powertrain

Summary: Transaxle wiring harness corrosion, which could cause the malfuntion indicator light (MIL) to illuminate and the transmission to shift into Second gear. *TT

Faulty automatic transmissions that shift erratically, gear down to "limp mode," are slow to shift in or out of Reverse, and self-destruct. This is an A604 (and its spin-offs) software and hardware problem that has bedevilled Chrysler owners for over a decade. Dozens of service bulletins address the problem and can be useful in small claims court. However, it's likely your filing won't go beyond the pretrial mediation stage: Chrysler reps are loath to defend the cases in front of independent garage testimony. **Warranty coverage:** Without a court threat, Chrysler usually denies any problem or refund program exists. If you have the assistance of your dealer's service manager, expect an offer of 50 percent (about $1,500) if your vehicle is less than 7 years old and has less than 160,000 km. File the case in small claims court and Chrysler will sweeten the offer considerably.

> I've just been told that I need my **FOURTH TRANSMISSION** on my '96 Town & Country minivan, with 132,000 miles [211,000 km] on it. I've driven many cars well past that mileage with only **ONE** transmission. The dealer asked Chrysler, who said they would not help me. My appeals to Chrysler's customer service department yielded me the same result...Chrysler split some of the costs with me on the previous rebuilt replacements....

1995–99 Breeze, Cirrus, Neon, Sebring convertible, Stratus, and minivans

Problem: Engine head gasket failure. **Warranty coverage:** Chrysler is quietly paying off owners after the warranty has expired up to 5 years/160,000 km if they threaten small claims action. Unfortunately, there is no written proof of this extended warranty. All we have is owner feedback confirming the free repairs. Nevertheless, Chrysler's technical service bulletin (TSB) #09-05-98 shows that 1995–99 models equipped with 2.0L and 2.4L engines have weak head gaskets that should be replaced with more durable ones. It's as close as you'll get to "smoking gun" proof the head gaskets are faulty, and so it can be useful in negotiating an out-of-court settlement.

I suggest you fight any payout of less than 100 percent with a small claims action, citing Ford's 7-year/160,000 km 3.8L engine extended warranty as a benchmark, along with Chrysler's 7-year AC warranty. (Is it fair the engine won't last as long as Ford's or Chrysler's own ACs?)

Chrysler/Jeep

1990–93 Dynasty, New Yorker, Fifth Avenue, and Imperial; 1991–92 Eagle Premier; 1991–93 minivans; 1989–93 Cherokee and Wagoneer

Problem: ABS brakes that fail or malfunction. **Warranty coverage:** Piggybacking a service campaign onto a recall, Chrysler extended the warranty to 10 years/160,000 km on a number of costly ABS components. Owners will also be reimbursed for previous ABS repairs—not applicable to calipers, pads/shoe linings, or other maintenance items. Two other ABS components, piston seals (excessive wear) and the pump motor (deterioration), will be repaired free of charge at any time during the life of the vehicle.

> Hi Phil. I thought I should let you know that Chrysler sent me a cheque to cover the full cost of the repair of my ABS brakes. I appreciate your book providing the copy of Safety Recall 685 as it sure makes things a whole lot easier when you have a document like that to refer to.
>
> Fred K.

SAFETY RECALL #685 — ANTI-LOCK BRAKE SYSTEM (ABS)

Dear Chrysler Canada Ltd. Vehicle Owner:

Chrysler Canada Ltd. has determined that a defect which relates to motor vehicle safety exists in some **1991 through 1993 Dodge Caravan/Grand Caravan, Plymouth Voyager/Grand Voyager and Chrysler Town and Country; late-1990 through 1993 Dodge Dynasty, Chrysler New Yorker, Fifth Avenue and Imperial; and 1991 and 1992 Eagle Premier vehicles equipped with an anti-lock brake system (ABS).**

The Problem is... **The ABS hydraulic control unit on your vehicle may experience excessive brake actuator piston seal wear and/or pump-motor deterioration.** If this occurs, the ABS function may be lost and reduced power assist may be experienced during braking. This may result in increased stopping distance that could result in an accident.

What you must do to ensure your safety... Owners of vehicles that experience any of the following symptoms should contact their dealers ***immediately*** to schedule a service appointment:

- Either the Brake System Warning Light or the Anti-lock Warning Light remains ***illuminated more than two minutes*** after starting the vehicle; ***or if either light comes on*** at any other time during vehicle operation;
- A ***substantial*** increase in ***brake*** pedal force is needed to stop the vehicle; or
- Any other ABS malfunction occurs.

NOTE: If your ABS brake system is operating properly and none of the above symptoms are present, no action is necessary at this time. However, if any of these symptoms appear in the future, contact your dealer immediately for a free repair. ***Keep this letter with your vehicle's other owner information in case you notice any of these conditions in the future.***

What Chrysler and your dealer will do... **Your Selling Dealer will test your vehicle's ABS for excessive piston seal wear and possible pump-motor deterioration.** If problems ***with these components*** are found at any time during the entire life of your vehicle, Chrysler will replace these components **free of charge**. The test will take about one hour to complete and you will not be charged for it. **Another one to two hours may be required if components must be replaced. However, additional time may be necessary depending on how dealer appointments are scheduled and processed.**

Extended Warranty...	In addition to this recall action, the warranty period on other ABS components in your vehicle is being extended to 10 years or 160,000 km (100,000 miles) whichever occurs first. This means that if any of these other ABS components fail within 10 years or 160,000 km (100,000 miles) your dealer will correct the problem free of charge. This extended warranty is limited to the same conditions defined in the original warranty and <u>does not include any base brake system components</u> (calipers, pad/shoe linings. etc.). Further, Chrysler will reimburse owners for any previous ABS components expenses incurred. Should you experience any difficulty in obtaining the recall service, please contact Chrysler Canada Customer Service at 1-800-465-2001 (English) or 1-800-387-9983 (French). We will take the necessary steps to ensure prompt servicing of your vehicle.

1993–99 Concorde, Intrepid, New Yorker, LHS, Vision, and Grand Cherokee

Problem: AC evaporator failure or malfunction. **Warranty coverage:** 7 years/ 115,000 km. Chrysler is free to limit this program to the evaporator and to the vehicles listed above; after all, they are simply modifying their *expressed* warranty. But you are just as free to plead the *implied* warranty (in your letter to Chrysler and small claims filing). Argue that this extension sets a benchmark for the warranty repairs on the entire AC as to what Chrysler considers reasonable durability under consumer protection statutes. Also, make the point that Chrysler is unfairly excluding other models using the same system with the same AC failures.

ADDENDUM TO BASIC WARRANTY

The following applies to 1993 through 1997 New Yorker, LHS, Concorde, Intrepid, Vision and Grand Cherokee vehicles equipped with factory-installed air conditioning:

The Basic Warranty coverage for the air conditioner evaporator has been extended to 7 years or 115,000 kilometres, whichever occurs first, from the vehicle's warranty start date.

This extended coverage applies to all owners of the vehicle. All of the other warranty terms apply to this extension.

You are looking at a rare piece of written evidence, showing Chrysler will pay for an AC failure—long after the original warranty has elapsed.

Dear Phil,

Thanks so much...you just saved me $1,200!

My 1997 Jeep Grand Cherokee was diagnosed with a badly leaking AC evaporator core. A local garage wanted $1,200 to replace the evaporator and dryer, and recharge the system. Upon reading your site I noticed the extended warranty for the Jeep's evaporator and called the dealer to verify.

Indeed it was covered so I took the car in to the local Chrysler dealer for the repair. After the repair was started the service advisor called to say the dryer must also be replaced at my expense to the tune of $338. I agreed to the charge and told them to continue.

After reflecting on the extra charge for the dryer, it didn't make sense to me why I had to pay for a part that must be replaced due to a covered part going bad so I called Chrysler customer service and explained my situation.

After a brief time on hold, the customer service rep told me "it was Chrysler's policy..." (I thought, "Oh boy, here it comes!") "...that if another part had to be replaced because of a covered component that Chrysler would cover that charge also." The customer rep also offered to call the local dealer and explain the coverage to them. Needless to say I was "jumping for joy." I picked up the car two days later...no charges and a "thanks for your business" from the service advisor.

What a success story! Please feel free to share my experience with your other readers and again thanks.

Scott Cline

Chrysler, Ford, and General Motors

All years, all models

Problem: Faulty paint jobs that cause paint to turn white and peel off horizontal panels. Out-of-court settlements proffered by all three automakers confirms they limit their "after-warranty assistance" to 6 years/100,000 km. In *Frank v. GM*, the Saskatchewan small claims court set a 15-year benchmark for paint finishes, and three other Canadian small claims judgments have extended the benchmark to seven years, second owners, and pickups. **Warranty coverage:** Automakers will offer a free paint job or partial compensation up to 6 years/no mileage limitation. Thereafter, all three manufacturers offer 50–75 percent refunds on the small claims courthouse steps.

Nancy Frith, a Nova Scotia owner of a 1992 Dodge Shadow, was awarded $2,000 for the paint delamination/peeling of her car. CBC Television interviewed her and you can watch the 5-minute program at: *cbc.ca/consumers/market/files/cars/paintpeel/lemonaid.html.*

> I wanted to let you and your readers know that the information you publish about Ford's paint failure problem is invaluable. Having read through your "how-to guide" on addressing this issue, I filed suit against Ford for the "latent" paint defect. The day prior to our court date, I received a settlement offer by phone for 75 percent of what I was initially asking for.
>
> This settlement was for a 9-year-old car. I truly believe that Ford hedges a bet that most people won't go to the extent of filing a lawsuit because they are intimidated or simply stop progress after they receive a firm no from Ford.
>
> I am angry at Ford for requiring me to hack through the nonsense and frequent denials and I am grateful for your site as it gave me the roadmap and confidence on how to proceed and win. I hope this inspires others to act.
>
> M.P.

All years, all models

Problem: Premature wearout of key front and rear brake components. **Warranty coverage:** Calipers and pads: Goodwill settlements confirm that brake calipers and pads that fail to last 2 years/40,000 km will be replaced for 50 percent of the repair cost; components not lasting 1 year/20,000 km will be replaced for free. Rotors: If they last less than 3 years/60,000 km, they will be replaced at half price; replacement is free up to 2 years/40,000 km. Interestingly, premature brake component wearout is quite common among all automobile manufacturers, although only Chrysler, Ford, and GM have standardized their policies for goodwill refunds.

Apparently, brake suppliers are using cheaper calipers, pads, and rotors that can't handle the heat generated by normal braking. Consequently, drivers find routine braking causes rotor warpage that produces excessive vibrations, shuddering, noise, and pulling to one side when braking.

All years, all models

Problem: Engine head gasket and intake manifold failures. All three Detroit automakers now have serious problems with poor-quality engine head gaskets and intake manifold gaskets failing between 60,0000 and 100,000 km. This serious engine problem is likely to be carried over to the 2004 models. Also, the engine may overheat, lose power, burn extra fuel, and possibly self-destruct. Under the best of circumstances, the repair will take a day and cost about $800. This engine defect afflicts Chrysler's 1995–99 models equipped with 4-cylinder engines, Ford's 1994–2001 cars and minivans and GM's 1996–2001 cars and minivans. **Warranty coverage:** If your dealer's service manager is on board, expect an offer of a 50 percent refund up to five years/100,000 km. If you threaten court action, the payout jumps to 75–100 percent up to 7 years/160,000 km.

> I just wanted to let you know that after contacting you back in January regarding our 2001 Chevy Venture head gasket problem, I have just received my judgment through the Canadian Arbitration Program.
>
> I used the sample complaint letter as well as the judgment you have posted in the *Ford Canada v. Dufour* court case. This combined with an avalanche of similar Chevy Venture complaints that are posted on the Internet helped us to win a $1,700 reimbursement of the $2,200 we were looking for.
>
> The reason for us not receiving the full amount is that the arbitrator stated that GM Canada would have only replaced one head gasket instead of replacing both as we had done, and that a dealer would of supplied us with a car free of charge and therefore did not allow us the car rental expense we incurred.
>
> We are still extremely happy with the results and thank you for your books and website, you have a fan for life.

Here's what an expert mechanic has to say about the GM intake manifold defect:

> Plastic manifold cracks around (very poorly designed placement) EGR return tube that is cooled in water jacket as it passes through intake housing. Over time hot exhaust gases passing through metal pipe cause pipe to get hot and coolant enters inside of engine when plastic intake housing finally cracks due to the heat.
>
> Engine swallows enough coolant after a hot run, owner comes out the next morning to an engine that will not crank. Customer is handy, puts a new battery and new starter motor in to no avail. Traditional diagnosis AT ANOTHER SHOP, pulled spark plugs and decided it was a head gasket after performing a pressure test (this would be a correct diagnosis in my view, but I was suspicious).
>
> Customer was adamant about not overheating this engine ever! He calls me up since now he has had no car since October but has been making the payments on it anyway. I try to confirm diagnosis and my pressure test has a lot of air hissing from throttle body, which prompts me to remove the throttle body and discover this engineering nightmare. This is like a $600 job if you catch it in time, but since the car sat, the engine is now rusted and frozen solid.

For more information see *www.consumeraffairs.com/automan/plastic.html.*

Ford

All models equipped with automatic transmissions

Problem: This problem is usually indicated by erratic shifting, delayed shifting, harsh engagement, and a tendency for the transmission to "hunt" for the proper gear. Barely noticeable at first, it will worsen progressively, until the transmission breaks down completely. **Warranty coverage:** Ford has repaired or replaced transmissions at no charge up to 7 years/160,000 km, no matter if the vehicle was bought new or used. The company's initial offer is usually 50 percent of the estimated $3,000 repair cost. Vehicles that have exceeded the above limitations will receive much less, if anything at all.

1995–98 Windstar

Problem: Defective front coil springs may suddenly break, puncturing the front tire, leading to loss of steering control. **Warranty coverage:** Under a "Safety Improvement Campaign," negotiated with NHTSA, Ford will replace *broken* coil springs at no charge up to 10 years/unlimited mileage (Transport Canada has no record of this free repair on its website). The company won't replace the springs until they have broken—if you're alive to submit a claim. 1997–98 models that are registered in rust-belt states and Canada have been recalled for the installation of a protective shield (called a spring catcher bracket in the Canadian recall) to prevent a broken spring from shredding the front tire.

General Motors

1997–2003 Venture, TranSport/Montana, and Silhouette

Problem: Roof paint delamination and peeling; rust perforation. **Warranty coverage:** GM will replace, repair, or repaint the roof for free up to 6 years/ 100,000 km.

GM: Roof Rust Perforation Bulletin No.: 02-08-67-006A

Date: December 2002

Subject: Roof perforation (replace roof)

Models:

1997–2003 Chevrolet Venture
1997–2003 Oldsmobile Silhouette
1997–2003 Pontiac TranSport/Montana

Important: Implementation of this service bulletin by "GM of Canada" dealers requires prior District Service Manager approval.

Condition: Some customers may comment that there is rust forming around the front or rear portion of the roof.

Cause: During production, the E-coating (ELPO primer) may have been missed in concealed areas of the front or rear portions of the outer roof panel.

Correction: Important: Partial repairs to the roof panel are not permitted.

1. Remove the headliner.
2. Replace the roof.
3. Reinstall the headliner.

Toyota

1997–2002 Toyota and Lexus vehicles with 2.2L 4-cylinder or 3.0L V6 engines

Problem: Sludge buildup may require a rebuilt engine. **Warranty coverage:** Toyota will repair or replace the engine at no charge up to 8 years/160,000 km (100,000 miles), whether you bought the vehicle new or used. Toyota has said owners will not be forced to show oil change receipts. Some dealers, apparently, haven't gotten the word.

> I also have a '98 Sienna. At 75,000 km it began to burn oil. I went to the local dealership and was advised that they would take the heads off and "clean" the engine, for $980. The dealership would offer no guarantee that it would work.
>
> We consulted another mechanic who told us they should change the rings and seals if they were going to take the heads off. When we asked the dealer about this they said that they weren't going to do that and were rather rude when we started to ask questions.
>
> We asked about warranty coverage and were told because we couldn't produce the receipts from one oil change that Toyota would not cover it.
>
> The service manager quietly told me that there was a "flaw" in the engine but would not elaborate. When I asked if there was a flaw why wouldn't Toyota warranty it, I was told "because you can't prove you did the oil change there is no use taking it to Toyota."
>
> In May 2002 we received the letter mentioned in your article, and we took our van into the dealer (after spending three weeks talking to them and producing all the oil change records except the missing one). The engine could not be rebuilt—it had to be replaced. The cost was $5,017.22, finally covered by this goodwill warranty.
>
> T.C., Toronto
>
> *Automotive News*, February 10, 1997

How Long Should a Part or Repair Last?

How do you know when a part or service doesn't last as long as it should and you should seek a partial refund for the failure? Sure, you have a gut feeling based on the use of the vehicle, how you maintained it, and the extent of work that was carried out. But you'll need more than emotion to win compensation from garages and automakers.

Automakers, mechanics, and the courts have their own benchmarks as to what's a reasonable period of time or amount of mileage one should expect a part or adjustment to last. I've prepared the following table to show what most automakers consider as reasonable durability, as expressed by their original and "goodwill" warranties.

ACCESSORIES

Air conditioner	7 years
Cellular phone	5 years
Cruise control	5 years/ 100,000 km
Power antenna	5 years
Power doors, windows	5 years
Radio	5 years

BODY

Paint (peeling)	7 years
Rust (perforations)	7 years
Rust (surface)	5 years
Sunroof	5 years
Water/wind/air leaks	5 years

BRAKE SYSTEM

Brake drum	120,000 km
Brake drum linings	35,000 km
Brake rotor	60,000 km
Disc brake calipers	30,000 km
Disc brake pads	30,000 km
Master cylinder, rebuild	100,000 km
Wheel cylinder, rebuild	80,000 km

ENGINE AND DRIVETRAIN

Constant velocity joint	6 years/ 160,000 km
Differential	7 years/ 160,000 km
Engine head gasket	7 years/ 160,000 km
Intake manifold gasket	7 years/ 160,000 km
Radiator	4 years/ 80,000 km
Transfer case	7 years/ 150,000 km
Transmission (auto.)	7 years/ 150,000 km
Transmission (man.)	10 years/ 200,000 km
Transmission oil cooler	5 years/ 100,000 km

EXHAUST SYSTEM

Catalytic converter	5 years/ 100,000 km or more
Muffler	2 years/ 40,000 km
Tailpipe	3 years/ 60,000 km

IGNITION SYSTEM

Cable set	60,000 km
Electronic module	5 years/ 80,000 km
Retiming	20,000 km
Spark plugs	20,000 km
Tune-up	20,000 km

SAFETY COMPONENTS

ABS brakes	7 years/ 160,000 km
ABS computer	10 years/ 160,000 km
Airbags	life of vehicle
Fuel gauge	5 years/ 100,000 km
Seat belts	life of vehicle

STEERING AND SUSPENSION

Alignment	1 year/ 20,000 km

Ball joints	80,000 km
Coil spring	10 years
Power steering	5 years/ 80,000 km
Shock absorber	3 years/ 60,000 km
Struts	5 years/ 80,000 km
Tires (radial)	5 years/ 80,000 km
Wheel bearing	3 years/ 60,000 km

VISIBILITY

Halogen/fog lights	3 years/ 60,000 km
Sealed beam	2 years/ 40,000 km
Windshield wiper motor	5 years/ 80,000 km

Much of the preceding guidelines were extrapolated from Chrysler and Ford payouts to thousands of dissatisfied customers over the past decade, in addition to Chrysler's original 7-year powertrain warranty applicable from 1991–95 and reapplied since 2001. Other sources for this chart were the Ford and GM transmission warranties outlined in their secret warranties; Ford, GM, and Toyota engine "goodwill" programs laid out in their internal service bulletins; and court judgments where judges have given their own guidelines as to what reasonable durability is.

Safety features generally have a lifetime warranty, with the exception of ABS, which Chrysler sees as having a 10-year lifespan.

Airbags are a different matter. Those that are deployed in an accident—and the personal injury and interior damage their deployment will likely have caused—are covered by your accident insurance policy. However, if there is a sudden deployment for no apparent reason, the automaker and dealer should be held jointly responsible for all injuries and damages caused by the airbag.

If your vehicle is equipped with a data recorder, you may be able to prove the automaker's liability by downloading the data. These small recorders are hidden under the seat or in the centre console and record data during the last five seconds before impact. This includes airbag performance. There will be more on these "black boxes" later in this section. Data-recorder evidence will likely lead to a more generous settlement from the two parties and prevent your insurance premiums from being jacked up. Inadvertent deployment may occur after passing over a bump in the road, slamming the car door, or, in some Chrysler minivans, simply putting the key in the ignition. This happens more often than you might imagine, judging by the hundreds of recalls and thousands of complaints recorded on the NHTSA website (*www.nhtsa.dot.gov/cars/problems/complain/Index.cfm*).

Finally, the manufacturer's emissions warranty is the primary guideline showing the expected durability of high-tech electronic and mechanical pollution control components, like powertrain control modules (PCM) and catalytic converters. Look first at your owner's manual for an indication of which parts on your vehicle are covered. If you come up with few specifics, ask the auto manufacturer for a list of specific components covered by the emissions warranty. If you're stonewalled, ask your local MP to get the info from

Transport Canada or Environment Canada, and invest in an ALLDATA Internet download or service bulletin DVD, where your car's emissions parts will be outlined.

Recall repairs

Vehicles are recalled for one of two reasons: they may be unsafe or they don't conform to federal and state pollution control regulations. Whatever the reason, recalls are a great way to get free repairs—if you know which ones apply to you and you have the patience of Job.

More than 450 million unsafe vehicles have been recalled by automakers for the free correction of safety-related defects since American recall legislation was passed in 1966 (equivalent Canadian laws were enacted in 1971). During that time, about one-third of the recalled vehicles never made it back to the dealership for repairs because owners were never informed, they gave up waiting for corrective parts, or they just didn't consider the defect to be that hazardous. This has been particularly evident with GM's handling of airbag defects, where owners waited over a year for parts to arrive, all the while fearful the airbags could suddenly explode in their face without provocation.

If you've moved or bought a used vehicle, it's smart to pay a visit to your local dealer, give him your address, and get a "report card" on which recalls, free-service campaigns, and warranties apply to your vehicle. Simply give the service advisor the identification number (VIN)—found on the dash just below the windshield on the driver's side, or on your insurance card—and have the number run through the automaker's computer system. Ask for a computer printout of the vehicle's history (have it faxed to you, if you're so equipped) and make sure you're listed in the automaker's computer as the new owner. This ensures that you'll receive notices of warranty extensions and emissions and safety recalls.

Still, don't expect to be welcomed with open arms when your vehicle develops a safety- or emissions-related problem that's not yet part of a recall campaign. Automakers and dealers generally take a restrictive view of what constitutes a safety or emissions defect and frequently charge for repairs that should be free under federal safety or emissions legislation. To counter this tendency, look at the following list of typical defects that are clearly safety related. If you experience similar problems, insist that the automaker fix the problem at no expense to yourself, including a car rental:

- airbag malfunctions
- corrosion affecting safe operation
- disconnected or stuck accelerators
- electrical shorts
- faulty windshield wipers
- fuel leaks
- problems with original axles, drive shafts, seats, seat recliners, or defrosters
- seat belt problems
- stalling or sudden acceleration
- sudden steering or brake loss

- suspension failures
- trailer coupling failures

In the U.S., recall campaigns force automakers to pay the entire cost of fixing a vehicle's safety-related defect for any vehicle purchased up to 8 years before the recall's announcement. A reasonable period beyond that time is usually a slam-dunk in small claims court. Recalls may be voluntary or ordered by the U.S. Department of Transportation. Canadian regulation has an added twist: Transport Canada can only order automakers to notify owners that their vehicles may be unsafe; it can't force them to correct the problem. Fortunately, most U.S.-ordered recalls are carried out in Canada, and when Transport Canada makes a defect determination on its own, automakers generally comply with an owner notification letter.

Voluntary safety recall campaigns, frequently called Special Service or Safety Improvement Campaigns, are a real problem, though. The notification of owners isn't monitored by the government; dealers and automakers routinely deny there's a recall, thereby dissuading most claimants; and the company's so-called fix, not authorized by any governing body, may not correct the hazard at all. Also, it may leave out many of the affected models or unreasonably exclude certain owners.

Safety defect information

If you wish to report a safety defect or want recall info, you may access Transport Canada's website at *www.tc.gc.ca/roadsafety/recalls/search_e.asp.* But be careful: Transport Canada's site misses many recall campaigns; compare its info with NHTSA's recall website. You can get recall information in French or English, as well as general information relating to road safety and importing a vehicle into Canada. Web surfers can now access the recall database for 1970–2004 model vehicles but, unlike NHTSA's website, owner complaints aren't listed, defect investigations aren't disclosed, voluntary warranty extensions (secret warranties) aren't shown, and service bulletin summaries aren't provided. You can also call Transport Canada at 1-800-333-0510 (toll-free within Canada) or 613-993-9851 (within the Ottawa region or outside Canada) to get additional information.

If you're not happy with Ottawa's treatment of your recall inquiry, try the U.S. government's NHTSA website. It's more complete than Transport Canada's site. (NHTSA's database is updated daily and covers vehicles built since 1952.) You can search the database for your vehicle or tires at *www.nhtsa.dot.gov/cars/problems/.* You'll get immediate access to four essential database categories applicable to your vehicle and model year: the latest recalls, current and closed safety investigations, defects reported by other owners, and a brief summary of technical service bulletins.

NHTSA's fax-back service provides the same info through a local line that can be accessed from Canada—although long-distance charges will apply. (Most calls take 5–10 minutes to complete.) The following local numbers get you into the automatic response service quickly and can be reached 24 hours a day: 202-366-0123 (202-366-7800 for the hearing impaired).

Black boxes: "snitches" and "saviours"?

Your car or truck is probably spying on you.

Data recorders the size of a VCR tape have been hidden under the seat or in the centre consoles of about 25 million Ford and GM vehicles sold in North America since the early 1990s.

Motorists weren't told at first that they were, in effect, unwitting test subjects used to amass accident data for American automakers' probing reports on malfunctioning ABS brakes and airbags. Now the disclosure is buried in the owner's manual, resulting in nearly two-thirds of people surveyed by an American insurance industry group still saying they knew nothing about them.

The data recorders operate in a similar fashion to flight data recorders used in airplanes: recording data during the last five seconds before impact, including the force of the collision, the airbag's performance, when the brakes were applied, engine and vehicle speed, gas pedal position, and whether the driver was wearing a seatbelt.

In addition to the "invasion of privacy" aspect of hiding recorders in customers' vehicles, Ford and GM have systematically hidden their collected data from U.S. and Canadian vehicle safety researchers investigating thousands of complaints relating to airbags that don't deploy when they should (or deploy when they shouldn't) and anti-lock brakes that don't brake.

This refusal to share data with customers and researchers is unfortunate, because the recorders are collecting critical information that could lead to better-functioning safety devices. In fact, experts say that highway safety could be vastly improved if black boxes that record information about car crashes were installed in all cars, just as similar devices are placed in all airplanes.

Fortunately, it has become impossible for automakers to hide recorder data now that Vitronix Corp. sells a $2,500 (U.S.) portable download device that accesses the data and stores it in any PC. It's presently marketed to accident reconstructionists, safety researchers, law enforcement agencies, and insurance companies.

Safety benefits

Enthusiastically promoted by government and law enforcement agencies around the world, these data recorders have actually had a positive effect in accident prevention: European studies show that drivers who know their vehicles are equipped with the device have 20–30 percent fewer accidents and drive more slowly.

The recorders are also sending people to jail, helping accident victims reap huge court awards, and prompting automaker recalls of unsafe vehicles. Last June, Edwin Matos of Pembroke Pines, Fla., was sentenced to 30 years in prison for killing two teenage girls, after crashing into their car at more than 100 mph. The recorder's speed data convicted him. Two months earlier, an Illinois police officer received a $10 million (U.S.) settlement after data showed the supposedly unconscious driver accelerated and braked in the moments before slamming into the officer's patrol car. Data recorders showed GM that its air bags were deploying inadvertently, forcing a recall in 1998 of more than 850,000 Cavaliers and Sunfires.

Parents of young drivers also have a keen interest in purchasing aftermarket data recorders to ensure their children are driving safely.

Three Steps to a Settlement

Step 1: Informal negotiations

If your vehicle was misrepresented, has major defects, or wasn't properly repaired under warranty, the first thing you should do is give the seller (the dealer and automaker or a private party) a written summary (by registered mail or fax) of the outstanding problems and stipulate a time period in which they will need to be corrected or your money will be refunded. Keep a copy for yourself, along with all your repair records. Be sure to check all of the sales and warranty documents you were given to see if they conform to provincial laws. Any errors, omissions, or violations can be used to get a settlement with the dealer in lieu of making a formal complaint.

At the beginning, try to work things out informally and, in your attempt to reach a settlement, keep in mind the cardinal rule: ask only for what is fair and don't try to make anyone look bad.

Speak in a calm, polite manner and try to avoid polarizing the issue. Talk about how "we can work together" on the problem. Let a compromise slowly emerge—don't come in with a hardline set of demands. Don't demand the settlement offer in writing, but make sure that you're accompanied by a friend or relative who can confirm the offer in court if it isn't honoured. Be prepared to act upon the offer without delay so your hesitancy won't be blamed for its withdrawal.

Dealer/service manager

Service managers have more power than you may have realized. They make the first determination of what work is covered under warranty or through post-warranty "goodwill" programs and are directly responsible to the dealer and manufacturer for that decision (dealers hate manufacturer audits that force them to pay back questionable warranty decisions). Service managers are paid to save the dealer and automaker money and to mollify irate clients—almost an impossible balancing act. Nevertheless, when a service manager agrees to extend warranty coverage, it's because you've raised solid issues that neither the dealer nor automaker can ignore. All the more reason to present your argument in a confident, forthright manner with your vehicle's service history and *Lemon-Aid*'s "How Long Should Parts/Repairs Last?" chart. Also bring as many technical service bulletins and owner complaint printouts as you can find, from websites like NHTSA's. It's not important that they apply directly to your problem; they establish parameters for giving out after-warranty assistance or "goodwill." Don't use your salesperson as a runner, since the sales staff are generally quite distant from the service staff and usually have less pull than you do. If the service manager can't or won't set things right, your next step is to convene a mini-summit with the service manager, the dealership principal, and the automaker's rep. By getting the automaker involved, you run less risk of having the dealer fob you off on the manufacturer and you can often get an agreement where the seller and automaker pay two-thirds of the repair cost.

Independent dealers and dealers who sell a brand of used vehicle that they don't sell new give you less latitude. You have to make the case that the vehicle's defects were present at the time of purchase or should have been known to the seller, or that the vehicle doesn't conform to the representations made when it was purchased. Emphasize that you intend to use the courts if necessary to obtain a refund—most independent sellers would rather settle than risk a lawsuit with all the attendant publicity. An independent estimate of the vehicle's defects and cost of repairs is essential if you want to convince the seller that you're serious in your claim and stand a good chance of winning your case in court. Come prepared with an estimated cost of repairs to challenge the dealer who agrees to pay half the repair costs and then jacks up the costs 100 percent so that you wind up paying the whole shot.

Step 2: Sending a registered letter, fax, or email

This is the step to take next if your claim is refused. Send the dealer and manufacturer a polite registered letter or fax that asks for compensation for repairs that have been done or need to be done; insurance costs while the vehicle is being repaired; towing charges; supplementary transportation costs like taxis and rented cars; and damages for inconvenience.

Specify five days (but allow 10) for either party to respond. If no satisfactory offer is made, file suit in small claims court. Make the manufacturer a party to the lawsuit, especially if the emissions warranty, a secret warranty extension, a safety recall campaign, or extensive chassis rusting is involved.

New Vehicle Complaint Letter/Fax/Email
Without Prejudice

Date:
Name and address of dealer:
Name and address of manufacturer:

Please be advised that I am not satisfied with my ____________ (indicate year, make, model, and serial number of vehicle). The vehicle was purchased on (indicate date) and currently indicates __________ km on the odometer. The vehicle presently exhibits the following defects:

1. Premature rusting
2. Paint peeling/discoloration
3. Water leaks
4. Other defects (explain)

(List previous attempts to repair the vehicle. Attach a copy of a report from an independent garage, showing cost of estimated repairs and confirming the manufacturer's responsibility.)

I hereby request that you correct these defects free of charge under the terms of the implied warranty provisions of provincial consumer protection statutes as applied in *Kravitz v. General Motors* (1979), I.S.C.R., and *Chabot v. Ford* (1983), 39 O.R. (2d).

If you do not correct the defects noted above to my satisfaction and within a reasonable length of time, I will be obliged to ask an independent garage to ___________________ (choose [a] estimate or [b] carry out) the repairs and claim the amount of $___________ (state the cost, if possible) by way of the courts without further notice or delay.

I have dealt with your company because of its competence and honesty. I close in the hope of hearing from you within five (5) days of receiving this letter, failing which I will exercise the alternatives available to me. Please govern yourself accordingly.

Sincerely,

(signed with telephone or fax number)

Used Vehicle Complaint Letter/Fax/Email
Without Prejudice

Date: ________________
Name: ________________

Please be advised that I am dissatisfied with my used vehicle, a (state model), for the following reasons:

1. ________________________________
2. ________________________________
3. ________________________________
4. ________________________________
5. ________________________________

In compliance with the provincial consumer protection laws and the "implied warranty" set down by the Supreme Court of Canada in *Donoghue v. Stevenson* and *Longpré v. St-Jacques Automobile*, I hereby request that these defects be repaired without charge.

This vehicle has not been reasonably durable and is, therefore, not as represented to me.

Should you fail to repair these defects in a satisfactory manner and within a reasonable period of time, I shall get an estimate of the repairs from an independent source and claim them in court, without further delay. I also reserve my right to claim up to $1 million for punitive damages, pursuant to the Supreme Court of Canada's February 22, 2002, ruling in *Whiten v. Pilot*.

I have dealt with your company because of its honesty, competence, and sincere regard for its clients. I am sure that my case is the exception and not the rule.

A positive response within the next five (5) days would be appreciated.

Sincerely,

(signed with telephone or fax number)

Secret Warranty Claim Letter/Fax/Email
Without Prejudice

Date: ______________
Name: ______________

Please be advised that I am dissatisfied with my vehicle, a ____________, bought from you on ___________.
It has had the following recurring problems that I believe are factory-related defects, as confirmed by internal service bulletins sent to dealers, and are covered by your "goodwill" policies:

1. ____
2. ____
3. ____

If your "goodwill" program has ended, I ask that my claim be accepted nevertheless, inasmuch as I was never informed of your policy while it was in effect and should not be penalized for not knowing it existed.

I hereby formally put you on notice under federal and provincial consumer protection statutes that your refusal to apply this extended warranty coverage in my case would be an unfair warranty practice within the purview of the above-cited laws.

Your actions also violate the "implied warranty" set down by the Supreme Court of Canada (*Donoghue v. Stevenson* and *Longpre v. St. Jacques Automobile*) and repeatedly reaffirmed by provincial consumer protection laws (*Lowe v. Chrysler, Dufour v. Ford du Canada*, and *Frank v. GM*).

I have enclosed several estimates (my bill) showing that this problem is factory related and will (has) cost $__________ to correct. I would appreciate your refunding me the estimated (paid) amount, failing which, I reserve the right to have the repair done elsewhere and claim reimbursement in court without further delay. I also reserve the right to claim up to $1 million for punitive damages, pursuant to the Supreme Court of Canada's February 22, 2002, ruling in *Whiten v. Pilot*.

A positive response within the next five (5) days would be appreciated.

Sincerely,

(signed with telephone or fax number)

Step 3: Mediation and arbitration

If the formality of a courtroom puts you off or you're not sure that your claim is all that solid and don't want to pay legal costs to find out, consider using mediation or arbitration. These services are sponsored by the Better Business Bureau, the Automobile Protection Association, the Canadian Automobile Association, the Canadian Automobile Manufacturers Vehicle Arbitration Program at *camvap.ca* (if you bought your vehicle new), small claims court (mediation is often a prerequisite to going to trial), and consumer mediation services set up by provincial and territorial governments.

CAMVAP arbitration hearings often result in a vehicle being repaired or replaced, as this Toronto owner of a 2001 Nissan Maxima discovered:

> At 10,000 km, my steering wheel would shimmy when the brakes were applied. I brought the car in to the dealership and they turned the rotors. The problem went away only to reappear at 18,000 km. At this time the dealership replaced the rotors. At 30,000 km, the problem returned.
>
> Once again the rotors were turned down and the problem went away. At 36,000 km the problem returned for the fourth time. This time around, Nissan Canada refused to fix the brakes. I got nowhere with Nissan Canada Customer Service. They were extremely rude and confrontational with me.
>
> I demanded CAMVAP arbitration.
>
> The result of the hearing was that Nissan had to reimburse my costs for brake repairs. They have also been given one last chance to fix the problem (they will replace pads and rotors). If, within the next 20,000 km, the brake problem reappears, upon examination of the car by my mechanic (indicating the car has not been driven hard), Nissan will have to buy the car back at the odometer reading at the hearing.

Getting outside help

Don't lose your case due to poor preparation. Ask government or independent consumer protection agencies to evaluate how well you're prepared before going to your first hearing. Also, use the Internet to ferret out additional facts and gather support (*www.lemonaidcars.com* is a good place to start).

Online services/Internet/websites

America Online and CompuServe are two online service providers with active consumer forums that use experts to answer consumer queries and to provide legal and technical advice. The Internet offers the same information using a worldwide database. If you or someone you know is able to create a website, you might consider using this site to attract attention to your plight and arm yourself for arbitration or court. You may wish to follow the example of some existing websites I've listed in Appendix I. A few of my favourites are: Chrysler Peeling Paint; Neon Enthusiasts (engine head gaskets); Chrysler, Plymouth, and Dodge Car Information; Taurus Transmission Victims; Toyota Celica Page; and the VW Lemon Page.

Classified ads

Use your local paper's classified section or *The Globe and Mail*'s "National Personals" column to gather data from others who may have experienced a problem similar to your own. This alerts others to the potential problem, helps build a base for a class action or group meeting with the automaker, and puts pressure on the dealer or manufacturer to settle. Sometimes the paper's news desk will assign someone to cover your story after your ad is published.

Federal and provincial consumer affairs

The wind left the sails of the consumer movement over a decade ago, leaving provincial consumer affairs offices understaffed and unsupported by the government. This has created a passive mindset among many staffers, who are tired of getting their heads kicked in by businesses and budget-cutters.

Consumer affairs offices can still help with investigation, mediation, and some litigation. Strong and effective consumer protection legislation has been left standing in most of the provinces, and resourceful consumers can use these laws in conjunction with media coverage to prod provincial consumer affairs offices into action. Furthermore, provincial bureaucrats aren't as well shielded from criticism as their federal counterparts. A call to your MPP or MLA, or to their executive assistant, can often get things rolling.

Federal consumer protection is a government-created PR myth. Don't expect the staffers in the reorganized Office of Consumer Affairs to be very helpful—they've been de-fanged and de-gummed through budget cuts and a succession of ineffective ministers. Although the beefed-up Competition Act has some bite with regards to misleading advertising and a number of other illegal business practices, the federal government has been more reactive than proactive in applying the law.

Nevertheless, you can lodge a formal complaint with Ottawa for misleading advertising, odometer tampering, or price fixing at *https:strategis.ic.gc.ca/sc_mrksv/competit/complaint/form.html*. An online complaint sent to the address above made Toyota cease its ACCESS price-fixing practices (though rumour has it, the company's dealers may try it on their own).

Invest in protest

You can have fun and put additional pressure on a seller or garage by putting a lemon sign on your car and parking it in front of the dealer or garage; creating a "lemon" website; or forming a self-help group like the Chrysler Lemon Owners Group (CLOG), the Ford Lemon Owners Group (FLOG), or the Ford Focus Lemon Owners Group (FFLOG).

> I went picketing last Saturday for two hours in front of the Suburban Motors dealership. I think I got a rise. The manager Michael Barron came charging out of the dealership yelling "what are you doing to me??!!" He also told me that it was very unprofessional, that I looked ridiculous.... Then he went away. An hour later, he walked to the end of his dealership, turned and started walking towards me with a camera in his face. He told me that I looked ridiculous and "won't the *Sun* be happy to see these pictures."
>
> He then sent out one of the carwashers to stand in front of me to block my sign (my back sign was still visible...). That carwasher stayed with me for 10 minutes waving a huge Ford Explorer sign in front of mine. So, that was my first day on the picket line. I plan to hit the dealership this Saturday and will tell you what happens.
>
> Ford Focus Lemon Group
> *flogbc@hotmail.com*

After forming your group, you can then have the occasional parade of creatively decorated cars visit area dealerships as the local media are convened. Just remember to keep your remarks pithy and factual, don't interfere with traffic or customers, and remain peaceful.

Use your website to gather data from others who may have experienced a problem similar to your own. This alerts others to the potential problem, helps build a base for a class action or group meeting with the automaker, and puts pressure on the dealer or manufacturer to settle. Sometimes the news media will assign someone to cover your story after the website is set up.

One other piece of advice from this consumer advocate with hundreds of picketing and mass demonstrations under his belt over the past 31 years: keep a sense of humour and never break off negotiations.

Finally, don't be scared off by threats that it's illegal to criticize a product or company. Unions, environmentalists, and consumer groups do it regularly ("informational" picketing) and the Supreme Court of Canada in *R. v. Guinard* reaffirmed this right in February 2002. In that judgment, an insurance policyholder posted a sign on his barn claiming the Commerce Insurance Company was unfairly refusing his claim. The municipality of St-Hyacinthe told him to take the sign down. He refused, maintaining that he had the right to state his opinion. The Supreme Court agreed.

This judgment means that consumer protests, signs, and websites that criticize the actions of corporations cannot be shut up or taken down simply because they say unpleasant things.

Safety and Performance Defects: Step-by-Step Resolution

Sudden acceleration, chronic stalling, and ABS and airbag failures

Incidents of sudden acceleration or chronic stalling are quite common. However, they are very difficult to diagnose and are treated quite differently by federal safety agencies. Sudden acceleration is considered to be a safety-related problem—stalling isn't. Never mind that a vehicle's sudden loss of power on a busy highway puts everyone's lives at risk. 2001–03 VW and Audi ignition coil failures are an example. The same problem exists with engine and transmission powertrain failures, which are only occasionally considered to be safety related. ABS and airbag failures are universally considered to be life-threatening defects. If your vehicle manifests any of these conditions, here's what you need to do:

1. Get independent witnesses to the fact that the problem exists. This includes verification by an independent mechanic, passenger accounts, downloaded data from your vehicle's data recorder and lots of Internet browsing using *www.lemonaidcars.com* and Google's browser as your primary tools. Notify the dealer/manufacturer by fax, email, or registered letter that you consider the problem to be a factory-induced, safety-related defect. Make sure you address your correspondence to the manufacturer's product liability or legal affairs department. At the dealership's service bay, make sure that every work order clearly states the problem, as well as the number of previous attempts to fix it. (This should result in you having a few complaint letters and a handful of work orders, confirming that this is an ongoing deficiency.) If the dealer won't give you a copy of the work order because the work is a warranty claim, ask for a copy of the order number "in case your estate wishes to file a claim, pursuant to an accident." (This will get the service manager's attention.) Leaving this kind of "paper trail" is crucial for any claim you may have later on because it shows your fear and persistence, and clearly indicates that the dealer and manufacturer had ample time to correct the defect. In California, for example, the state's recently revamped Lemon Law requires that car owners clearly show they made two attempts to have a safety defect corrected (other states require three or four attempts) before the court will grant a refund, order the car taken back, or impose punitive damages.
2. Note on the work order that you expect the problem to be diagnosed and corrected under the emissions warranty or a "goodwill" program. It also wouldn't hurt to add the phrase on the work order or in your claim letters that any deaths, injuries, or damage caused by the defect will be the dealer's and manufacturer's responsibility since this work order (or letter, fax, or email) constitutes you putting them on "formal notice."

3. If the dealer does the necessary repairs at little or no cost to you, send a follow-up confirmation that you appreciate the assistance. Also, emphasize that you'll be back if the problem reappears, even if the warranty has expired, because the repair renews your warranty rights applicable to that defect. In other words, the warranty clock is set back to its original position. Understand that you won't likely get a copy of the repair bill, either, because dealers don't like to admit that there was a serious defect present. Keep in mind, however, that you can get your complete vehicle file from the dealer and manufacturer by issuing a subpoena (cost: about $25), if the case goes to small claims or a higher court. This request has produced many out-of-court settlements when the internal documents show extensive work was carried out to correct the problem.
4. If the problem persists, send a letter, fax, or email to the dealer and manufacturer saying so, look for ALLDATA service bulletins to confirm your vehicle's defects are factory related, and call Transport Canada or NHTSA or log onto NHTSA's website to report the failure. Also, call the Nader-founded Center for Auto Safety in Washington, D.C., (Tel.: 202-328-7700) for a lawyer referral and an information sheet covering the problem. For tire complaints, also notify researchers at the Strategic Safety website (*www.strategicsafety.com*).
5. Now come two crucial questions: repair the defect now or later; use the dealer or an independent? Generally it's smart to use an independent garage if you know the dealer isn't pushing for free corrective repairs from the manufacturer; weeks or months have passed without any resolution of your claim; the dealer keeps repeating it's a maintenance item; and you know an independent mechanic who will give you a detailed work order showing the defect is factory related and not due to poor maintenance. Don't mention that a court case may ensue, since this will scare the dickens out of your only independent witness. An added bonus is that the repair charges will be about half of what a dealer would demand. Incidentally, if the automaker later denies warranty "goodwill" because you used an independent repairer, use the argument that the defect's safety implications required emergency repairs, carried out by whomever could see you first.
6. Dashboard-mounted warning lights usually come on prior to airbags suddenly deploying, ABS brakes failing, or engine glitches causing the vehicle to stall out. (Sudden acceleration usually occurs without warning.) Automakers consider these lights to be critical safety warnings and generally advise drivers to *immediately* have the vehicle serviced to correct the problem (advice found in the owner's manual) when any of the above lights come on. This bolsters the argument that your life was threatened, emergency repairs were required, and your request for another vehicle or a complete refund isn't out of line.
7. Sudden acceleration can have multiple causes, isn't easy to duplicate, and is often blamed on the driver mistaking the accelerator for the brakes or failing to perform proper maintenance. Yet NHTSA data shows that with the 1992–2000 Explorer, for example, a faulty cruise control or PCV valve

and poorly mounted pedals are the most likely causes of the Explorer's sudden acceleration. So how do you satisfy the burden of proof showing the problem exists and is the automaker's responsibility? Use the legal doctrine called "the balance of probabilities" by eliminating all of the possible dodges the dealer or manufacturer may employ. Show that proper maintenance has been carried out, you're a safe driver, and the incident occurs frequently and without warning.

8. If any of the above defects causes an accident, the airbag fails to deploy, or you're injured by its deployment, ask your insurance company to have the vehicle towed to a neutral location and clearly state that neither the dealer nor automaker should touch the vehicle until your insurance company and Transport Canada have completed their investigation. Also, get as many witnesses as possible and immediately go to the hospital for a check-up, even if you're feeling okay. You may be injured and not know it because the adrenalin coursing through your veins is masking your injuries. Plus, a hospital exam will easily confirm that your injuries are accident related, which is essential in court or for future settlement negotiations.
9. Peruse NHTSA's online accident database to find reports of other accidents caused by the same failure.
10. Don't let your insurance company settle the case if you're sure the accident was caused by a mechanical failure. Even if an engineering analysis fails to directly implicate the manufacturer or dealer, you can always plead the aforementioned balance of probabilities. If the insurance company settles, your insurance premiums will probably be increased.

Defective tires

Tire companies are far easier to deal with than automobile manufacturers because, under the legal doctrine of *res ipsa loquitor* (liability is shown by the failure), tires aren't supposed to fail. It's for this reason that tire companies try to avoid liability by imputing blame to someone or something else, like punctures, impact damage, overloading, over-inflating, or under-inflating. If you have a premature tire failure, consider the 10 steps outlined previously, and include the following.

1. Access NHTSA and Strategic Safety websites on the Internet (see Appendix I) for current data on which tires are failure prone and which companies are under investigation, conducting recalls, or carrying out "silent recalls."
2. Keep the tire. If the tiremaker says an analysis must be done, permit only a portion of the tire to be taken away.
3. Plead the balance of probabilities, using friends and family to refute the tire company's contention that you caused the failure.
4. Ask for damages that are adequate for the replacement of all the tires on your vehicle, including mounting costs.
5. Include in your damage claim any repairs needed to fix body damage caused by the tire's failure.

Paint and body defects

The following settlement advice applies mainly to paint defects, but you can use these tips for any other vehicle defect that you believe is the automaker's or dealer's responsibility. If you're not sure that the problem is a factory-related deficiency or a maintenance item, have it checked out by an independent garage or get a technical service bulletin summary for your vehicle. The summary may include specific bulletins relating to the diagnosis, correction, and ordering of upgraded parts needed to fix your problem.

1. If you believe the paint problem is factory related, take your vehicle to the dealer and ask for a written, signed estimate of what the service manager feels needs to be done. When you're handed the estimate, ask if the paint job can be covered by some "goodwill" assistance. (Ford's euphemism for this secret warranty is "Owner Notification Program" or "Owner Dialogue Program," GM's term is "Special Policy," and Chrysler simply calls it "Owner Satisfaction Notice" or "goodwill." Don't use the term "secret warranty" yet; you'll just make everyone angry and evasive.)
2. Your request will probably be met with a refusal, an offer to repaint the vehicle for half the cost, or (if you're lucky) an agreement to repaint the vehicle free of charge. If you accept half the costs, make sure that it's based on the original estimate you have in hand, since some dealers jack up their estimates so that your 50 percent is really 100 percent of the true cost.
3. If the dealer or automaker has already refused your claim and the repair hasn't been done yet, get an additional estimate from an independent garage that shows the problem is factory related.
4. Again, if the repair has yet to be done, mail or fax a registered claim to the automaker (send a copy to the dealer), claiming the average of both estimates. If the repair has been done at your expense, mail or fax a registered claim with a copy of your bill.
5. If you don't receive a satisfactory response within a week, deposit a copy of the estimate or paid bill and claim letter/fax before the small claims court and await a trial date. This means that the automaker/dealer will have to appear, no lawyer is required, and costs should be minimal (under $100). Usually, an informal pretrial mediation hearing with the two parties and a court clerk will be scheduled in a few months, followed by a trial a few weeks later (the time varies among different regions). Most cases are settled at the mediation stage.

Things that you can do to help your case: collect photographs, maintenance work orders, previous work orders dealing with your problem, and technical service bulletins; and speak to an independent expert (the garage or body shop that did the estimate or repair is best, but you can also use a local teacher who teaches automotive repair). Remember, service bulletins can be helpful, but they aren't critical to a successful claim. Concentrate on what you can prove happened to your vehicle, not what may have occurred as outlined in the TSB.

Other situations

- If the vehicle has just been repainted but the dealer says that "goodwill" coverage was denied by the automaker, pay for the repair with a certified cheque and write "under protest" on the cheque. Remember, though, that if the dealer does the repair, you won't have an independent expert who can affirm that the problem was factory related or that it was a result of premature wearout. Plus, the dealer can say that you or the environment caused the paint problem. In these cases, technical service bulletins can make or break your case.
- If the dealer/automaker offers a partial repair or refund, take it. Then sue for the rest. Remember, if a partial repair has been done under warranty, it counts as an admission of responsibility, no matter what "goodwill" euphemism is used. Also, the repaired component/body panel should be just as durable as if it were new. Hence, the clock starts ticking from the beginning until you reach the original warranty parameter—again, no matter what the dealer's repair warranty limit says.
- It's a lot easier to get the automaker to pay to replace a defective part than it is to be compensated for a missed day of work or a ruined vacation. Manufacturers hate to pay for consequential expenses—apart from towing bills—because they can't control the amount of the refund. Fortunately, Canadian courts have taken the position that all expenses (damages) flowing from a problem covered by a warranty or service bulletin are the manufacturer's/dealer's responsibility under negligence and product liability provisions found in provincial consumer protection statutes, common law jurisprudence, Quebec civil law, and federal consumer protection legislation. Nevertheless, don't risk a fair settlement for some outlandish claim of "emotional distress," "pain and suffering," etc. If you have invoices to prove actual consequential damages, then use them. If not, don't be greedy.

Very seldom do automakers contest these paint claims before small claims court, opting instead to settle once the court claim is bounced from their customer relations people to their legal affairs department. At that time, you'll probably be offered an out-of-court settlement for 50–75 percent of your claim.

Stand fast and make reference to the service bulletins you intend to subpoena in order to publicly contest in court the unfair nature of this "secret warranty" program. (Automaker lawyers cringe at the idea of trying to explain why consumers aren't made aware of these bulletins.) One hundred percent restitution will probably follow.

Three good examples of favourable paint judgments are *Shields v. General Motors of Canada*, *Bentley v. Dave Wheaton Pontiac Buick GMC Ltd. and General Motors of Canada*, and, the most recent, *Maureen Frank v. General Motors of Canada Limited.*

Shields v. General Motors of Canada, No. 1398/96, Ontario Court (General Division), Oshawa Small Claims Court, 33 King Street West, Oshawa,

Ontario L1H 1A1, July 24, 1997, Robert Zochodne, Deputy Judge. The owner of a 1991 Pontiac Grand Prix purchased the vehicle used with over 100,000 km on its odometer. Commencing in 1995, the paint began to bubble and then flake and eventually peel off. Deputy Judge Robert Zochodne awarded the plaintiff $1,205.72 and struck down every one of GM's environmental/acid rain/UV rays arguments. Other important aspects of this 12-page judgment that GM did not appeal:

1. The judge admitted many of the technical service bulletins referred to in *Lemon-Aid* as proof of GM's negligence.
2. Although the vehicle had 156,000 km when the case went to court, GM still offered to pay 50 percent of the paint repairs if the plaintiff dropped his suit.
3. Deputy Judge Zochodne ruled that the failure to protect the paint from the damaging effects of UV rays is akin to engineering a car that won't start in cold weather. In essence, vehicles must be built to withstand the rigours of the environment.
4. Here's an interesting twist: the original warranty covered defects that were present at the time it was in effect. The judge, taking statements found in the GM technical service bulletins, ruled the UV problem was factory related, and therefore it existed during the warranty period and thereby represented a latent defect that appeared once the warranty expired.
5. The subsequent purchaser was not prevented from making the warranty claim, even though the warranty had long since expired from a time and mileage standpoint and he was the second owner.

Bentley v. Dave Wheaton Pontiac Buick GMC Ltd. and General Motors of Canada, Victoria Registry No. 24779, British Columbia Small Claims Court, December 1, 1998, Judge Higinbotham. This small claims judgment builds upon the Ontario *Shields v. General Motors of Canada* decision and cites other jurisprudence as to how long paint should last on a car. If you're wondering why Ford and Chrysler haven't been hit by similar judgments, remember that they usually settle.

Maureen Frank v. General Motors of Canada Limited, No. SC#12 (2001), Saskatchewan Provincial Court, Saskatoon, Saskatchewan, October 17, 2001, Provincial Court Judge H.G. Dirauf.

> On June 23, 1997, the Plaintiff bought a 1996 Chevrolet Corsica from a General Motors dealership. At the time the odometer showed 33,172 km. The vehicle still had some factory warranty. The car had been a lease car and had no previous accidents.
>
> During June of 2000, the Plaintiff noticed that some of the paint was peeling off from the car and she took it to a General Motors

dealership in Saskatoon and to the General Motors dealership in North Battleford where she purchased the car. While there were some discussions with the GM dealership about the peeling paint, nothing came of it and the Plaintiff now brings this action claiming the cost of a new paint job.

During 1999, the Plaintiff was involved in a minor collision causing damage to the left rear door. This damage was repaired. During this repair some scratches to the left front door previously done by vandals were also repaired.

The Plaintiff's witness, Frank Nemeth, is a qualified auto body repairman with some 26 years of experience. He testified that the peeling paint was a factory defect and that it was necessary to completely strip the car and repaint it. He diagnosed the cause of the peeling paint as a separation of the primer surface or colour coat from the electrocoat primer. In his opinion no primer surfacer was applied at all. He testified that once the peeling starts, it will continue. He has seen this problem on General Motors vehicles. The defect is called delamination.

Mr. Nemeth stated that a paint job should last at least 10 years. In my opinion most people in Saskatchewan grow up with cars and are familiar with cars. I think it is common knowledge that the original paint on cars normally lasts in excess of 15 years and that rust becomes a problem before the paint fails. In any event, paint peeling off, as it did on the Plaintiff's vehicle, is not common. I find that the paint on a new car put on by the factory should last at least 15 years.

Counsel for the defendant submitted that any award I should make should be reduced because of betterment. While betterment was discussed at trial, I am not persuaded that any award for damages should be reduced in this case. See *Scheeler v. C.M. Holdings Inc.* (1997) 183 Sask R (Q.B.) and *Nan v. Black Pine Manufacturing Ltd.* (1991) 5WWR172 (B.C.C.A.). I have reviewed *Pitch Snyder, Damages for Breach of Contract*, 2nd edition, pages 2-14.3 to 2-22 and read *Betterment Before Canadian Common Law Courts* by J. Berryman, (1993) 72 *Canadian Bar Review.*

It is clear that the onus to show and calculate any betterment is on the Defendant. He has not done so. In any event, I doubt that any betterment in this case would be significant.

The repair cost given by Mr. Nemeth of Superior Auto Body Ltd., Exhibit P-7 shows the repair cost (including minor dents) to be $3,679.90. I reduce the sum by $267.52 for the repair cost of the dents.

There will be judgment for the Plaintiff in the amount of $3,412.38 plus costs of $81.29.

Some of the important aspects of the *Frank* judgment are:

1. The judge accepted that the automaker was responsible, even though the car was bought used. The subsequent purchaser was not prevented from making the warranty claim, even though the warranty had long since expired from a time and mileage standpoint and she was the second owner.
2. The judge stressed that the provincial warranty can kick in anytime the automaker's warranty has expired or isn't applied.
3. By awarding full compensation to the plaintiff, the judge didn't feel there was a significant "betterment" or improvement added to the car that would warrant reducing the amount of the award.
4. The judge decided that the paint delamination was a factory defect.
5. The judge also concluded that, without this factory defect, a paint job should last up to 15 years.
6. GM offered to pay $700 of the paint repairs if the plaintiff dropped the suit; the judge awarded five times that amount.
7. Maureen Frank won this case despite having to confront GM lawyer Ken Ready, a lawyer who has argued other paint cases for GM and Chrysler.

Other paint/rust cases

Martin v. Honda Canada Inc., March 17, 1986, Ontario Small Claims Court (Scarborough), Judge Sigurdson. The original owner of a 1981 Honda Civic sought compensation for the premature "bubbling, pitting, cracking of the paint and rusting of the Civic after five years of ownership." Judge Sigurdson agreed and ordered Honda to pay the owner $1,163.95.

Thauberger v. Simon Fraser Sales and Mazda Motors, 3 B.C.L.R., 193. This Mazda owner sued for damages caused by the premature rusting of his 1977 Mazda GLC. The court awarded him $1,000. Thauberger had previously sued General Motors for a prematurely rusted Blazer truck and was also awarded $1,000 in the same court. Both judges ruled that the defects could not be excluded from the automaker's express warranty or from the implied warranty granted by ss. 20, 20(b) of the B.C. Sale of Goods Act.

Whittaker v. Ford Motor Company (1979), 24 O.R. (2d), 344. A new Ford developed serious corrosion problems in spite of having been rustproofed by the dealer. The court ruled that the dealer, not Ford, was liable for the damage for having sold the rustproofing product at the time of purchase. This is an important judgment to use when a rustproofer or paint protector goes out of business or refuses to pay a claim, since the decision holds the dealer jointly responsible.

See also:

- *Danson v. Chateau Ford* (1976) C.P., Quebec Small Claims Court, No. 32-00001898-757, Judge Lande.
- *Doyle v. Vital Automotive Systems*, May 16, 1977, Ontario Small Claims Court (Toronto), Judge Turner.

- *Lacroix v. Ford*, April 1980, Ontario Small Claims Court (Toronto), Judge Tierney.
- *Marinovich v. Riverside Chrysler*, April 1, 1987, District Court of Ontario, No. 1030/85, Judge Stortini.

Going to Court

When to sue

If the seller you've been negotiating with agrees to make things right, give him or her a deadline and then have an independent garage check the repairs. If no offer is made within 10 working days, file suit in court. Make the manufacturer a party to the lawsuit only if the original, unexpired warranty was transferred to you; your claim falls under the emissions warranty, a TSB, a secret warranty extension, or a safety recall campaign; or there is extensive chassis rusting due to poor engineering.

Choosing the right court

You must decide what remedy to pursue; that is, whether you want a partial refund or a cancellation of the sale. To determine the refund amount, add the estimated cost of repairing existing mechanical defects to the cost of prior repairs. Don't exaggerate your losses or claim for repairs that are considered routine maintenance.

A suit for cancellation of sale involves practical problems. The court requires that the vehicle be "tendered" or taken back to the seller at the time the lawsuit is filed. This means you are without transportation for as long as the case continues, unless you purchase another vehicle in the interim. If you lose the case, you must then take back the old vehicle and pay storage fees. You could go from having no vehicle to having two, one of which is a clunker.

Generally, if the cost of repairs or the sales contract amount falls within the small claims court limit (discussed later), file the case there to keep costs to a minimum and get a speedy hearing. Small claims court judgments aren't easily appealed, lawyers aren't necessary, filing fees are minimal (about $125), and cases are usually heard within a few months.

> Mr. Edmonston, I emailed you earlier in the year seeking help on my small claims case against Ford. I'm happy to report that I won my case and received a $1,900 settlement cheque from Ford in the mail yesterday! As you may recall I have a 1991 Explorer that has a significant paint peel problem.
>
> I followed all the steps recommended by your website—ultimately I ended up in small claims court. Ford had indicated in court documents that they were going to send a representative to the hearing, but nobody showed. The judge made a quick ruling in my favour and I was out the door. I didn't even get a chance to show the load of material I had brought to make my case.
>
> Mark G.

If the damages exceed the small claims court limit and there's no way to reduce them, you'll have to go to a higher court—where costs quickly add up and delays of a few years or more are commonplace.

Small claims courts

There are small claims courts in most counties of every province, and you can make a claim in the county where the problem happened or where the defendant lives and conducts business. The first step is to make sure that your claim doesn't exceed the dollar limit of the court. (The limits differ from province to province.) Then, you should go to the small claims court office and ask for a claim form. Instructions on how to fill it out accompany the form. Remember, you must identify the defendant correctly. It's a practice of some dishonest firms to change a company's name to escape liability; for example, it would be impossible to sue Joe's Garage (1999) if your contract is with Joe's Garage Inc. (1984).

At this point, it would be a smart idea to hire a lawyer or a paralegal for a brief walk-through of small claims procedures to ensure that you've prepared your case properly and that you know what objections will likely be raised by the other side. If you'd like a lawyer to do all the work for you, there are a number of law firms around the country that specialize in small claims litigation. Small claims doesn't means small legal fees, however. In Toronto, some law offices charge a flat fee of $1,000 for the basic small claims lawsuit and trial.

Remember that you're entitled to bring to court any evidence relevant to your case, including written documents, such as a bill of sale or receipt, contract, or letter. If your car has developed severe rust problems, bring a photograph (signed and dated by the photographer) to court. You may also have witnesses testify in court. It's important to discuss a witness's testimony prior to the court date. If a witness can't attend the court date, he or she can write a report and sign it for representation in court. This situation usually applies to an expert witness, such as an independent mechanic who has evaluated your car's problems.

If you lose your case in spite of all your preparation and research, some small claims court statutes allow cases to be retried, at a nominal cost, in exceptional circumstances. If a new witness has come forward, additional evidence has been discovered, or key documents (that were previously not available) have become accessible, apply for a retrial. In Ontario, this little-known provision is Rule 18.4 (1).B.

Key Court Decisions

The following Canadian and U.S. lawsuits and judgments cover typical problems that are likely to arise. Use them as leverage when negotiating a settlement or as a reference should your claim go to trial. Legal principles applying to Canadian and American law are similar; however, Quebec court decisions may be based on legal principles that don't apply outside that province.

Additional court judgments can be found in the legal reference section of your city's main public library or at a nearby university law library. Ask the librarian for help in choosing the legal phrases that best describe your claim.

Two useful Internet sites for legal research are Lexis/Nexis (*www.lexis-nexis.com*) and Findlaw (*www.findlaw.com*). Their main drawback, though, is you may need to subscribe or use a lawyer's subscription to access jurisprudence and other areas of the sites.

An excellent reference book that will give you plenty of tips on filing, pleading, and collecting your judgment is Judge Marvin Zuker's *Ontario Small Claims Court Practice 2002–2003*, Carswell, 2002. Judge Zuker's book is easily understood by non-lawyers and uses court decisions from across Canada to help you effectively plead your case in almost any Canadian court. Interestingly, small claims court is quickly becoming a misnomer, now that Alberta allows claims up to a limit of $25,000 and most other provinces permit $10,000 claims.

Product Liability

Almost three decades ago, the Supreme Court of Canada in *Kravitz v. GM* clearly affirmed that automakers and their dealers are jointly liable for the replacement or repair of a vehicle if independent testimony shows it is afflicted by factory-related defects that compromise its safety or performance. The existence of a secret warranty extension or technical service bulletins also help prove that the vehicle's problems are the automaker's responsibility. For example, in *Lowe v. Fairview Chrysler* (see pages 128–129), technical service bulletins were instrumental in showing an Ontario small claims court judge that Chrysler had a history of automatic transmission failures since 1989!

In addition to replacing or repairing the vehicle, an automaker can also be held responsible for any damages arising from the defect. This means that loss of wages, supplementary transportation costs, and damages for personal inconvenience can be awarded. However, in the States, product liability damage awards often exceed millions of dollars, while Canadian courts are far less generous.

Before settling any claim with GM or any other automaker, download the latest information from dissatisfied customers who've banded together and set up their own self-help websites. Follow the links at *www.lemonaidcars.com.*

Implied Warranty

This is that "other" warranty they never tell you about. Judges use the *implied*, or legal warranty whenever the manufacturer's *expressed* warranty has expired and the vehicle's manufacturing defects remain uncorrected.

Engine intake manifold and head gasket failures (Ford Windstar)

Engine intake manifold and head gasket failures are the defects *du jour* among Detroit Big Three automakers. Some of the models afflicted are Chrysler's

1995–99 4-cylinder-equipped cars; GM's 1994–97 Saturn, Cavalier, Sunbird, and Sunfire; 1996–2003 cars and minivans; and Ford's 1994–2001 models. Although many of the above vehicles are eligible for after-warranty "goodwill" repairs, owners are often forced to threaten small claims action before being compensated.

In the following judgment, Ford was forced to reimburse the cost of engine head gasket repairs carried out on a 1996 Windstar 3.8L engine—a vehicle not covered by the automaker's Owner Notification Program, which cut off assistance after the '95 model year.

Use this letter as a guide for any complaint relating to premature engine failure and cite the judgment in your claim.

COUR DU QUÉBEC
Division petites créances

QUÉBEC
DISTRICT DE HULL

NO: 550-32-008335-009

Hull, le 10 avril 2001

SOUS LA PRESIDENCE DE:
L'HONORABLE PIERRE CHEVALIER
Juge de la Cour du Québec

BASTIEN DUFOUR
Partie requérante,
-c.-
FORD DU CANADA LTÉE, 7800, route Transcanadienne à Pointe-Claire (Québec) H9H 1C6
Partie intimée

JUGEMENT

Les parties essentielles de la requête se lisent comme suit :

I am hereby claiming from Ford Canada expenses and collateral expenses incurred for the repair of the 3.8 litter engine of a Ford Windstar GL 1996, VIN 2FMDA5147TBA95586.

The said engine had to have the head gasket and thermostat replaced on 02 June 2000, after a total of 118,892 kms indicated on the vehicle odometer. This repair was deemed necessary by my hometown Ford dealership (Mont-Bleu Ford in Gatineau, Que.) after I observed inadequate performance of the interior heating system and abnormal engine coolant temperature indications, and after a leak down test performed by the Mont-Bleu Ford dealership.

I consider such a defect to be abnormal as components such as an engine should have a life expectancy of at least 160,000 kms of 7 years without major repairs such as head gasket repair or replacement.

2

I have enclosed a copy of my bill showing that this problem is factory related and has cost $1364.35 to correct; this amount includes the cost for the engine head gasket repair and appropriate provincial and federal sales taxes

L'ensemble de la preuve satisfait le Tribunal par prépondérance de preuve que la détérioration impliquée est survenue prématurément par rapport à un bien identique et que cette détérioration n'est pas due à un défaut d'entretien.

L'article 1729 C.c.Q. stipule qu'en cas de vente par un vendeur professionnel, l'existence d'un vice au moment de la vente est présumée, lorsque la détérioration du bien survient prématurément par rapport à des biens identiques. De plus, les intimés n'ont pas repoussé la présomption en établissant que le défaut serait dû à une mauvaise utilisation du bien par l'acheteur. Selon l'art. 1730, le fabricant est soumis à cette même garantie.

Vu les articles 1729 et 1730 du Code civile du Québec, le Tribunal fait droit à la réclamation et condamne la partie intimée à payer à la partie requérante la somme de 1 364,35 $ avec intérêts au taux légal de 5% depuis la requête, soit le 19 septembre 2000 et les frais de 72 $.

PIERRE CHEVALIER
Juge de la Cour du Québec

Use the above judgment as leverage in your negotiations to get compensation for premature engine or automatic transmission repairs involving any automaker. If you have to go to court, cite the decision in your filing.

Automatic transmission failures (Chrysler)

Lowe v. Fairview Chrysler-Dodge Limited and Chrysler Canada Limited, May 14, 1996, Ontario Court (General Division), Burlington Small Claims Court, No. 1224/95. The following judgment, in the plaintiff's favour, raises important legal principles relative to Chrysler:

- Internal dealer service bulletins are admissible in court to prove that a problem exists and certain parts should be checked out.

- If a problem is reported prior to a warranty's expiration, warranty coverage for the problematic component(s) is automatically carried over after the warranty ends.
- It's not up to the car owner to tell the dealer/automaker what the specific problem is.
- Repairs carried out by an independent garage can be refunded if the dealer/automaker unfairly refuses to apply the warranty.
- The dealer/automaker cannot dispute the cost of the independent repair if they fail to cross-examine the independent repairer.
- Auto owners can ask for and win compensation for their inconvenience, which in this judgment amounted to $150.
- Court awards quickly add up. Although the plaintiff was given $1,985.94, with the addition of court costs and prejudgment interest, plus costs of inconvenience fixed at $150, the final award amounted to $2,266.04.

New-Vehicle Defects

Bagnell's Cleaners v. Eastern Automobile Ltd. (1991), 111 N.S.R. (2nd), No. 51, 303 A.P.R., No. 51 (T.D.). This Nova Scotia company found that the new van it purchased had serious engine, transmission, and radiator defects. The dealer pleaded unsuccessfully that the sales contract excluded all other warranties except for those contained in the contract. The court held that there was a fundamental breach of the implied warranty and that the van's performance differed substantially from what the purchaser had been led to expect. An exclusionary clause could not protect the seller, who failed to live up to a fundamental term of the contract.

Burridge v. City Motor, 10 Nfld. & P.E.I.R., No. 451. This Newfoundland resident complained repeatedly of his new car's defects during the warranty period, and stated that he hadn't used his car for 204 days after spending almost $1,500 for repairs. The judge awarded all repair costs and cancelled the sale.

Davis v. Chrysler Canada Ltd. (1977), 26 N.S.R. (2nd), No. 410 (T.D.). The owner of a new $28,000 diesel truck found that a faulty steering assembly prevented him from carrying on his business. The court ordered that the sale be cancelled and that $10,000 in monthly payments be reimbursed. There was insufficient evidence to award compensation for business losses.

Fox v. Wilson Motors and GM, February 9, 1989, Court of Queen's Bench, New Brunswick, No. F/C/308/87. A trucker's new tractor-trailer had repeated engine malfunctions. He was awarded damages for loss of income, excessive fuel consumption, and telephone charges under the provincial Sale of Goods Act.

Gibbons v. Trapp Motors Ltd. (1970), 9 D.L.R. (3rd), No. 742 (B.C.S.C.). The court ordered the dealer to take back a new car that had numerous defects and required 32 hours of repairs. The refund was reduced by mileage driven.

Johnson v. Northway Chevrolet Oldsmobile (1993), 108 Sask. R., No. 138 (Q.B.). The court ordered the dealer to take back a new car that had been brought in for repairs on 14 different occasions. Two years after purchase, the buyer initiated a lawsuit for the purchase price of the car and for general damages. General damages were awarded.

Julien v. GM of Canada (1991), 116 N.B.R. (2nd), No. 80. The plaintiff's new diesel truck produced excessive engine noise. The dealer claimed that the problem was caused by the owner's engine alterations. The plaintiff was awarded the $5,000 cost of repairing the engine through an independent dealer.

Kravitz v. General Motors, January 1979, Supreme Court of Canada, I.R.C.S., No. 393. This owner of a new Oldsmobile was never able to have it properly repaired under warranty. When the warranty period was over, General Motors and the dealer refused to do further free work or to give him another vehicle. The presiding judge awarded the car owner damages and a refund of the purchase price. This Quebec precedent is based on articles 1522–1530 of the Quebec Civil Code (hidden defects), but it applies in common-law provinces as well. The Supreme Court ruled that both the dealer and the manufacturer can be held jointly or separately responsible and that the manufacturer's warranty does not negate the implied legal warranty of fitness.

Magna Management Ltd. v. Volkswagen Canada Inc., May 27, 1988, Vancouver (B.C.C.A.), No. CA006037. This precedent-setting case allowed the plaintiff to keep his new $48,325 VW while awarding him $37,101—three years after the car was purchased. The problems were centred on poor engine performance. The jury accepted the plaintiff's view that the car was practically worthless with its inherent defects.

Maughan v. Silver's Garage Ltd., Nova Scotia Supreme Court, 6 B.L.R., No. 303, N.S.C. (2nd), No. 278. The plaintiff leased a defective backhoe. The manufacturer had to reimburse the plaintiff's losses because the warranty wasn't honoured. The Court rejected the manufacturer's contention that the contract's exclusion clause protected the company from lawsuits for damages resulting from a latent defect.

Murphy v. Penney Motors Ltd. (1979), 23 Nfld. & P.E.I.R., No. 152, 61 A.P.R., No. 152 (Nfld. T.D.). This Newfoundland trucker found that his vehicle's engine problems took his new trailer off the road for 129 days during a 7-month period. The judge awarded all repair costs, as well as compensation for business losses, and cancelled the sale.

Murray v. Sperry Rand Corp., Ontario Supreme Court, 5 B.L.R., No. 284. The seller, dealer, and manufacturer were all held liable for breach of warranty when a forage harvester did not perform as advertised in the sales brochure or as promised by the sales agent. The plaintiff was given his money back and reimbursed for his economic loss, based on the amount his harvesting usually earned. The court held that the advertising was a warranty.

Oliver v. Courtesy Chrysler (1983) Ltd. (1992), 11 B.C.A.C., No. 169. This new car had numerous defects over a 3-year period, which the dealer attempted to fix to no avail. The plaintiff put the car in storage and sued the dealer for the purchase price. The court ruled that the car wasn't roadworthy and that the plaintiff couldn't be blamed for putting it in storage rather than selling it and purchasing another vehicle. The purchase price was refunded minus $1,500 for each year the plaintiff used the car.

Olshaski Farms Ltd. v. Skene Farm Equipment Ltd., January 9, 1987, Alberta Court of Queen's Bench, 49 Alta. L.R. (2nd), No. 249. The plaintiff's Massey-Ferguson combine caught fire after the manufacturer had sent two notices to dealers informing them of a defect that could cause a fire. The judge ruled under the Sale of Goods Act that the balance of probabilities indicated that the manufacturing defect caused the fire, even though there was no direct evidence proving that the defect existed.

Western Pacific Tank Lines Ltd. v. Brentwood Dodge, June 2, 1975, B.C.S.C., No. 30945-74, Judge Meredith. The court awarded the plaintiff $8,600 and cancelled the sale of a new Chrysler New Yorker with the following defects: badly adjusted doors, water leaks into the interior, and electrical short circuits.

Leasing

Ford Motor Credit v. Bothwell, December 3, 1979, Ontario County Court (Middlesex), No. 9226-T, Judge Macnab. The defendant leased a 1977 Ford truck that had frequent engine problems, characterized by stalling and hard starting. After complaining for one year and driving 35,000 km (22,000 miles), the defendant cancelled the lease. Ford Credit sued for the money owing on the lease. Judge Macnab cancelled the lease and ordered Ford Credit to repay 70 percent of the amount paid during the leasing period. Ford Credit was also ordered to refund repair costs, even though the corporation claimed that it should not be held responsible for Ford's failure to honour its warranty.

Salvador v. Setay Motors/Queenstown Chev-Olds, Hamilton Small Claims Court, Case No.1621/95. Robert Salvador, an Ontario consumer advocate and founder of the Consumer Action Group (CAG), was awarded $2,000 plus costs from Queenstown Leasing. The court found that the company should have tried harder to sell the leased vehicle, and at a higher price, when the "open lease" expired.

Salvador gave Queenstown a list of offers from independent buyers when he returned the vehicle, but they were never contacted. Instead, the leasing agency auctioned off the van to the highest bidder. You guessed it—Queenstown Leasing.

This judgment can also be helpful in cases where a repossessed vehicle is sold or auctioned off for far less than what it's worth, or where the seller is in a conflict of interest by being the buyer as well. Copies of this judgment can be obtained by calling the Ontario Consumer Action Group at 416-587-0070.

Schryvers v. Richport Ford Sales, May 18, 1993, B.C.S.C., No. C917060, Justice Tysoe. The court awarded $17,578.47, plus costs, to a couple who paid thousands of dollars more in unfair and hidden leasing charges than if they had simply purchased their Ford Explorer and Escort. The court found that this price difference constituted a deceptive, unconscionable act or practice, in contravention of the Trade Practices Act, R.S.B.C. 1979, c. 406.

Judge Tysoe concluded that the total of the general damages awarded to the Schryvers for both vehicles would be $11,578.47. He then proceeded to give the following reasons for awarding an additional $6,000 in punitive damages:

> Little wonder Richport Ford had a contest for the salesperson who could persuade the most customers to acquire their vehicles by way of a lease transaction. I consider the actions of Richport Ford to be sufficiently flagrant and high-handed to warrant an award of punitive damages.
>
> There must be a disincentive to suppliers in respect of intentionally deceptive trade practices. If no punitive damages are awarded for intentional violations of the legislation, suppliers will continue to conduct their businesses in a manner that involves deceptive trade practices because they will have nothing to lose. In this case I believe that the appropriate amount of punitive damages is the extra profit Richport Ford endeavoured to make as a result of its deceptive acts. I therefore award punitive damages against Richport Ford in the amount of $6,000.

See also:

- *Barber v. Inland Truck Sales*, 11 D.L.R. (3rd), No. 469.
- *Canadian-Dominion Leasing v. Suburban Super Drug Ltd.* (1966), 56 D.L.R. (2nd), No. 43.
- *Neilson v. Atlantic Rentals Ltd.* (1974), 8 N.B.R. (2d), No. 594.
- *Volvo Canada v. Fox*, December 13, 1979, New Brunswick Court of Queen's Bench, No. 1698/77/C, Judge Stevenson.
- *Western Tractor v. Dyck*, 7 D.L.R. (3rd), No. 535.

Repairs

Faulty diagnosis

Babcock v. Servacar (1970), 1 O.R., No. 125. A motorist took a car he was planning to buy to an Ottawa Esso Diagnostic Clinic to determine its condition. He did the recommended repairs, and then had serious problems with the car while on vacation. The judge ruled that the clinic would have to pay for the repairs and reimburse the diagnostic cost as well. The court held that the garage's advertising claims gave rise to a contractual warranty that promised to root out the car's defects before it was purchased.

Davies v. Alberta Motor Association, August 13, 1991, Alberta Provincial Court, Civil Division, No. P9090106097, Judge Moore. The plaintiff had a used 1985 Nissan Pulsar NX checked out by the AMA's Vehicle Inspection Service prior to buying it. The car passed with flying colours. A month later, the clutch was replaced and numerous electrical problems ensued. At that time, another garage discovered that the car had been involved in a major accident, had a bent frame and a leaking radiator, and was unsafe to drive. The court awarded the plaintiff $1,578.40 plus three years of interest. The judge held that the AMA set itself out as an expert and should have spotted the car's defects. The AMA's defence—that it was not responsible for errors—was thrown out. The court held that a disclaimer clause could not protect the association from a fundamental breach of contract.

False Advertising

Vehicle not as ordered

Whether you're buying a new or used vehicle, the seller can't misrepresent the vehicle through a lie or a failure to disclose important information. Anything that varies from what one would commonly expect, or from the seller's representation, must be disclosed prior to signing the contract. Typical scenarios are odometer turnbacks, accident damage, used or leased cars sold as new, new vehicles that are the wrong colour and the wrong model year, or vehicles that lack promised options or standard features.

Goldie v. Golden Ears Motors (1980) Ltd, Port Coquitlam, June 27, 2000, British Columbia Small Claims Court, Case No. CO8287, Justice Warren. In a well-written eight-page judgment, the court awarded plaintiff Goldie $5,000 for engine repairs on a 1990 Ford F-150 pickup in addition to $236 court costs. The dealer was found to have misrepresented the mileage and sold a used vehicle that didn't meet Section 8.01 of the provincial motor vehicle regulations (unsafe tires, defective exhaust, and headlights).

In rejecting the seller's defence that he disclosed all information "to the best of his knowledge and belief," as stipulated in the sales contract, Justice Warren stated:

> The words "to the best of your knowledge and belief" do not allow someone to be wilfully blind to defects or to provide incorrect information. I find as a fact that the business made no effort to fulfill its duty to comply with the requirements of this form.... The defendant has been reckless in its actions. More likely, it has actively deceived the claimant into entering into this contract. I find the conduct of the defendant has been reprehensible throughout the dealings with the claimant.

This judgment closes a loophole that sellers have used to justify their misrepresentation and it allows for cancellation of the sale and damages if the vehicle doesn't meet highway safety regulations.

MacDonald v. Equilease Co. Ltd., January 18, 1979, Ontario Supreme Court, Judge O'Driscoll. The plaintiff leased a truck that was misrepresented as having an axle stronger than it really was. The court awarded the plaintiff damages for repairs and set aside the lease.

Seich v. Festival Ford Sales Ltd. (1978), 6 Alta. L.R. (2nd), No. 262. The plaintiff bought a used truck from the defendant after being assured that it had a new motor and transmission. It didn't, and the court awarded the plaintiff $6,400.

Bilodeau v. Sud Auto, Quebec Court of Appeal, No. 09-000751-73, Judge Tremblay. This appeals court cancelled the contract and held that a car can't be sold as new or as a demonstrator if it has ever been rented, leased, sold, or titled to anyone other than the dealer.

Chenel v. Bel Automobile (1981) Inc., August 27, 1976, Quebec Superior Court (Quebec), Judge Desmeules. The plaintiff didn't receive his new Ford truck with the Jacob brakes essential to transporting sand in hilly regions. The Court awarded the plaintiff $27,000, representing the purchase price of the vehicle less the money he earned while using the truck.

Lasky v. Royal City Chrysler Plymouth, February 18, 1987, Ontario High Court of Justice, 59 O.R. (2nd), No. 323. The plaintiff bought a 4-cylinder 1983 Dodge 600 that was represented by the salesman as being a 6-cylinder model. After putting 40,000 km on the vehicle over a 22-month period, the buyer was given her money back, without interest, under the provincial Business Practices Act.

Rourke v. Gilmore, January 16, 1928, (Ontario Weekly Notes, vol. XXXIII, p. 292). Before discovering that his new car was really used, the plaintiff drove it for over a year. For this reason the contract couldn't be cancelled. However, the appeals court instead awarded damages for $500, which was quite a sum in 1928!

Secret warranties

It's common practice for manufacturers to secretly extend their warranties to cover components with a high failure rate. Customers who complain vigorously get extended warranty compensation in the form of "goodwill" adjustments.

Desjardins v. Canadian Honda and Du Portage Mercury, February 20, 1981, Quebec Small Claims Court (Hull), No. 550-32-000933-801, Judge Dagenais. The plaintiff bought a used Honda "as is" from a Ford dealership.

One year later, the head gasket blew and was repaired free of charge by a Honda dealer under a secret warranty extension and the emissions warranty. Ten months later, the head gasket blew again and was repaired without charge. A few months later, the part failed again and was repaired a third time at Honda's expense.

The plaintiff brought suit 22 months after purchase for expenses relating to motor adjustments, rental cars, and general inconvenience. Judge Dagenais forced both Honda Canada and the Ford dealer to pay $281 in damages for having made and sold such a defect-ridden vehicle.

François Chong v. Marine Drive Imported Cars Ltd. and Honda Canada Inc., May 17, 1994, British Columbia Provincial Small Claims Court, No. 92-06760, Judge C. L. Bagnall. Mr. Chong is the first owner of a 1983 Honda Accord with 134,000 km on the odometer. He's had seven engine camshafts replaced—four under Honda "goodwill" programs, one where he paid part of the repairs, one via a small claims court judgment. Please note that the Honda no longer has serious engine problems.

In his ruling, Judge Bagnall agreed with Chong and ordered Honda and the dealer to each pay half of the $835.81 repair bill, for the following reasons:

> The defendants assert that the warranty which was part of the contract for purchase of the car encompassed the entirety of their obligation to the claimant, and that it expired in February 1985. The replacements of the camshaft after that date were paid for wholly or in part by Honda as a "goodwill gesture." The time has come for these gestures to cease, according to the witness for Honda. As well, he pointed out to me that the most recent replacement of the camshaft was paid for by Honda and that, therefore, the work would not be covered by Honda's usual warranty of 12 months from date of repair. Mr. Wall, who testified for Honda, told me there was no question that this situation with Mr. Chong's engine was an unusual state of affairs. He said that a camshaft properly maintained can last anywhere from 24,000 to 500,000 km. He could not offer any suggestion as to why the car keeps having this problem.
>
> The claimant has convinced me that the problems he is having with rapid breakdown of camshafts in his car is due to a defect, which was present in the engine at the time that he purchased the car. The problem first arose during the warranty period and in my view has never been properly identified nor repaired.

Damages (Punitive)

Punitive damages (also known as exemplary damages) allow the plaintiff to get compensation that exceeds his or her losses, as a deterrent to those who carry out dishonest or negligent practices. These kinds of judgments, common in the U.S., sometimes reach hundreds of millions of dollars.

Punitive damages are rarely awarded in Canadian courts and are almost never used against automakers. When they are given out, it's usually for sums less than $100,000. The most recent award, in *Prebushewski v. Dodge City Auto (1985) Ltd. and Chrysler Canada Ltd.* (2001 SKQB 537; Q.B. No. 1215) was for $25,000 and was handed down December 6, 2001, in Saskatoon, Saskatchewan. It followed testimony from Chrysler's expert witness that the company was aware of many cases where daytime running lights shorted and caused 1996 Ram pickups to catch fire. The plaintiff's truck had burned to the ground and Chrysler refused the owner's claim, in spite of its knowledge that fires were commonplace.

Angered by Chrysler's stonewalling, Justice Rothery rendered the following judgment:

> Not only did Chrysler know about the problems of the defective daytime running light modules, it did not advise the plaintiff of this. It simply chose to ignore the plaintiff's requests for compensation and told her to seek recovery from her insurance company. Chrysler had replaced thousands of these modules since 1988. But it had also made a business decision to neither advise its customers of the problem nor to recall the vehicles to replace the modules. While the cost would have been about $250 to replace each module, there were at least one million customers. Chrysler was not prepared to spend $250 million even though it knew what the defective module might do.
>
> Counsel for the defendants argues that this matter had to be resolved by litigation because the plaintiff and the defendants simply had a difference of opinion on whether the plaintiff should be compensated by the defendants. Had the defendants some dispute as to the cause of the fire, that may have been sufficient to prove that they had not wilfully violated this part of the Act. They did not. They knew about the defective daytime running light module. They did nothing to replace the burned truck for the plaintiff. They offered the plaintiff no compensation for her loss. Counsel's position that the definition of the return of the purchase price is an arguable point is not sufficient to negate the defendants' violation of this part of the Act. I find the violation of the defendants to be wilful. Thus, I find that exemplary damages are appropriate on the facts of this case.
>
> In this case, the quantum ought to be sufficiently high as to correct the defendants' behaviour. In particular, Chrysler's corporate policy to place profits ahead of the potential danger to its customer's safety and personal property must be punished. And when such corporate policy includes a refusal to comply with the provisions of the Act and a refusal to provide any relief to the plaintiff, I find an award of $25,000 for exemplary damages to be appropriate. I therefore order Chrysler and Dodge City to pay:

1. Damages in the sum of $41,969.83
2. Exemplary damages in the sum of $25,000
3. Party and party costs

Vlchek v. Koshel (1988), 44 C.C.L.T. 314, B.C.S.C., No. B842974. The plaintiff was seriously injured when she was thrown from a Honda all-terrain cycle on which she had been riding as a passenger. The Court allowed for punitive damages because the manufacturer was well aware of the injuries likely to be caused by the cycle. Specifically, the Court ruled that there is no firm and inflexible principle of law stipulating that punitive or exemplary damages must be denied unless the defendant's acts are specifically directed against the plaintiff. The Court may apply punitive damages "where the defendant's conduct has been indiscriminate of focus, but reckless or malicious in its character. Intent to injure the plaintiff need not be present, so long as intent to do the injurious act can be shown."

See also:
- *Granek v. Reiter*, Ont. Ct. (Gen. Div.), No. 35/741.
- *Morrison v. Sharp,* Ont. Ct. (Gen. Div.), No. 43/548.
- *Schryvers v. Richport Ford Sales*, May 18, 1993, B.C.S.C., No. C917060, Judge Tysoe.
- *Varleg v. Angeloni*, B.C.S.C., No. 41/301.

Provincial business practices acts cover false, misleading, or deceptive representations, and allow for punitive damages should the unfair practice toward the consumer amount to an unconscionable representation. (See C.E.D. (3d) s. 76, pp. 140–45.) "Unconscionable" is defined as "where the consumer is not reasonably able to protect his or her interest because of physical infirmity, ignorance, illiteracy, or inability to understand the language of an agreement or similar factors."

- Exemplary damages are justified where compensatory damages are insufficient to deter and punish. See *Walker et al. v. CFTO Ltd. et al.* (1978), 59 O.R. (2nd), No. 104 (Ont. C.A.).
- Exemplary damages can be awarded in cases where the defendant's conduct was "cavalier." See *Ronald Elwyn Lister Ltd. et al. v. Dayton Tire Canada Ltd.* (1985), 52 O.R. (2nd), No. 89 (Ont. C.A.).
- The primary purpose of exemplary damages is to prevent the defendant and all others from doing similar wrongs. See *Fleming v. Spracklin* (1921).
- Disregard of the public's interest, lack of preventive measures, and a callous attitude all merit exemplary damages. See *Coughlin v. Kuntz* (1989), 2 C.C.L.T. (2nd) (B.C.C.A.).
- Punitive damages can be awarded for mental distress. See *Ribeiro v. Canadian Imperial Bank of Commerce* (1992), Ontario Reports 13 (3rd) and *Brown v. Waterloo Regional Board of Comissioners of Police* (1992), 37 O.R. (2nd).

In the States, punitive damage awards have been particularly generous. Do you remember the Alabama fellow who won a multi-million dollar damages award because his new BMW had been repainted before he bought it and the seller didn't tell him so? The case was *BMW of North America, Inc. v. Gore*, 517 U.S. 559, 116 S. Ct. 1589 (1996). In *Gore*, the Supreme Court cut the damages award and established standards for jury awards of punitive damages. Nevertheless, million-dollar awards are still quite common. For example, an Oregon dealer learned that a $1 million punitive damages award was not excessive under *Gore* and under Oregon law.

The Oregon Supreme Court determined that the standard it set forth in *Oberg v. Honda Motor Company*, 888 P.2d 8 (1996), on remand from the Supreme Court, survived the Supreme Court's subsequent ruling in *Gore*. The court held that the jury's $1 million punitive damages award, 87 times larger than the plaintiff's compensatory damages in *Parrott v. Carr Chevrolet, Inc.*, (2001 Ore. LEXIS 1, January 11, 2001) wasn't excessive. In that case, Mark Parrott sued Carr Chevrolet, Inc. over a used 1983 Chevrolet Suburban under Oregon's Unlawful Trade Practices Act. The jury awarded Parrott $11,496 in compensatory damages and $1 million in punitive damages because the dealer failed to disclose collision damage to a new car buyer.

See also:

- *Grabinski v. Blue Springs Ford Sales, Inc.*, 2000 U.S. App. LEXIS 2073 (8th Cir. W.D. MO, February 16, 2000).

Now that we know how to get the best deal and protect our rights, let's take a look in Part Three at which cars and minivans give us the most for our money.

NEW-VEHICLE RATINGS

3

Poor Mercedes?

As *The Wall Street Journal* recently reported, when Mercedes sets sales records in North America, their cars get crappier, as witnessed by the further downturn in their quality for 2001. New Mercedes owners are bonding with their dealer's service managers like never before, dealing with malfunctioning electronic key fobs, bitchy "Command" systems and faulty door hinges. Based on a century of engineering leadership and excellence, Mercedes cars are supposed to be "Engineered like no other car in the world." Hard to swallow that line these days, isn't it?

John Rogers, car columnist
The Crank, Issue No. 8
www.straight-six.com

What Makes a Good Car or Minivan?

Your new car or minivan should first live up to the promises made by the manufacturer and dealer. It *must* be safe, crashworthy, reasonably durable (lasting at least 10 years), cost no more than about $800 a year to maintain, and provide you with a fair resale value a few years down the road. Parts should be reasonably priced and easily available, and servicing shouldn't be hard to find or performed incompetently by a dealer network afflicted by a "what, me worry?" mindset.

Toyota's 2004 Sienna minivan is an even better buy with this year's refinements.

Lemon-Aid guides try to publish up-to-date photographs of each new model rated, but some automakers disagree with our ratings and refuse to co-operate with us in any manner whatsoever, including sending us recent photos and current technical specifications. We regret any errors or omissions that may result. Our independence is more important than a book with pretty pictures.

Under no circumstances are dealers or manufacturers solicited for free "test" vehicles, as is the case with most auto columnists and some consumer groups. When a test vehicle is needed, it is rented from a major rental agency or borrowed from its owner. I've adopted this practice from my early experience as a consumer reporter, over 30 years ago. At that time, Nissan asked me to test-drive their new 1974 240Z—no strings attached. I took the car for a week, had it examined by an independent garage, spoke with satisfied and dissatisfied owners, and accessed internal service bulletins. I came to the conclusion that the car's faulty brakes made it unsafe to drive, and said so in my report. Nissan sued me for $4 million, fixed the brakes through a "product improvement campaign," and dropped the lawsuit two years later. I never went back for another car.

Definitions of Terms

Ratings

This edition makes use of owner complaints, confidential technical service bulletins (TSBs), and test-drives to expose serious factory-related defects, design deficiencies, or servicing glitches. It should be noted that customer complaints alone do not make a scientific sampling, and that's why they are used in conjunction with other sources of information. On the other hand, owner complaints combined with inside information found in dealer service bulletins are a good starting point to cut through the automakers' hyperbole and get a glimpse of reality. Since ratings can change dramatically from one year to the next, depending upon the manufacturer's warranty performance, you will want to keep abreast of these changes between editions by logging onto *www.lemonaidcars.com.*

This guide emphasizes important new features that add to a vehicle's safety, reliability, road performance, and comfort, and points out those changes that are merely gadgets and styling revisions. Also noted are important improvements to be made in the future, or the dropping of a model line. In addition to the "Recommended" or "Not Recommended" rating, each vehicle's strong and weak points are summarized.

Unlike most auto guides, *Lemon-Aid* isn't bedazzled by high-tech wizardry. Three decades of consumer advocacy in the auto industry have taught me that complex components are usually quite troublesome during their first few years on the market. Complexity drives up ownership costs, reduces overall reliability, and puts extra stress on such expensive major parts as the powertrain, fuel system, and emissions components.

Depreciation is the biggest—and most often ignored—expense you encounter when you trade in your vehicle, or when an accident forces you to buy another vehicle before the depreciated loss can be amortized. Most new cars depreciate a whopping 30–40 percent during the first two years of ownership. Although minivans, vans, trucks, and sport-utilities lose their value at a much slower rate, you still lose money. The best way to use depreciation rates to your advantage is to choose a vehicle listed as being both reliable and economical to own and keep it for a minimum of eight years. Alternatively, you may buy insurance to protect yourself from depreciation's bite.

During a car or minivan's first year on the market it takes about six months to acquire enough information for a fair-minded evaluation, unless it's a hybrid that has been in service under another name or has only been re-badged. Most new cars hit the market before all of the bugs have been worked out, so it would be irresponsible to recommend them before they've been owner-driven, or before the quality of service from the dealer and manufacturer has been customer-tested. The Chrysler Neon is a case in point. Hailed as "Car of the Year" by the motoring press in 1995, it's now noted mostly for its costly engine, AC, electrical, and body deficiencies. Ford's Windstar has followed a similar path.

Recommended

This rating indicates a best buy. This category includes new vehicles that combine a high level of crashworthiness with good road performance, few safety-related complaints, decent reliability, and better-than-average resale value. Servicing must be readily available, and parts inexpensive and easy to find.

A vehicle may lose its "Recommended" rating from the previous edition of *Lemon-Aid* whenever complaints registered by the National Highway Traffic Safety Administration (NHTSA) increase, its price becomes unreasonable, or its warranty performance falters.

Above Average

Vehicles in this class exhibit quality construction, durability, and safety features as standard equipment. They may have expensive parts and servicing, safety-related complaints, an unreasonably high price tag, or only satisfactory warranty performance, one or all of which may have disqualified them from the "Recommended" category.

Average

Vehicles in this group have some deficiencies or flaws that make them a second choice. In many cases, certain components are prone to premature wear or breakdown, or some other positive aspect of long-term ownership is lacking. An "Average" rating can also be attributed to such factors as substandard assembly quality, lack of a solid long-term reliability record, a substantial number of safety-related complaints, or some flaw in the parts and service network.

Below Average

This rating category denotes a vehicle that may have had a poor safety or reliability record, but where improvements have been made with regard to durability and/or safety features. Ensure you get an extended warranty with a vehicle in this category.

Not Recommended

Buy at your own risk. Substandard crashworthiness, poor overall reliability and safety, inadequate road performance, and poor dealer service, among other factors, can make owning one of these vehicles a traumatic and expensive experience. It doesn't necessarily follow that every single vehicle produced in a "Not Recommended" model line will have exactly the same reliability shortcomings, but chances of having trouble are higher than normal.

Vehicles that have not been on the road long enough to assess, or that are sold in such small numbers that owner feedback is insufficient, are "Not Recommended" or left unrated.

Cost analysis/best alternatives

Fall prices for the 2004 models haven't risen much this year due to the ramping up of generous discounts and rebates by GM and Chrysler. Only a few automakers, like BMW with its redesigned 5 Series, are bucking this trend with substantial price increases (around 6 percent).

BMW's redesigned 5 Series.

Each model's cost is analyzed in light of cheaper earlier models eligible for substantial rebates, PDI and destination charges, insurance costs, parts costs, depreciation, and fuel consumption. According to the Canadian Automobile Association (CAA), annual maintenance costs average about $800. A listing of competing recommended new models is also included; good used alternatives can be found in *Lemon-Aid Used Cars and Minivans 2004*.

Quality/reliability/safety

Lemon-Aid bases its quality and reliability evaluations on owner comments, confidential manufacturer service bulletins, and government reports from NHTSA safety complaint files, among other sources. This year's edition also draws on the knowledge and expertise of professionals working in the automotive marketplace, including mechanics and fleet owners. The aim is to have a wide range of unbiased (and irrefutable) data on quality, reliability, durability, and ownership costs. Allowances are made for the number of vehicles sold versus the number of complaints, as well as for the seriousness of problems reported and the average number of problems reported by each owner.

TSBs listed in this section give the most probable cause of factory-related defects on 2003 models that will likely be carried over to the 2004 version. TSBs are reliable in this way because manufacturers depend on the dealer corrections outlined in their bulletins until a permanent, cost-effective engineering solution is found at the factory, which often takes several model years with lots of experimentation.

A/T - Firm Shifts/Shudder/Lit SES Lamp

Bulletin No.: 02-07-30-039B **Date:** April 2003
Firm transmission shifts, shudder/chuggle, transmission won't downshift on deceleration, Service Engine Soon light illuminated, DTC P0742 set (Perform diagnostics and replace TCC PWM solenoid)

2003 Buick Century, LeSabre, Park Avenue, Regal, Rendezvous
2003 Cadillac DeVille, Seville
2003 Chevrolet Cavalier, Impala, Malibu, Monte Carlo, Venture
2003 Oldsmobile Alero, Aurora, Silhouette
2003 Pontiac Aztek, Bonneville, Grand Prix, Grand Am, Montana, Sunfire

with 4T65E Transaxle (RPO's MN3, MN7, M15, M76), 4T40E/4T45E Transaxle (RPO's MN4 or MN5) or 4T80E Transaxle (RPO MH1) with transaxle manufacturing Julian dates between 219 and 3048

Condition: In addition to the above concerns, customers with 4TBOE transaxles may also comment on a "bump" feeling after decelerating and then applying light throttle at speeds of 40–76 km/h (25–47 mph). No SES light will illuminate and the technician will find no DTCs stored for this condition.

Most bulletins listed in this edition come from American sources and often differ from Canadian bulletins only where part numbers are concerned. Nevertheless, the problems and defects they treat are exactly the same on both sides of the border. Some vehicles have more TSBs than others, but this doesn't necessarily mean they're lemons. It may be that the listed problems affect only a small number of vehicles, or are minor and easily corrected. TSBs should also be used to verify that a problem was correctly diagnosed, the correct upgraded replacement part was used, and the billed labour time was fair.

As you read through the quality and reliability ratings (safety is more of a mixed bag) you'll quickly discover that most Japanese and South Korean manufacturers are far ahead of Chrysler, Ford, and GM in maintaining a high level of quality control in their vehicles. What once was a small-car phenomenon has spread to all divisions of cars and trucks, say both *Consumer Reports* and J.D. Power and Associates.

Warranty performance

A manufacturer's warranty is a legal commitment. It promises that the vehicle will perform in the normal and customary manner for which it is designed. If a part malfunctions or fails (not owing to owner negligence or poor maintenance), the dealer must fix, repair, or replace the defective part or parts and bill the automaker or warranty company for all of the part and labour costs.

Warranties are an important factor in *Lemon-Aid*'s ratings, and they are judged by how fairly they're applied—not by what's promised. Unfortunately, it has been our experience that most automakers cheat on their warranty obligations and inflate the cost of scheduled maintenance work.

Most new vehicle warranties fall into two categories: bumper-to-bumper for a period of three to five years, and powertrain for up to 7 years/115,000 km. Automakers sometimes charge an additional $50–$100 fee for repairs requested by purchasers of used vehicles with unexpired base warranties. For snowbirds, the federal and provincial governments can charge GST and sales tax on warranty and non-warranty repairs done south of the border. Beware. Also, keep in mind that some automakers, like Honda, may not honour your warranty if a vehicle is purchased in Canada and registered in the United States.

Road performance

The main factors considered in this rating are acceleration and torque, transmission operation, routine handling, emergency handling, steering, and braking.

Every vehicle must at minimum be able to merge safely onto a highway and have adequate passing power for two-lane roads. Steering feel and handling should inspire confidence. The suspension ought to provide a reasonably well-controlled ride on most road surfaces. Ideally, the passenger compartment will be roomy enough to accommodate passengers comfortably on extended trips. The noise level should not become tiresome and annoying. As a rule, handling and ride comfort are inversely proportional—good handling requires a stiff suspension, which pounds the kidneys. Variations from this pattern are reflected in the ratings.

Cost

We list the manufacturer's suggested retail price (MSRP) in effect at press time and applicable to standard models, that price's negotiability, the range of the dealer's markup, and the vehicle's estimated residual value over the next five years (particularly helpful when leasing). Undoubtedly, the MSRP will be a bit higher when the fall prices are announced. However, ask for a copy of the manufacturer's notice to the dealer of the MSRP increase. If the dealer refuses, you can confirm the MSRP figure by accessing each manufacturer's website.

To help you negotiate the best price, this edition indicates those MSRP prices that are firm and those that are negotiable, and gives the approximate price markup percentage in parentheses (including assorted extra fees).

Destination charges and the pre-delivery inspection (PDI) are a "backdoor" into your wallet that can cost you almost $1,300 (in the case of a 2004 Toyota Sienna). If you get tired of haggling with the dealer, agree to pay no more than 2 percent of a vehicle's MSRP for these extra charges. Also, don't fall for the $99–$475 "administration fee" scam, unless the bottom-line price is so tempting that it won't make much difference. Principle is one thing; not losing an attractive deal is another.

Technical data

Note that towing capacities differ depending on the kind of powertrain/suspension package or towing package you buy. Remember that there's a difference between how a vehicle is rated for cargo capacity or payload and how heavy a boat or trailer it can pull. Do not purchase any new vehicle without receiving from the dealer very clear information, in writing, about a vehicle's towing capacity and the kind of special equipment you need to meet your requirements. Have the towing capacity and necessary equipment written into the contract.

Safety features/crashworthiness

Some of the main features weighed in the safety ratings are a model's crashworthiness and the availability of seatbelt pretensioners, de-powered airbags, airbag disablers, adjustable brake and accelerator pedals, integrated child safety seats, effective head restraints, traction control, and front and rearward visibility.

Side airbags won't be considered a safety plus until real-world crash findings prove their worth and confirm that they're not a danger to children, women, seniors, or small-statured adults, as a number of safety researchers have postulated.

Crash protection figures are taken from NHTSA's New Car Assessment Program. Vehicles are crashed into a fixed barrier, head-on, at 57 km/h (35 mph), in order to evaluate the effects of the consequent forces exerted on the specially constructed dummies placed in the two front seats. The latest results—although they may be several years old—have been included in the ratings.

NHTSA shows a vehicle's level of crashworthiness by the likelihood, expressed as a percentage, of the belted occupants being seriously injured. The higher the number of stars, the greater the protection:

NHTSA Front Collision Ratings

*****	— a 10 percent or less chance of serious injury
****	— an 11 to 20 percent chance of serious injury
***	— a 21 to 35 percent chance of serious injury
**	— a 36 to 45 percent chance of serious injury
*	— a 46 percent or greater chance of serious injury

Mercedes doesn't always agree with NHTSA crash results.

NHTSA tests don't necessarily provide an accurate picture of how a given model will perform in every accident; test figures are only valid if they're used to compare vehicles that are of the same size (i.e., compact, mid-size, or large).

Mercedes-Benz, proud of its reputation for building crashworthy vehicles of all sorts, hasn't always fared well in NHTSA head-on collisions, and questions the validity of the ratings. The company claims that its own crash data shows that most frontal collisions occur at an angle (offset), and that that's the kind of test wherein its vehicles excel. The Insurance Institute for Highway Safety (IIHS) sides with Mercedes, and crash-tests at an angle and at 64 km/h (40 mph). NHTSA doesn't do front offset testing, but it has tested many 1997–2003 cars for side-impact protection.

NHTSA Side Collision Ratings

***** — a 5 percent or less chance of serious injury
**** — a 6 to 10 percent chance of serious injury
*** — an 11 to 20 percent chance of serious injury
** — a 21 to 25 percent chance of serious injury
* — a 26 percent or greater chance of serious injury

AMERICAN VEHICLES

DaimlerChrysler

Daimler-Benz's $36-billion buyout of Chrysler in 1998 was widely hailed as a giant leap forward in the cost-efficient manufacture and development of cars and trucks (*Lemon-Aid* said it wasn't). Instead, it has become a case study in deception, corporate mismanagement, and greed.

Daimler promised that the merger was a coming together of equals when it was really a "shotgun marriage" where the American and European cultures continually clashed until Daimler forced out most of the top American executives.

But don't feel sorry for Chrysler's head honchos: the top 30 executives divided nearly $500 million in cash, stock, and severance pay. Chrysler Chairman Robert Eaton got the lion's share with $3.7 million in cash and $66.2 million in stock. Chrysler vice chairman Bob Lutz got $1.3 million in cash and $25.7 million in stock.

And Chrysler's car and truck owners got royally screwed with poor-quality vehicles, a shorter warranty, and expensive servicing.

Daimler's Chrysler takeover hasn't improved the American automaker's profits, product mix, or quality control. The company is losing money hand over fist and will launch only a small number of 2004 model vehicles (Pacifica, Crossfire, and the Durango).

The Pacifica is an overpriced "kinda" minivan and "kinda" SUV that comes to the "crossover" party late and is an unimpressive performer. Its low-volume Crossfire luxury sports coupe is stylish, but expensive, and it targets a specialty car segment that does nothing for Chrysler's bottom line. Look for brutal discounting of both vehicles.

You'd expect Chrysler to put lots of Mercedes content into its 2004 cars and trucks, but that's not the case, except for the SLK-inspired, rear-drive, two-passenger Crossfire coupe and Sprinter full-sized van. Surprisingly, the company's 2005 and later small cars will be more Mitsubishi- and Hyundai-bred than Mercedes-sourced.

Having bought a controlling interest in "almost on the skids" Mitsubishi, Chrysler is offering more small cars in Canada without increasing costs by piggybacking Mitsubishi models on its existing Chrysler dealer body. This places an additional burden upon Chrysler dealers who have more parts to buy and store, more mechanics to train, and more floor space taken up by less-profitable models that haven't done that well in the States and have yet to prove themselves in Canada (old Colts don't count).

Chrysler's vehicles *do* look good. Whether it's the bold and quirky styling of its sports cars or PT Cruiser, or simply the sleek "cab forward" styling of the 300M, Charger R/T, Intrepid, or Concorde large sedans, the automaker's vehicles seldom go unnoticed. And when it comes to convenience and savings, you can't beat the company's minivans for providing oodles of performance and convenience features at prices that usually beat the competition.

But what good is smart styling or folding seats if your vehicle's powertrain is changed more often than your oil, the paint turns chalky-white, or the ABS brakes are literally "hit or miss?"

Daimler's myopic cost-cutting management is currently squeezing dealers and suppliers by cutting their profit margins to the point where quality and reliability are lost. The prevailing attitude is that customers will ignore the factory-related defects as they snap up bargain-priced vehicles.

But, instead, buyers are staying away in droves.

Consumer advocates, owners, and exasperated dealers (speaking ever so softly to *Lemon-Aid*) recite the same litany of quality shortcomings found year after year throughout the automaker's lineup of vehicles—like failure-prone engine head gaskets, automatic transmissions that "limp" home, or shift erratically, ABS brakes that don't brake, and inoperative air conditioners, that can cost as much as $1,500 to fix. Owners will also point out Chrysler's subpar finish and build quality that produces excessive noise, vibration, and harshness, in addition to water/air leaks and paint delamination.

There are no signs whatsoever that Chrysler's poor quality control has been addressed in its 2003 and 2004 models—judging by the automaker's own service bulletins and press statements. The company's warranty performance (the manner in which it handles customer complaints) is presently stagnating, after initially improving several years ago, following angry car owner protests that led to CLOGs (Chrysler Lemon Owners Groups) sprouting up in New Brunswick, Alberta, and British Columbia.

A longer powertrain warranty doesn't make up for poor quality, that's for sure, but Chrysler's 7-year/115,000 km powertrain warranty will take the $3,000 sting out of the average transmission repair and pay the $1,000 needed to replace a leaking engine head gasket.

Generous rebates and financing

After announcing that it was abandoning its costly rebate programs, Chrysler has made an about-face this year and is matching GM's sales incentives throughout its lineup. Additionally, Chrysler intends to keep prices down by rendering optional many features that were once standard. For the average buyer, this turnaround in Chrysler's strategy means vehicles will become substantially less costly by mid-2004, making patience more of a virtue than ever before. The only question remaining will be how effective is a good warranty when applied to a bad product?

VIPER SRT-10 ★★★★★

Depreciation? A new 1998 Viper Coupe GTS sold for $94,380. Today, it's worth about $55,000. Almost as good as investing in Nortel and Corel.

RATING: Recommended. **Strong points:** Good acceleration and handling; slow depreciation; and only two minor safety-related complaints logged in the past three years. **Weak points:** Poor fuel economy; and passenger comfort is compromised by a hard ride, excessive wind noise, and limited storage room. **New for 2004**: Just an additional colour.

OVERVIEW: This $108,920 (RT/10: $104,500), mid-sized, two-door, rear-drive roadster breaks all the marketing rules—and wins. Its awesome 500-hp 8.2L V10 engine, 6-speed manual gearbox, and "in your face" styling aren't equalled by any vehicle in its class. It features de-powered airbags and an airbag cut-off switch, though side airbags and traction-control system aren't available. Service bulletins and owner comments paint a positive picture of the Viper's overall dependability. Some bulletins address poor idling and engine service light malfunctions. 2003s were revised from the ground up. Using "500-500-500" as its slogan, the Viper engine now has 500 horsepower, torque, and displacement (50 more than the 2002). The new engine got a lower intake manifold to accommodate a flatter hood, the convertible was re-styled, the coupe was dropped, and sidepipes were added. **Best alternatives:** A base Porsche 911.

SX 2.0, SRT-4 

RATING: Average. **Strong points:** Plenty of interior space; good handling; a thrifty engine; impressive turbocharged power; and a 7-year powertrain warranty. **Weak points:** Base engine lacks power in the upper gear ranges; a reputation for early powertrain breakdowns; turbo's long-term reliability is still to be determined; imprecise manual shifter; mediocre braking; and excessive engine, road, and body noise. Log onto *www.neons.org* for updated reports from owners. **New for 2004:** Nothing significant.

OVERVIEW: This Neon, now sold in Canada as the SX 2.0, is Chrysler's homegrown small car. It's presently simply treading water until the Hyundai- and Mitsubishi-sourced lineup arrives sometime next year. The SX 2.0 is roomy and reasonably powered for urban use, and recent refinements have given it a softer, quieter ride while enhancing the car's handling and powertrain performance. Interior room easily accommodates six-footers.

Chrysler has tried to make these low-end cars more appealing by loading up on features that are usually only seen with higher-priced vehicles. One of those higher-priced entries is the $26,950 SRT-4 high-performance version, equipped with a turbocharged 215-hp 2.4L 4-cylinder engine hooked to a manual 4-speed transmission, a hood scoop, 17-inch wheels, a revised suspension, and four-wheel disc brakes. The turbocharged SRT-4 should arrive in showrooms before the end of October.

Cost analysis/best alternatives: Buy an identical 2003 version if the car's sufficiently discounted. Other cars worth looking at are the Honda Civic, Hyundai Accent and Elantra, Mazda Protegé, Suzuki Swift, and Toyota Echo and Corolla. If you're in the market for a sport coupe, check out the Hyundai Tiburon. **Options:** Consider getting the height adjustment for the steering wheel (for short drivers). Optional front side airbags, traction control, and ABS aren't the proven lifesavers they pretend to be and may simply add extra cost and complexity to maintenance repairs. **Rebates:** $1,500–$2,500 by mid-year on most models, except for the high-performance SRT-4. **Delivery/PDI:** $850. **Depreciation:** About average. **Insurance cost:** Average. **Parts supply/cost:** Average. **Annual maintenance cost:** Higher than average. **Warranty:** Bumper-to-bumper 3 years/60,000 km; powertrain 7 years/115,000 km; rust perforation 5 years/160,000 km. **Supplementary warranty:** A bumper-to-bumper extended warranty would be advisable, even though this takes away the car's price advantage. **Highway/city fuel economy:** 7–9.5L/100 km with the base 2.0L engine.

Quality/Reliability/Safety

Pro: Apparently, the 4-cylinder engine's faulty head gaskets, present since 1995, have been remedied with improved gaskets on 2001–04 Neons. Interestingly, although 2003 bulletins show some problems with engine and automatic transmission operation, owner complaints don't reflect this as a major problem. Owners complain that the front windshield pillars create a huge blind spot. Helpful front adjustable shoulder belt anchors.

Con: Quality control: Mediocre. Although quality control has improved a bit, it's still no match for Japanese and South Korean competitors. That's why Chrysler is looking to Mitsubishi as its small-car supplier. **Reliability:** Engine and transmission performance problems continue to impinge a bit upon the car's overall reliability. **Warranty performance:** Mediocre at best. **Owner-reported problems:** A perusal of owner comments confirms that 2003s have had a dramatic drop in owner complaints compared with previous years' models. Problems mentioned relate to automatic transmission failures on the highway; the premature replacement of the brake rotors and power-steering pump; and that the head restraint support may poke through the seatback. **Service bulletin problems:** Engine won't start; "limping" automatic transmission;" delayed transmission engagement; harsh 4–3 downshift; right side cowl water leak; engine malfunction indicator light (MIL) remains lit; possible loss of engine mount bolt torque; and defective powertrain control module (PCM). **NHTSA safety complaints/safety:** Airbags fail to deploy.

Road Performance

Pro: Better than average acceleration with plenty of low-end torque. The 2.0L engine's 132 horses and the Neon's low weight give it acceptable performance in lower gear ranges, but restrict it to mainly urban use. The 4-speed automatic provides relatively smooth performance, though it saps power in the upper ranges. **Emergency handling:** Better than average. **Steering:** Precise and easy to control on smooth roads.

Con: Acceleration/torque: Base engine runs out of steam in high gear and is buzzy from 4000 rpm on up. It runs particularly roughly after 5000 rpm. This is especially irritating because horsepower and torque peak at 5000 and 6000 rpm. Even with the manual transmission, highway passing requires downshifting from Fifth gear to Third, with the 2.0L engine crying all the way. **Transmission:** The manual transmission is harsh and noisy—no comparison with Honda and Toyota vehicles. The manual gearbox also requires lots of downshifting when going over small hills. **Routine handling:** Definitely improved, but the jittery ride becomes fairly rough when traversing anything but the smoothest roads. Ride deteriorates and the suspension bottoms with a vengeance with a full load. Turbo's stiffer suspension may be too firm for some. **Braking:** Disc/drum; worse than average.

Comfort/Convenience

Pro: Driving position: Good driving position; the bucket seats are comfortable; and there's plenty of headroom. **Controls and displays:** Easy-to-operate controls and clear gauges; easy-to-use and effective heating and ventilation system. **Climate control:** Everything's within easy reach and easily read. **Entry/exit:** Good up front; large door openings for long legs. **Interior space/comfort F/R:** The Neon seats five adults and has a spacious interior. Practical cloth bucket seats. Seats are firmer and more comfortable than one would expect. Plenty of rear legroom and elbow room. **Cargo space:** Lots of little storage areas, including split-folding seatbacks that allow for extra-long cargo. Upgraded radio antenna reduces wind noise.

Con: Standard equipment: Low-quality interior appointments; rear visibility compromised by high rear parcel shelf; gauges lose contrast in dim light with headlights on; only front windows are power assisted, and hard-to-operate manually cranked rear windows don't roll down completely; door-mounted power window switches press uncomfortably against the driver's leg; rear door shape makes for difficult entry and exit; hard-to-access rear seat doesn't provide enough thigh support; inadequate rear headroom and toe space; short, flimsy trunk lid restricts trunk access. **Trunk/liftover:** Narrow trunk with a low sill; trunk hinges eat up lots of storage space, as well as possibly damage your luggage; folding rear seatbacks can't be locked, allowing anyone access to the trunk's contents. **Quietness:** Plenty of engine boom and growling; automatic transmission whine; and tire thumping.

SX 2.0, SRT-4

List Price (negotiable)	**Residual Values** (months)			
	24	**36**	**48**	**60**
Base: $15,195 (13%)	$9,500	$7,500	$6,000	$4,500
Sport: $18,195 (14%)	$11,500	$9,500	$7,500	$6,000
R/T: $20,995 (15%)	$13,000	$11,000	$9,000	$7,000
SRT-4: $26,950 (18%)	$16,000	$14,000	$12,000	$10,000

Technical Data

Powertrain (front-drive)
Engines: 2.0L 4-cyl. (132 hp)
• 2.4L 4-cyl. (150 hp)
• 2.4L 4-cyl. Turbo (215 hp)
Transmissions: 5-speed man.
• 4-speed auto.

Dimensions/Capacity
Passengers: 2/3
Height/length/width: 53/171.8/67.5 in.
Headroom F/R: 39.6/36.5 in.
Legroom F/R: 42.5/35.1 in.
Wheelbase: 104 in.
Turning circle: 35.4 ft.
Cargo volume: 42.6 cu. ft.
Tow limit: 1,000 lb.; 2,000 lb. (manual)
Fuel tank: 47L/reg.
Weight: 2,400 lb.

Safety Features/Crashworthiness

	Std.	**Opt.**
Anti-lock brakes	■	■
Seat belt pretensioners F/R	—	—
Side airbags	—	—

Traction control	—	—
Head restraints F/R	*	*
Visibility F/R	***	*
Crash protection (front) D/P	****	****
Crash protection (side) F/R	***	***
Crash protection (offset)	**	—

SEBRING, STRATUS ★★★★

The Sebring convertible is a good buy.

RATING: Above Average. **Strong points:** Good V6 performance; comfortable ride; plenty of interior room; a strong 7-year powertrain warranty; and few owner complaints. The convertible is especially attractive due to its reasonable price and slow depreciation. **Weak points:** Mediocre 4-cylinder engine performance; lots of engine and road noise; and subpar warranty performance. **New for 2004:** Minor interior changes.

OVERVIEW: These coupes, sedans, and convertibles are good buys mainly because they've had fewer new-model "teething" problems than other Chrysler-built vehicles. Sebring is a reasonably priced luxury model equipped with standard amenities, including AC, bucket seats, and a tilt steering wheel, while the Stratus fills the sporty coupe niche with standard tinted glass and an awesome sound system.

These front-drives use powertrains and platforms from Mitsubishi's Eclipse and Galant and also share most safety features and mechanical components, including standard dual airbags and a 150-hp 2.4L 4-banger along with an optional 200-hp 2.7L V6. In 2003, convertibles got a manual gearbox; and coupes received a new front- and rear- end treatment.

Cost analysis/best alternatives: The Sebring and Stratus are presently in a sales slump, so they should be heavily discounted. You should consider a heavily discounted 2003 version instead of this year's model. Other models in contention: the Honda Accord, Hyundai Elantra or Sonata, or Toyota Solara and Camry. **Options:** You'll want the V6 engine for better all-around performance, without much of a fuel penalty. Be wary of the untested curtain side airbags and failure-prone ABS brakes. **Rebates:** $2,500 rebates and zero percent financing on all models. **Delivery/PDI:** $895. **Depreciation:** A bit slower than average. **Insurance cost:** Average. **Parts supply/cost:** Reasonably priced parts are easily found. **Annual maintenance cost:** Average. **Warranty:** Bumper-to-bumper 3 years/60,000 km; powertrain 7 years/115,000 km; rust perforation 5 years/160,000 km. **Supplementary warranty:** An extended warranty isn't needed. **Highway/city fuel economy:** 7.3–11.2L/100 km with the base 4-cylinder; 8–12L/100 km with the V6.

Quality/Reliability/Safety

Pro: Quality control: Quality control is unusually good; body construction and assembly are solid, but with a few water leaks; body flexing and door creaking on convertible models. **Reliability:** Better than average. **Warranty performance:** Average.

Con: Owner-reported problems: Automatic transmission, steering, and suspension glitches; poor engine idling, rear brake grinding, and water leaks into the vehicle; early replacement of brake rotors. **Service bulletin problems:** Mis-machined crankshaft oil passage; delayed or temporary loss of transmission engagement after initial startup; harsh 4-3 downshift; transmission goes into "limp" mode; defective power control module; power windows click when changing directions; front-end pop or clunk sound; rear brake drum clunk; window fogging; driver's window automatic-down function operates only intermittently; wind noise from top of windshield; horn operating on its own calls for a new airbag cover; erratic fuel gauge operation; lower steering column clicking; diagnostic tips for lower centre dash panel squeaking or rattling; and poor AC operation (see below).

AC - Erratic Operation

NUMBER: 24-006-02 **DATE:** July 22, 2002

OVERVIEW: This bulletin provides diagnostic information for AC and heater performance complaints.

MODELS: 1998–2003 (LH) LHS/300M/Concorde/Intrepid
2001–03 (JR) Sebring Convertible/Sebring Sedan/Stratus Sedan

NOTE: PERFORM CUSTOMER SATISFACTION RECALL NO. 857,

SYMPTOM/CONDITION: Erratic operation of the AC and heater systems including:

^Lack of cold air

^Lack of hot air

^Unrequested mode change - Automatic Temperature Control (ATC) only

^No control of mode or temperature control

^Tapping noise from blend door

NHTSA safety complaints/safety: Sudden, unintended acceleration, engine overheating and stalling, and steering failure.

Road Performance

Pro: Emergency handling: Better than average. **Steering:** Steering is responsive and light. **Acceleration/torque:** Adequate, but far from sporty acceleration (0–100 km/h: 10 seconds with the V6). The base 4-cylinder is thrifty and adequate for most driving tasks. It works well with both the manual and automatic transmission. **Transmission:** Smooth and quiet operation in all gear ranges. **Routine handling:** Handling is exceptional with either engine, and there's little body lean when cornering under speed. The ride deteriorates only slightly on bad surfaces due to the car's compliant suspension. Jarring is reduced as the weight of passengers is added.

Con: The base 2.4L engine loses power when mated to the 4-speed automatic and, when pushed, it's noisy and less responsive than the V6. Some front-end plow in turns. The car's large turning circle makes parking a chore. **Braking:** Disc/disc; unimpressive (100–0 km/h: 43 m).

Comfort/Convenience

Pro: Standard equipment: Well-appointed with many standard features. **Controls and displays:** Controls and gauges are easily reached and seen. **Entry/exit:** Large doors facilitate easy entry and exit. **Interior space/comfort F/R:** The rear seating area is more adult-friendly than most sport coupes. Three tall passengers can sit in the rear for short trips, but the seating is really designed for two. **Cargo space:** Limited trunk cargo area can be extended by folding the split seatback. **Trunk/liftover:** Average trunk space with a low liftover. Rear seatback and remote trunk release can be locked.

Con: Driving position: Front seats need additional lateral support, and the seatback bulge is annoying to some drivers; the tilt steering wheel is set too low (almost in the driver's lap); short drivers may have difficulty with forward vision, even with the seat raised to its maximum setting; tall drivers will find forward visibility limited. **Climate control:** The climate control system is hampered by a noisy fan and uneven air distribution; the rear defroster leaves some of the glass untouched; with rear-quarter windows that won't open and the optional rear deck-lid spoiler cutting rear visibility, it's not hard to feel a bit claustrophobic. **Quietness:** Excessive tire/road noise intrudes into the passenger compartment.

Sebring, Stratus

List Price (very negotiable)	**Residual Values** (months)			
	24	**36**	**48**	**60**
Sebring 4d: $24,115 (17%)	$16,000	$15,000	$13,500	$12,000
Stratus: $23,380 (17%)	$16,000	$15,000	$13,500	$12,000
Sebring Cvt.: $39,195 (24%)	$24,000	$20,000	$16,000	$12,000

Technical Data

Powertrain (front-drive)
Engines: 2.4L 4-cyl. (150 hp)
• 2.7 L V6 (200 hp)
Transmissions: 5-speed man.
• 4-speed auto.

Dimensions/Capacity
Passengers: 2/3
Height/length/width:
54.9/190.7/70.6 in.
HeadroomHeadroom F/R: 37.6/36.5 in.
Legroom F/R: 43.3/35 in.
Wheelbase: 106; 108 in.
Turning circle: 33.6 ft.
Cargo volume: 13.1 cu. ft.
Tow limit: 1,000 lb.
Fuel tank: 61L/reg.
Weight: 3,012 lb.

Safety Features/Crashworthiness

	Std.	**Opt.**
Anti-lock brakes	■	■
Seat belt pretensioners F/R	—	—
Side airbags	—	■
Traction control	—	—
Sebring		
Head restraints F/R	***	**
4d	*****	*****
Visibility F/R	****	*****
Crash protection (front) D/P	****	****
4d	*****	*****
Convertible	***	***
Crash protection (side) F/R	*****	*****
4d	*****	*****
Convertible	***	***
Crash protection (offset)	***	—
Stratus		
Head restraints F/R	****	***
Visibility F/R	*****	*****
Crash protection (front) D/P	****	****
4d	*****	*****
Crash protection (side) F/R	***	*****
4d	***	***
Crash protection (offset)	***	—

300M, CONCORDE, INTREPID ★★

RATING: *Concorde and Intrepid:* Below Average. These cars look better than they run. *300M:* Below Average. You get a nicely packaged array of generic components that aren't as durable as they are attractive. Owners will see traditional Chrysler-type problems compromising reliability and safety. **Strong points:** The 7-year powertrain warranty has worked well over the past three years. *Concorde and Intrepid:* Nice riding; easy entry/exit; and lots of passenger and cargo room. *300M:* Good acceleration and handling; plenty of passenger and cargo room; and reasonable price. **Weak points:** *Concorde and Intrepid:* Excessive road and wind noise; vague steering; limited rear visibility caused by the cars' high rear end; impractical trunk; and poor reliability. *300M:* Rough-running engine and transmission; excessive road and wind noise; five-passenger capacity; limited rear legroom, a narrow rear windshield reduces rear visibility and carries a smaller trunk; mediocre fit and finish; and questionable reliability. **New for 2004:** Intrepid's SXT gets a 6-hp boost to 250, while the 300M adds satellite navigation and improved speakers.

OVERVIEW: *Concorde and Intrepid:* These full-sized cars share the same chassis and offer most of the same standard and optional features. Both cars have loads of passenger space and many standard features that usually sell as options, such as four-wheel disc brakes and an independent rear suspension. The Concorde is marketed to the more conservative buyer. The Intrepid, the more popular model, is the entry-level version. Base models are equipped with a 2.7L V6 aluminum engine that delivers 200 hp. Higher-line variants get a more powerful 232–250-hp 3.5L V6.

300M: This model is the Concorde's near-luxury clone. Although they use the same front-drive platform as the Concorde, their bodies are shorter and they're styled differently. In fact, the 300M is the shortest of Chrysler's mid-sized sedans. Both cars are powered by a 255-hp 3.5L V6 and mated to Chrysler's AutoStick semi-automatic transmission. The 300M Special comes with an upgraded 3.5L V6, enhanced steering and suspension, and better-performing Michelin Pilot Sport tires on alloy wheels.

For 2003, a higher-output version of the 3.5L V6 boosted Intrepid horsepower to 250, Concorde horsepower to 250, and the 300M's horsepower to 255. Optional side airbags were also offered. All of the above vehicles will be totally redesigned for the 2005 model year as they are transformed into rear-drives, using some of Mercedes' E-Class components and marketed to a higher-class clientele.

Cost analysis/best alternatives: Look for heavy discounting on the 2003 and 2004 versions as these cars make room for 2005 rear-drive replacements in mid-2004. If you opt to wait, stay away from the first six months' production. If you want something from Chrysler in the same price range, but more reliable, and perkier, try a V6-equipped Stratus or Sebring. For first-class quality, an Acura TL, Audi A6, Infiniti G35, Lexus ES 300, Nissan's Altima or Maxima, or a fully loaded Toyota Camry or Avalon should be your first choice. **Options:** You'd be smart to get a V6 engine. The harsh sports suspension, however, is a waste of money. Opt for the handling package and you get crisper steering response, better brakes and tires, and stiffer springs that eliminate much of the body roll. The large rear windshield may need tinting to protect rear-seat passengers from the sun. Optional front side airbags are unproven. **Rebates:** All four models will have $2,500+ rebates coming into 2004. **Delivery/PDI:** $850–$995. **Depreciation:** Average. The Intrepid suffers from a low level of owner loyalty—industry stats show that about nine out of 10 Intrepid owners walk away from the model when considering their next vehicle purchase. **Insurance cost:** Higher than average. **Parts supply/cost:** Reasonably priced and easily found. **Annual maintenance cost:** Much higher than average. **Warranty:** Bumper-to-bumper 3 years/60,000 km; powertrain 7 years/ 115,000 km; rust perforation 5 years/160,000 km. **Supplementary warranty:** A bumper-to-bumper extended warranty is essential, judging from the serious problems reported on previous models. **Highway/city fuel economy:** 7.2–11.3L/100 km with the 2.7L engine; 7.6–12.5L/100 km with the 3.2L engine; 8.0–13L/100 km with the 3.5L engine.

Quality/Reliability/Safety

Pro: Plastic front fenders are dent- and corrosion-resistant. Platinum-tipped spark plugs are expected to last 160,000 km. Improved headlights really illuminate the road, and the upgraded defroster now clears the entire windshield.

Con: Quality control: Below average. It's doubtful Chrysler will put much new money into correcting quality control deficiencies, since these vehicles are nearing the end of their model run. **Reliability:** Poor overall. **Warranty performance:** Average. The 7-year powertrain warranty has proven itself. **Owner-reported problems:** Hard starts due to faulty fuel system, fuel sensor light stays lit, vehicle power loss explained by dealer as due to crankshaft being misaligned with the camshaft; glitches in the computerized transmission's shift timing, which cause driveability problems (stalling, hard starts, and surging).

Amazingly, the 4-speed LE42 automatic transmission—a spin-off of Chrysler's failure-prone A604—appears still to be just as problem-plagued, with reports of driveline shudder during 3–4 shifts and frequent transmission defaults into Second gear (limp-in mode). Other owner-reported defects include premature brake wear and brake malfunctions, electrical problems, and sloppy body construction. For example, owners complain of water and air leaks into the interior; uneven fit and finish; poor-quality trim items that break or easily fall off; exposed screw heads; faulty door hinges that make the doors rattle and render them hard to open; windows that stick, come off their tracks, or are misaligned; and power-window motor failures. **Service bulletin problems:** Engine stumbling or misfire; defective power control module; delayed or temporary loss of transmission engagement after initial startup; transmission goes into "limp" mode; harsh 4-3 downshift; headliner sag or rattle; front brake noise or pulsation; poor AC performance; rear strut squeaks; wind noise from sunroof or B-pillar when driving; fuel tank slow to fill, note that this has been a chronic problem affecting five model years (see below).

Fuel System - Fuel Tank Slow to Fill

NUMBER: 14-001-03 **DATE:** Jan. 24, 2003

OVERVIEW: This bulletin involves correcting any or all of the following items as necessary:

^Kinked/plugged fuel tank vent lines

^Replacing the fuel tank control valve

^Replacing the Leak Detection Pump (LDP) filter

^Unplugging or replacing the fuel tank fill tube assembly

2000-2004 (LH) LHS/300M/Concorde/Intrepid

SYMPTOM/CONDITION: The fuel tank is slow to fill because of lack of venting.

NHTSA safety complaints/safety: Sudden brake failure. Surging when stopped. Rear visibility compromised by narrow rear windows in the 300M. Headlights don't work properly.

Road Performance

Pro: Emergency handling: Steering is a bit vague and ponderous with some tire squealing, but no worse than the competition. **Acceleration/torque:** The 3.5L V6 engine provides plenty of low-end torque and acceleration. **Routine handling:** Handling and steering response as good as, or better than, the Taurus, and the optional AutoStick clutchless manual transmission available with the Intrepid sports package improves overall performance even more. Independent suspension also maximizes control and provides lots of suspension travel so that you don't get bumped around on rough roads. The ride doesn't deteriorate as the load is increased. The best balance between performance and ride is found with the mid-level touring suspension. It's not as harsh as the optional sport suspension, and it's firmer than the standard settings. **Braking:** Disc/disc; good braking performance with standard brakes during tests, however, owners report that premature brake rotor and pad wear leads to unacceptably long braking distance.

Con: The traction control system is noisy when activated. The smaller, 2.7L V6 engine is inadequate to handle a fully loaded Concorde or Intrepid. The 3.5L V6 isn't as smooth a performer as the Lexus GS 300 or the Acura TL power plants. **Transmission:** The 4-speed automatic transmission shifts in and out of Overdrive with a jolt. A considerable amount of body roll occurs in tight manoeuvring.

Comfort/Convenience

Pro: Standard equipment: Loaded with standard features. A sleeker, more aerodynamic body than the Taurus/Sable, the Honda Accord, and Chevrolet's Lumina. Excellent front visibility due to the large windshield and low front end. **Controls and displays:** Very user-friendly. Analogue instruments are clearly laid out. **Climate control:** Automatic climate-control system is much improved. Efficient and easy to adjust. **Entry/exit:** Excellent front and rear access. The user-friendly interior features passenger grab handles for easy access. **Interior space/comfort F/R:** Extended front seat tracks for long-legged drivers or simply for people wanting to sit away from the airbag. Comfortable front bucket seats and rear seat sits three abreast. **Cargo space:** Lots of storage space, including map pockets in the door.

Con: Driving position: Front seat lacks lateral support and the adjustable lumbar support is uncomfortable. Instrumentation illumination could be brighter and more distinct on the LHS and 300M. **Truck/liftover:** The trunk has a high deck lid, making for difficult loading and unloading, and there's no inside access by folding down the rear seat, as in the Camry. **Quietness:** All four vehicles have excessive engine, road, and wind noise that comes mainly from the tires and poor sealing around the doors and windows.

300M, Concorde, Intrepid

List Price (very negotiable)	**Residual Values** (months)			
	24	**36**	**48**	**60**
Intrepid SE: $25,615 (19%)	$16,500	$13,500	$11,500	$9,500
Concorde LX: $30,775 (20%)	$18,000	$15,000	$13,000	$11,000
300M: $40,900 (23%)	$25,000	$20,000	$18,000	$15,000

Technical Data

Powertrain (front-drive)
Engines: 2.7L V6 (200 hp)
• 3.5L V6 (232 hp)
• 3.5L V6 (250 hp)
Transmissions: 5-speed man.
• 4-speed auto.

Headroom F/R: 38.3/37.2 in.; 38.3/37.7 in.
Legroom F/R: 42.2/41.6 in.; 42.2/39.1 in.
Wheelbase: 113 in.
Turning circle: 35.4 ft.; 37.6 ft.

Dimensions/Capacity (Concorde; 300M)
Passengers: 3/3; 2/3
Height/length/width: 55.9/207.7/74.4 in.; 56/197.8/82.7 in.
Cargo volume: 18.7 cu. ft.; 16.8 cu. ft.
Tow limit: 2,000 lb.
Fuel tank: 68L/reg.; 65L/reg.
Weight: 3,550 lb.; 3,567 lb.

Safety Features/Crashworthiness

	Std.	Opt.
Anti-lock brakes	■	■
Seat belt pretensioners F/R	—	—
Side airbags	—	■
Traction control	■	■
Head restraints F/R		
Concorde and Intrepid	*****	**
300M	**	**
Visibility F/R		
Concorde and Intreprid	*****	*****
300M	*****	*
Crash protection (front) D/P	****	*****
Concorde	****	****
300M	***	****
Crash protection (side) F/R	****	***
Concorde	****	***
300M	****	***
Crash protection (offset)	***	—

PT CRUISER ★★★★

RATING: Above Average. This Neon spin-off's popularity is waning. Cobbled together with Neon parts and engineering, the PT Cruiser is essentially a fuel- and space-efficient hatchback mini-minivan that uses the same nostalgic hot-rod flair that was so successful with the Prowler. **Strong points:** Excellent fuel economy (regular fuel); nimble handling around town; good braking; lots of interior space; easy access; versatile cargo area; many thoughtful interior amenities; and slow depreciation. **Weak points:** Lethargic base engine (especially with the automatic transmission); downshifts roughly; poor crashworthiness rating (driver); mediocre highway performance and handling; a firm ride; a tacky-looking interior; lots of engine, wind and road noise; and limited rear visibility. Log on to *www.ptcruiser.org* for the latest owners' comments. **New for 2004:** A less expensive turbo option and a Dream Cruiser Series 3 that includes a standard 220-hp High Output Turbo engine, a distinctive Midnight Blue over Bright Silver two-tone paint scheme and blue-tinted glass. Cost: $35,340.

OVERVIEW: The Cruiser is ideal for buyers tired of the trucklike handling of rear-drive minivans, and who want good passenger/cargo flexibility. True, its bold styling is an attention-getter, but the basic configuration has been used in other countries for years by Asian and European automakers and in North America with the bland-looking Mitsubishi Expo LRV/Eagle Summit, popular in the early '90s.

The Cruiser comes with a nice assortment of new parts that include attractive chrome door handles, a four-spoke steering wheel, and a cue-ball shifter for the 5-speed manual that sits atop a chrome stalk with a vinyl boot. It carries a Dodge Stratus 150-hp 2.4L 16-valve engine, although in international markets, a 1.6L and 2.0L 4-cylinder will be offered.

The 2003s got a 215-hp turbocharged version, a new AutoStick optional automatic transmission, larger 17-inch tires, four-wheel disc brakes (with the turbo), and a firmer suspension. A convertible arrived late in the model year.

Cost analysis/best alternatives: Cruiser sales aren't very good these days, as the car's unique styling has worn thin after three years. Dealers are more than eager to haggle. In a pinch, you may wish to consider these wagons: VW's Jetta wagon and Passat GLS 4, the Subaru Legacy Outback Limited, or the Pontiac Vibe or Matrix. Sport-utilities worth a gander are the Subaru Forester, Honda CR-V EX, GM Tracker, Hyundai Santa Fe, and Suzuki Grand Vitara. **Options:** Try to get a manual transmission and you'll have a much more reliable and better-performing car. The optional front folding rear seatback allows objects up to 2.4 m (8.0 ft.) long to fit inside the vehicle. The same option permits the back of the front passenger seat to be used as a table for the driver. Sunroof design causes excessive interior wind noise. Stay away from the side airbag and ABS brakes—neither safety item is worth the extra risk it poses. Although the disc brakes have been improved, Chrysler's large number of ABS failures is worrisome. It would be a good idea to replace the rear drum brakes with discs; however, that option is only available with the expensive, and still unproven, four-wheel ABS option coupled with electronic traction control. **Rebates:** If sales don't improve, look for $1,000+ rebates and discounting on both model year vehicles. **Delivery/PDI:** $810. **Depreciation:** Practically non-existent. Expect 10 percent depreciation per year to kick in after the second year. **Insurance cost:** Higher than average, because many insurance companies impose the higher rates they use for trucks. **Parts supply/cost:**

Average, since many parts come from the Neon generic parts bin. Body parts are another matter. Expect long delays and high costs. **Annual maintenance cost:** Average. **Warranty:** Bumper-to-bumper 3 years/60,000 km; powertrain 7 years/115,000 km; rust perforation 5 years/160,000 km. **Supplementary warranty:** A toss-up; only needed if you plan to exceed base warranty. **Highway/city fuel economy:** 8.3–11.7L/100 km.

Quality/Reliability/Safety

Pro: Reliability: Better than average. With the 7-year warranty working well and the Neon's improvements, many of the generic deficiencies have been cleaned up or fixed under warranty. **Quality control:** Very few problems reported by owners. **Warranty performance:** Average. Apparently, DaimlerChrysler is much more sensitive to Cruiser complaints than to complaints on models that will soon get the axe like the Concorde, 300M, etc.

Con: Owner-reported problems: The automatic transmission, AC, and brakes have been the most troublesome components. Other problems include excessive oil consumption due to a faulty valve cover gasket; annoying wind noise when driving with the rear window or sunroof open; moisture between clearcoat and paint turns the hood a chalky white colour; water leaks through the side passenger window; and drivetrain whine. **Service bulletin problems:** Moisture accumulation in headlights; high-speed engine surging; white powder collects on airbag cover.

Rear Bumper Fascia Wavy/Warped

NUMBER: 13-002-03 **DATE:** Feb. 14, 2003

OVERVIEW: This bulletin involves replacing the rear fascia stud plates, energy absorbers and fascia (paint fascia as required).

200 (PT) PT Cruiser

SYMPTOM/CONDITION: The top horizontal surface of the rear fascia may appear wavy.

This cosmetic problem is covered under the warranty and entails at least three hours of labour.

NHTSA safety complaints/safety: Sudden, unintended acceleration; gas pedal went to floor with no acceleration; airbags deployed for no reason, or failed to deploy; vehicle suddenly shifts into First gear while cruising; sudden brake lock-up; headrests are too high, block vision; optima battery leaks acid; white powder leaks from airbag; and speedometer is hard to read.

Road Performance

Pro: Steering: Acceptable, with good road feedback. **Acceleration/torque:** Blistering acceleration with the turbocharged engine. Competent acceleration with a manual transmission coupled to the base engine. **Transmission:** Precise manual shifter. Optional automatic gearbox works quite well. The AutoStick is no big deal. **Routine handling:** Getting around town is easy. **Braking:** Disc/drum; acceptable ABS braking when it functions as it should.

Con: Acceleration/torque: The 2.4L 150-hp, 4-cylinder engine is not very smooth running when matched to the automatic transmission; it struggles when going uphill or merging with freeway traffic. This requires frequent downshifting and lots of patience—accelerating to 100 km/h takes about 9 seconds. Automatic transmission doesn't have much low-end torque, forcing early kickdown shifting and deft manipulation of the accelerator pedal. Solution? Get a manual tranny or pay extra for a turbocharger and cross your fingers when the warranty expires. **Emergency handling:** Slow and sloppy. The turning diameter seems excessive for such a short vehicle. Hard cornering produces an unsteady, wobbly ride due to the car's height.

Comfort/Convenience

Pro: Standard equipment: Very well appointed with lots of innovative convenience features. Standard AM/FM/cassette stereo with six speakers. Front passengers have two 12-volt power plug-ins. **Controls and displays:** User-friendly rotary ventilation and heater controls. Centre dash area has buttons for the defogger, rear wiper/washer, and (optional) traction control on/off button. Easy access to all gauges and controls. **Climate control:** Efficient system that's easily understood and accessed. **Entry/exit:** Getting in and out is quite easy because of the Cruiser's tall roof, big doors, and elevated seats. The rear doors in particular open about 10 degrees more than in most cars. **Interior space/comfort F/R:** Slightly smaller than the Neon in wheelbase and overall length, the Cruiser's about 8 cm (3 in.) shorter than the base minivan. But its additional 25 cm (10 in.) of height over the Neon provides lots of room for tall passengers in front and rear. Seating is midway between a Caravan and a Neon, also making access quite easy. Tall, chair-like seats provide lots of legroom, and there's also generous headroom, although the Cruiser is a bit narrower than the Neon. The front and rear seats have a reasonable amount of bolster to keep occupants from sliding side to side. **Cargo space:** Storage spaces include two large front door pockets, an underseat drawer, and a small glove box. Removable seats and a flat floor put the PT into the truck class for government taxonomists. The folding front passenger seat is a nice touch. Rear seats can be easily folded, flipped forward, and unlatched for quick removal. They have easy-to-carry handles, and wheels for rolling away. The seats aren't light, though, with the 29 kg (65 lb.) right rear seat weighing about twice as much as the left. Another innovative feature is the car's rigid cargo-hold cover that can double as a tailgate-party table. **Trunk/liftover:** The rear hatch door has a very low 62 cm (25 in.) liftover height, and is easy to open and close.

Con: Controls and displays: Left side of the tachometer not easily seen. Front power window switches are found at the top of the dashboard; rear window switches are inconveniently located on the back of the centre console. **Quietness:** Some drivetrain; wind, tire, and road noise intrudes into the interior.

PT Cruiser

List Price (soft)	**Residual Values** (months)			
	24	**36**	**48**	**60**
Classic: $24,360 (16%)	$18,000	$16,000	$14,000	$12,000
Limited: $28,800 (18%)	$22,000	$19,000	$16,000	$14,000

Technical Data

Powertrain (front-drive)
Engines: 2.4L 4-cyl. (150 hp)
• 2.4L 4-cyl. turbo (180 hp)
• 2.4L 4-cyl. turbo (220 hp)
Transmissions: 5-speed man.
• 4-speed auto.
Dimension/Capacity (base)
Passengers: 2/3
Height/length/width:
63/168.8/67.1 in.
Headroom F/R: 40.4/39.6 in.
Legroom F/R: 41/40.8 in.
Wheelbase: 103 in.
Turning circle: 42 ft.
Cargo volume: 64 cu. ft.
Tow limit: 1,000 lb.
Fuel tank: 57L/reg.
Weight: 3,300 lb.

Safety Features/Crashworthiness

	Std.	**Opt.**
Anti-lock brakes (4W)	—	■
Seat belt pretensioners F/R	—	■
Side airbags	—	■
Traction control	—	■
Head restraints F/R	*****	*****
Visibility F/R	****	***
Crash protection (front) D/P	****	****
Crash protection (side) F/R	****	*****
Crash protection (offset)	—	—

CARAVAN, GRAND CARAVAN, TOWN & COUNTRY ★★★

RATING: Average. These minivans are better built and have better warranties than the failure-prone Ford Freestar/Windstar. Still, early automatic transmission burnout is a major concern. **Strong points:** Very reasonably priced and available with AWD (though it's not recommended). You get a comfortable ride; excellent braking; lots of innovative convenience features; user-friendly instruments and controls; two side sliding doors; easy entry/exit; and plenty of interior room. **Weak points:** Poor 4-cylinder acceleration and mediocre handling with the extended versions, though the ride is smoother over bumps. A sad history of chronic powertrain, AC, ABS, and body defects, that's exacerbated by the automaker's hard-nosed attitude in interpreting its after-warranty-assistance obligations. No power doors on the regular length models. Get used to a cacophony of rattles, squeals, moans, and groans, caused by the vehicle's poor construction and subpar components. Crashworthiness has declined a bit, though it's not bad; side curtain airbags aren't available. Fuel economy isn't as good as advertised. **New for 2004**: Voyager has been dropped in the States; a tire-pressure warning monitor, enhanced audio and new keyless entry options are new this year.

OVERVIEW: These versatile minivans return with a wide array of standard and optional features that include AWD, anti-lock brakes, child safety seats integrated into the seatbacks, flush design door handles, and front windshield wiper/washer controls located on the steering column lever for easier use. Childproof locks are standard and the front bucket seats incorporate vertically adjustable head restraints. The Town & Country, a luxury version of the Caravan, comes equipped with a 3.8L V6 and standard luxury features that make the vehicle more fashionable for upscale buyers.

Cost analysis/best alternatives: The redesigned 2004 Toyota Sienna and Nissan Quest should be your first choice. Honda's latest Odyssey isn't as refined, but its older models have a bit better performance and reliability record than Toyota. Mazda's 2002–03 MPV also is a good used alternative. GM front- and rear-drive minivans are also credible alternatives, Some full-sized GM or Chrysler rear-drive cargo vans, ripe for conversion, might be a more affordable and practical buy if you intend to haul a full passenger load, do some regular heavy hauling, are physically challenged, use lots of accessories, or take frequent motoring excursions. Don't splurge on a luxury Chrysler minivan: its upscale Town & Country may cost up to $15,000 more than a base Caravan, yet be worth only a few thousand more after five years on the market. **Options:** As you increase body length, you lose manoeuvrability, but gain ride comfort, except with the Sport Touring Group suspension, which makes for a firmer ride. Don't even consider the 4-cylinder engine—it has no place in a minivan, The 3.3L V6 is a better choice for most city-driving situations, but don't hesitate to get the 3.8L if you're planning lots of highway travel or carrying four or more passengers. Since its introduction, it's been relatively trouble-free, plus it's more economical on the highway than the 3.3L, which strains to maintain speed. Try to get 16-inch wheels and four-wheel disc brakes for a relatively cheap handling and safety upgrade. The sliding side doors make it easy to load and unload children, install a child safety seat in the middle, or remove the rear seat. On the downside, they expose kids to traffic and are a costly, failure-prone option. Child safety seats integrated into the rear seatbacks are convenient and reasonably priced, but Chrysler's versions have had a history of tightening up excessively or not tightening enough, allowing the child to slip out. Try the seat with your child before buying it. Other important features to consider are the optional defroster, power mirrors, power door locks, and power driver's seat (if you're shorter than 5' 9" or expect to have different drivers using the minivan). You may wish to pass on the tinted windshields; they seriously reduce visibility. Town & Country buyers should pass on the optional all-wheel drive coupled with four-wheel disc brakes (instead of the standard rear drums). Chrysler's large number of ABS failures is worrisome. Ditch the failure-prone Goodyear original equipment tires and remember that a night drive is a prerequisite to check out headlight illumination, called inadequate by many. **Rebates:** 2003 and 2004 models will get $2,500–$3,000 rebates and zero percent financing throughout the year. **Delivery/PDI:** $995. **Depreciation:** Slightly slower than average, but not as slow as pickups, SUVs, or the PT Cruiser. **Insurance cost:** Higher than average, but about average for a minivan. **Parts supply/cost:** Higher than average, especially for paint, AC, transmission, and ABS components, which are covered under a number of "goodwill" warranty programs and several recall campaigns. Chrysler says that its 3.3L and 3.8L engines won't require tune-ups before 160,000 km. Prepare to be disappointed; many owners have had to tune up their minivans way before then. **Annual maintenance cost:** Repair costs are average during the warranty period. **Warranty:** Base warranty is inadequate, if you plan to keep your minivan more than five years. Bumper-to-bumper 3 years/60,000 km; powertrain 7 years/115,000 km; rust perforation 5 years/160,000 km. **Supplementary warranty:** A wise buy. Bargain the price down to about one-third the $1,500 asking price. **Highway/city fuel economy:** *Caravan:* 9–13L/100 km; *Grand Caravan AWD:* 10–14L/100 km; *Town & Country:* 9–14L/100 km; *Town & Country AWD:* 10–15L/100 km.

Quality/Reliability/Safety

Pro: At least Chrysler's new warranty will go a long way toward keeping engine and transmission repair costs down. Dual airbags include knee bolsters to prevent front occupants from sliding under the seat belts. Side-impact protection has been increased with steel beams in door panels. An innovative engine compartment layout also makes for a larger "crumple zone" in the event of collision. Remote-control power door locks can be programmed to lock when the vehicle is put in gear. Chrysler has developed a mechanism that releases the power door locks and turns on the interior lights when the airbag is deployed.

Con: **Quality control:** You are *not* buyng a Mercedes minivan, thankfully, because Mercedes quality is also on the downswing. 2004 Chrysler minivans will likely have similar powertrain, electrical system, brake, suspension and body deficiencies to previous versions. Rear outward vision is reduced by the high head restaints. *Consumer Reports* rates only the extended versions as worse than average, yet *Lemon-Aid* readers report that the base versions are just as failure-prone. This isn't surprising—quality control has been below average since these vehicles were first launched almost 20 years ago, and, surprisingly, got much worse after the '90 model year introduced "limp-prone" transmissions. **Reliability:** Somewhat improved, but inferior to the competition. **Warranty performance:** Fair. **Owner-reported problems:** Owners say powertrain failures make them afraid to drive these minivans. Their complaints also focus on electrical glitches, erratic AC performance, and early brake wear.

Premature wearout was reported on these parts: cooling system, clutch, front suspension components, wheel bearings, air conditioning, and body parts (trim; weather stripping becomes loose and falls off; plastic pieces rattle and break easily). In spite of improvements over the years, the front brakes need constant attention: if not to replace the pads or warped rotors, then to silence the excessive squeaks when braking. Premature wearout of the front brake rotors within two years or 30,000 km is a common Chrysler minivan failing that also has serious safety implications. **Service bulletin problems:** Oil leaks at oil adapter; rear-brake rubbing sound; rear brake drum snow ingestion; sliding door won't open; troubleshooting water leaks (see following).

Water Leaks/Crack in Upper Body Seam Sealer

NUMBER: 23-001-03 **DATE:** Jan. 17, 2003

SUBJECT: Upper body seam seal water leak

OVERVIEW: This bulletin involves repairing a crack/split in the upper body seam sealer.

2003 (RS) Town & Country/Caravan/Voyager

SYMPTOM/CONDITION: Customer may describe a water leak and/or visible crack/split in the upper body seam sealer.

This cosmetic problem is covered under the warranty and entails at least three hours of labour.

NHTSA safety complaints/safety: NHTSA has recorded numerous complaints of airbags failing to deploy in an accident or deploying unexpectedly—when passing over a bump in the road, or simply turning on the vehicle. Chrysler continues to downplay the safety implications of its minivan safety defects, whether in the case of ABS failures, inadvertent airbag deployments, or sudden transmission breakdowns. Interestingly, Chrysler's five-star side-impact crash rating was earned without the assistance of optional side airbags, whereas Ford's Windstar needed extra side airbags to garner five stars. Seat belts are another recurring problem: they may become unhooked from the floor anchor; buckles jam or suddenly release; and the child safety seat harness easily pulls out or over-retracts, trapping children. Other problems: Automatic transmission won't shift or suddenly drops into Reverse; exposed electrical wires under the front seats; excessive steering vibration; seat belts unlatch themselves; missing suspension bolt caused the right side to collapse; wiper blades stick together.

Road Performance

Pro: Acceleration/torque: The most versatile, though a bit underpowered, powertrain for short-distance commuting is the 3.3L V6 engine coupled with the 4-speed automatic transmission (0–100 km/h: in about 10.8 seconds with a 3.3L-equipped Grand Caravan). Chrysler's top-of-the-line 3.8L engine is a better choice if you intend to do a lot of highway cruising: it's smooth and quiet with lots of much-needed low-end torque. The AWD transfers 90 percent of the engine power to the front wheels during normal driving conditions. It's easy to use and performs well. As the front wheels lose traction, the rear wheels get additional power until traction has been stabilized or the 55/45 percent front-to-rear limit is reached. **Routine handling:** The regular-sized models are the closest thing to a passenger car when it comes to ride and hand-ling. The redesigned chassis and improved steering provide a comfortable, no-surprise ride. Stiff springs greatly improve handling and comfort. Manoeuvrability around town is easy. **Emergency handling:** Slow, but acceptable. **Steering:** Generally precise and predictable. **Braking:** Acceptable ABS braking when it functions as it should.

Con: Transmission: The 41TE 4-speed automatic transmission with Overdrive shifts slowly and imprecisely. Excessive transmission whine. It's hard to use engine compression to slow down by "gearing down." Downshifting from the electronic gearbox provides practically no braking effect. The AWD option is overrated and not worth the fuel penalty for most driving situations. The stretched wheelbase version gives less-than-nimble handling. A large turning radius and long nose can make parking difficult. **Steering:** Power steering is vague and over-assisted as speed increases. **Braking:** Brake pedal feels mushy and the brakes tend to heat up after repeated applications, causing considerable loss of effectiveness (fade) and warping of the front discs. The ABS system has proved unreliable on older vans, and repair costs are astronomical.

Comfort/Convenience

Pro: Standard equipment: This is where Chrysler minivans shine. They offer plenty of standard comfort and versatility features. **Driving position:** Drivers are treated to a carlike driving position. Good overall view of the road. **Controls and displays:** The instrument panel features easy-to-read gauges and warning lights, and radio and heater controls are set close to the driver. The location of the turn signal indicators is particularly well thought out—they're in the lower portion of your field of vision. Lots of cupholders and interior reading lights. **Climate control:** Adequate. Triple-zone AC allows for different temperature settings for the driver and rear-seat passenger; overhead heating and ventilation ducts to the rear seat are well placed. **Entry/exit:** Easy. The step-up height isn't too high for most people. Another nice touch: grab handles on the rear hatch and sliding door. There's convenient "walk through" access to the rear seating area. **Interior space/comfort F/R:** Minivan doesn't mean "mini" in terms of passenger space. The aerodynamic exterior design, increased window area, and lower sills make for an attractive, roomier-feeling vehicle, and the interior is large and versatile with excellent outward visibility. Chrysler has copied the Windstar in providing theatre-type seating; each row is set a little higher than the row in front, giving most passengers a better view. A fairly high roofline means that a six-foot-tall passenger will sit comfortably in the back seat. An optional "Convert-a-Bed" package is available with the seven-passenger seating configuration. **Cargo space:** Plenty and practical. This is where the dual sliding doors come in handy. Rear seatbacks fold down, and removing the centre and rear seats is a "snap," thanks to the addition of little wheels on the base of the rear seat. Snap them down and you can roll the seat anywhere. Lifting them is a chore, though. **Trunk/liftover:** Easy to load and unload 4' x 8' sheets of plywood, thanks to the wide doors and low floor. Courtesy lights in front and on the liftgate are an added convenience. Interestingly, with the rear seat removed, the regular-sized Caravan provides more cargo space than the Grand Caravan. **Quietness:** Body soundproofing effectively keeps out road and wind noise.

Con: Controls and displays: Driver is faced with a sea of switches, dials, and buttons on the dash/door panel/console. Centre console is way too low, console storage bin for small objects is sometimes hard to open and close, and the panel dimmer is hidden behind the steering wheel. **Entry/exit:** Both sliding side doors are unwieldy to slide. Rear hatch door is hard to close. **Interior space/comfort F/R:** Tall drivers will find legroom a bit tight without an adequate left footrest. Three adults will find the third-row bench seat a bit cramped and the head restraints set too low. **Cargo space:** Removable seats, which weigh 23–45 kg (50–100 lb.) aren't as easy to remove as Chrysler would have you believe. **Quietness:** At first, interior noise is very low, then it gets progressively louder as these vehicles age and the body rattles, groans, and moans increase.

Caravan, Grand Caravan, Town & Country

List Price (soft)	**Residual Values** (months)			
	24	**36**	**48**	**60**
Base Caravan: $27,620 (16%)	$19,000	$16,000	$14,000	$11,000
G. Caravan: $34,090 (20%)	$19,000	$16,000	$14,000	$11,500
Town & Country LXI: $44,095 (25%)	$29,000	$26,000	$23,000	$20,000

Technical Data

Powertrain (front-drive)
Engines: 3.3L V6 (180 hp)
• 3.8L V6 (215 hp)
Transmission: 4-speed auto.
Dimension/Capacity (base)
Passengers: 2/2/3
Height/length/width:

Legroom F: 41.2/R1: 36.6/R2: 35.8 in.
Wheelbase: 113.3 in.
Turning circle: 39.5 ft.
Cargo volume: 146.7 cu. ft.
GVWR: 5,800 lb.
Payload: N/A
Tow limit: 2,700–3,500 lb.

68.5/186.3/76.8 in.
Headroom F: 39.8/R1: 40.1/R2: 38.1 in.

Fuel tank: 76L/reg.
Weight: 3,985 lb.

Safety Features/Crashworthiness

	Std.	Opt.
Anti-lock brakes (4W)	—	■
Seat belt pretensioners F/R	■	—
Side airbags	■	■
Traction control	■	■
Head restraints F/R	**	*
G. Caravan	****	***
Town & Country	***	**/*
Visibility F/R	****	****
Crash protection (front) D/P		
Caravan	****	****
G. Caravan	****	****
Town & Country	****	****
Crash protection (side) F/R		
Caravan	****	****
G. Caravan	*****	*****
Town & Country	*****	*****
Crash protection (offset)	***	—

Ford

Fix Or Repair Daily (FORD)

Ford needs repairing badly. It's in even worse shape than DaimlerChrysler, having lost $6 billion (U.S.) in 2001 and 2002.

Except for its re-engineered F-150 truck, very few new Ford models will arrive before the 2005 model year and a number of them will be dropped, like the venerable rear-drive Crown Victoria (the only sedan with a six-passenger bench seat), and the quality-challenged Taurus, and Windstar. Most surprising of all, Ford's Thunderbird, a model resurrected in August 2001, will be dropped after the 2005 model year. Ford says it never intended the car to be in regular production.

Bull.

Isn't it obvious? No one wants to admit they have almost bankrupted the company through unwanted products and gross mismanagement. Disgraced ex-CEO Jac Nasser's recruits are fleeing the company faster than President Bush's speechwriters; the Blue Oval dealer program has been killed by dealer protests and successful lawsuits after burning through millions of dollars; the company's "Ford 2000" attempt to centralize its global business has spun out of control and been abandoned; and the Premier Automotive Group, which sells many of the company's European cars, is floundering, having lost its independence and the confidence of Ford's senior management in Dearborn.

The Jaguar X-Type is a declawed pussycat that has a Taurus V6 Duratec engine block and styling that is more Taurus than "Jag." That explains its imprecise automatic shift linkage and quality lapses. One car columnist commented: "I have to duck to see most traffic lights if I am at the front of the line. The view out the top of the windshield is optically distorted, shrinking text on road signs vertically and causing my eyes to work harder as they try to adjust…" Source: *www.europeancarweb.com/longtermers/0209ec_jaguar.*

What could Ford have been thinking when they bought this turkey called Land Rover, a company that nearly bankrupted BMW? Actually, Ford was thinking globally, knowing that it couldn't sell luxury cars throughout the world with just its Lincoln models. Therefore, Ford acquired those companies that already had a worldwide luxury cachet. Land Rover became part of Ford's Premier Automotive Group, joining Aston Martin, Jaguar, Lincoln, and Volvo. All money losers.

But what really is scary is Ford's cutbacks on warranty repairs. The automaker is testing new warranty repair procedures that could result in shorter authorized flat-rate repair times and lower payments to dealers. When Ford last overhauled its warranty repairs guidelines in 2000, labour times were cut by 24 percent. Ford internal documents confirm that warranty costs were cut almost 40 percent from 2002 through 2003—the same years *Lemon-Aid* readers complained most bitterly about denied warranty claims for engine intake manifolds/head gaskets and automatic transmissions.

On top of this, Ford has managed to profoundly tick off its suppliers, dealer body, and customers. Parts suppliers have been berated by Ford management for simply following the company's insane cost-cutting orders. The dealer body no longer trusts the company that has shut down Mercury, cut warranty payouts with its Blue Oval program, and made dealers beg for after-warranty assistance for customers screwed by the automaker's abysmal quality control.

Poor quality: Ford's weakest link

Ford quality control is still quite poor as senior parts managers are shuffled and reshuffled faster than a Las Vegas card deck.

In spite of the initial sales success of its Escape and Focus and a much-improved Explorer 4X4, Ford is going downhill fast according to virtually every yardstick used to measure automotive success: profit, market share, quality, productivity, morale, bonuses, credit rating, public image, and relations with car owners, dealers, investors, and shareholders.

Engine and transmission defects

Ford's engine and automatic transmission deficiencies are broadly based and have existed for over a decade, judging by NHTSA complaints and confidential Ford internal documents cited by *www.blueoval.com:*

> According to Ford, the [Focus] rear axles "may exhibit rear axle shaft and/or axle bearing premature wear. This is caused by excessive load, temperature, and inadequate lubrication."
>
> Ford will replace the axle bearings under dealer operation code "030505A" under the provisions of the bumper-to-bumper warranty.
>
> This is considered a "silent recall" where upon Ford may NOT notify owners of this condition. Owners should contact Pat Hoye (313) 248-8336 at Ford World Headquarters.

Premature brake wear and brake failures are other areas where Ford needs immediate improvement and a more generous policy in handling customer complaints. The quality of body components also remains far below Japanese and European standards.

FOCUS ★

RATING: Not Recommended. Performance-wise, this is a pretty good vehicle for urban use, but it's so poorly made, you risk both your life and your wallet if you drive one. **Strong points:** Powerful 170-hp optional 4-cylinder engine; excellent handling and road holding; impressive fuel economy; plenty of interior space for occupants and cargo (especially in the wagon); well appointed for an entry-level vehicle; user-friendly control layout; and now covered by a 5-year warranty. **Weak points:** Mediocre acceleration with base power plants;

excessive engine and road noise; difficult rear-seat access with the hatchback version; and a large number of safety- and performance-related complaints that aren't likely to be fixed under the newly extended warranty. **New for 2004:** Nothing significant.

OVERVIEW: Hailed as Europe's 1999 Car of the Year (shows you how much they know), Ford's sleek Focus came to North America as an uplevel, premium small car, positioning itself between the entry-level four-door Escort and Contour. Two inches taller and almost seven inches longer than the Escort, and embodying Ford's "new edge" styling (less aero, more creases) are three body styles: a two-door hatchback in sporty ZX3 trim; LX, SE, and upscale ZTS four-door sedans; and a four-door SE wagon. The Escort's base engine, a 110-hp 2.0L 4-banger, is carried over to the Focus LX and SE, while the 130-hp twin-cam 2.0L (also used on the Escort ZX2 coupe) is standard on the ZTS and ZX3 and optional on the SE. Either engine can be hooked to a manual or an optional 4-speed automatic transmission. There's also a high-performance, 170-hp SVT Focus that's essentially a hatchback equipped with a sport suspension and 17-inch wheels. The ZX5, a four-door hatchback looks like a shortened version of the Focus wagon.

Cost analysis/best alternatives: Ford probably won't increase the 2004's price by more than a few hundred dollars. If so, that model year represents the better buy because you'll stand a better chance of avoiding some of the car's quality glitches from 2003. Other viable alternatives include the Honda Civic, Hyundai's Accent, the Mazda Protegé, or the Toyota Echo. **Options:** Go for the $300 upgraded, twin-cam engine; the extra horsepower is sorely needed. Stay away from the cruise control: it's been failure-prone and still may malfunction, despite recall repairs. Don't go for the side airbags until more is known about their overall effect. **Rebates:** $1,500 and zero percent financing. **Delivery/PDI:** $895. **Depreciation:** Average. **Insurance cost:** Average. **Parts supply/cost:** Average. **Annual maintenance cost:** Too early to tell, but pervasive quality problems aren't a good sign. **Warranty:** Bumper-to-bumper 5 years/100,000 km; rust perforation 5 years/unlimited km. **Supplementary warranty:** A good idea for the powertrain. **Highway/city fuel economy:** 6.5–9.4L/100 km with a manual transmission; 6.9-9.4L/100 km with the automatic.

Quality/Reliability/Safety

Pro: Warranty performance: Ford Canada is attending to the Focus' quality problems under warranty, but once the warranty expires, plan on regular meetings with your bank's loan officer.

Con: Quality control: Abysmal. A disturbingly large number of life-threatening defects have been reported on previous year models, yet few service bulletins offering remedies have been found. **Reliability:** Worse than average, particularly in view of the many stalling and hard starting complaints.

Owner-reported problems: Chronic stalling; hard starts due to defective mass airflow system (MAS), fuel pump, or fuel rail pressure sensor; transmission suddenly downshifts, axle and transmission failures; cruise control disengages and Transmission Trouble light comes on; transmission makes a "shuddering" sound while in Reverse; PCM fails repeatedly; frequent replacement of the brake rotors and calipers; water leaks into the trunk compartment; inoperative power door locks; trunk won't close or fails to open; long delay for wheel delivery; and a wobbling rear end due to excessive suspension vibrations. **Service bulletin problems:** Engine exhibits a rolling, rough idle following a cold start; whine, moan, whistling, humming, drumming, or rattling noise emanating from the right halfshaft support bearing; excessive front brake pad wear, front brake creep or groan noise, front brake dust, and rotor wear concern; fuel level sensor circuit malfunction; coolant leak from cracked coolant degas bottle; back glass window cracks for no apparent reason; instrument panel squeak and rattle; rear suspension rattling and chucking noise; front end creaking and crunching while turning; fog lamp assembly condensation or water accumulation; MIL stays lit; poorly calibrated stability assist; intermittent wiper failure; and an inoperative horn. **NHTSA safety complaints/ safety:** Airbags failed to deploy; complete brake failure; SVT engine surges and misses upon acceleration; constant stalling, especially in cold weather; vehicle (wagon) shakes violently when driven 100 km/h with the windows down; second-row middle seat belt buckle unlatched when brakes were applied; rear window suddenly shattered (a TSB covers this); trunk opened on its own while driving; and speedometer reading washes out in sunlight.

Road Performance

Pro: Acceleration/torque: Very good with the 170-hp SVT engine. **Emergency handling:** Competent and predictable. Good tire grip and little body lean under power when going into corners. Automatic transmission is smooth and quiet; however, it does compromise performance a bit and long-term durability is still undecided. **Steering:** Precise with good road feedback. **Routine handling:** Excellent. **Braking:** Good, but some brake fade.

Con: Acceleration/torque: Not good with 110–130-hp engines. Expect a ho-hum 0-100km/h time of about 9.6 seconds. **Transmission:** Manual's short gearing requires frequent shifting.

Comfort/Convenience

Pro: Standard equipment: Very well appointed for an entry-level Ford. Height-adjustable driver's seat provides good outward visibility. **Driving position:** Comfortable, with good all-around visibility. **Controls and displays:** Easy to find and very user-friendly. **Climate control:** AC works well and is easily adjusted for almost any setting. **Interior space/comfort F/R:** Passenger comfort has been enhanced with the high roofline, raised seating position, and tall, wide doors. Passengers sit comfortably high with lots of headroom and an acceptable amount of legroom (for a small car). **Entry/exit:** No problem; the large front doors are helpful, but the rear doors could open wider. **Cargo space:** Best in class. The wagon version, for example, has plenty of cargo space and rear seats that fold or flip flat for additional room. **Trunk/liftover:** Low, making for easy luggage loading and unloading. Plus, trunk lid hinges are luggage friendly.

Con: Controls and displays: Could use a tachometer to keep the engine on track. Crank handle for seat adjustments is both primitive and annoying. Air vents could be more conveniently placed. **Quietness:** Excessive engine and wind noise at higher speeds.

Focus

List Price (very negotiable)	**Residual Values** (months)			
	24	**36**	**48**	**60**
Focus LX: $16,475 (15%)	$13,000	$11,000	$9,500	$7,000
ZX3: $17,750 (16%)	$14,500	$12,500	$11,000	$8,000
ZX5: $21,560 (16%)	$16,500	$13,500	$12,000	$9,500
ZTW Wagon: $22,405 (16%)	$17,500	$14,500	$13,000	$10,500

Technical Data

Powertrain (front-drive)
Engines: 2.0L 4-cyl. (110 hp)
• 2.0L DOHC 4 (130 hp)
• 2.0L DOHC 4 (170 hp)
Transmissions: 5-speed man.
• 4-speed auto.
Dimensions/Capacity
Passengers: 2/3
Height/length/width:
56.3/168.1/66.9 in.
Headroom F/R: 39.3/38.7 in.
Legroom F/R: 43.1/37.6 in.
Wheelbase: 103 in.
Turning circle: 36. ft.
Cargo volume: 13 cu. ft.
Tow limit: 1,000 lb.
Fuel tank: 49L/reg.
Weight: 2,551 lb.

Safety Features/Crashworthiness

	Std.	**Opt.**
Anti-lock brakes	■	■
Seat belt pretensioners F/R	■	—
Side airbags	■	■
Traction control	—	■
Head restraints F/R	****	****
Visibility F/R	*****	*****
Crash protection (front) D/P	****	****
4d	*****	****
Crash protection (side) F/R	****	*
4d	***	****
Crash protection (offset)	*****	—

MUSTANG ★★★★

RATING: Above Average. With the Firebird and Camaro gone, the Mustang is your only reasonably priced American sports car. **Strong points:** Fast acceleration; impressive handling, braking, and resale value. Better-than-average crashworthiness and overall reliability. **Weak points:** Insufficient rear-seat room; limited cargo space; fishtails easily when accelerating upon a wet surface or cornering under moderate speed; and has a history of chronic stalling. **New for 2004:** Nothing, but there will be an all-new 2005 model.

OVERVIEW: Ford has renewed its commitment to the Mustang through a "retro" re-styling, using sharp exterior creases that harken back to the pony car's early years. It carries a base 3.8L V6 and offers an optional 4.6L V8 along with a high performance, limited edition Cobra variation that delivers 60 more horses than the stock 4.6L V8. The single- and twin-cam V8 options make the rear-drive Mustang a powerful street machine that has far more brawn than finesse (one of its failings when compared with the recommended Camaro and Firebird). V6 models are an acceptable compromise, even though the engine fails to deliver the gobs of power expected of a Mustang by most performance enthusiasts. Base models come equipped with a host of luxury and convenience items. Furthermore, the price is hard to beat, and resale value stays relatively high through the fifth year of ownership.

This is definitely not a family car. But for those who want a sturdy and stylish second car, or who don't need room in the back or standard ABS, the Mustang is a less risky buy than most of the Big Three sports cars.

Cost analysis/best alternatives: Get a cheaper 2003 model discounted in the $1,500–$2,000 range, or wait for the reworked 2005 version. Other cars worth considering are the Acura RSX, Mazda's Miata, and the Toyota Celica. **Options:** Traction control (consider this rear-drive's tendency to "spin out" when pushed) and an anti-theft system that includes an engine immobilizer. **Rebates:** $3,000. **Delivery/PDI:** $950. **Depreciation:** Much slower than average. **Insurance cost:** Way higher than average. **Parts supply/cost:** Inexpensive and easily found among independent suppliers. **Annual maintenance cost:** Lower than average: most mechanics will be able to fix a Mustang, although computer module powertrain glitches will drive them nuts. **Warranty:** Bumper-to-bumper 3 years/60,000 km; rust perforation 5 years/ unlimited km. **Supplementary warranty:** Not necessary; put your money into handling options and theft protection. **Highway/city fuel economy:** 7–12L/100 km with the 3.8L; 9–14.5L/100 km with the 4.6L.

Quality/Reliability/Safety

Pro: Quality control: Average. As the Mustang returns relatively unchanged year after year, one can expect fewer problems to crop up. This presumes, however, that Ford has the inclination and the money to correct the old deficiencies and pay suppliers for better quality parts on the newer vehicles—two presumptions that don't always pan out with Ford. Be wary of the Cobra's chronic stalling. **Reliability:** Good. Body assembly has improved of late, with fewer rattles and a more solid feeling. **Warranty performance:** Few warranty complaints.

Con: Owner-reported problems: The automatic transmission, fuel system, front suspension, brakes, electrical system, steering components, and fit and finish remain the primary weak spots. Sudden stalling with all models, but the Cobra is particularly affected. Noisy automatic transmission; AC failure; side windows leak when it rains; the engine computer can be temperamental; and electrical problems are common. Keep an eye on the 4.6L engine: excessive oil burning and head gasket failures may be in the offing. Body assembly quality is not up to the level of Japanese vehicles; numerous complaints of convertible, door, and trunk water leaks. **Service bulletin problems:** Hard to shift manual transmission; noise from the door trim panel when vehicle passes over bumps; popping noise when car is turning into or out of an incline; condensation within the fog lamp assembly; drift or pull to the right or left; and driveline clunk (Cobras). **NHTSA safety complaints/safety:** Fire ignited in the wiring harness under dash area; airbags failed to deploy; chronic stalling when coasting, braking, or when clutch is depressed; sudden, unintended acceleration; engine surging; Cobra pedal fractures easily; complete brake failure; seat belt doesn't retract properly; Goodyear tire blowout; serpentine belt came off, causing loss of power steering and brakes; wheel lug nuts fell off; sudden acceleration; sudden loss of steering when making a left-hand turn; car left on an incline with transmission in Park and motor shut off, rolled down after 10 minutes and hit a tree; airbags failed to deploy in a frontal collision; emergency brake ratchet assembly broke, making mechanism inoperable; gas spills out of fuel tank due to clamps not sufficiently tightened; left front wheel fell off when the lower control arm and ball joint became loose; defective transmission spider gear; stuck gas pedal; seat ratchet assembly teeth broke; seat belt continually tightens up when worn; engine produces an annoying "harmonic vibration"; a light rear end makes the car dangerously unstable on wet roads or when cornering at moderate speeds.

Road Performance

Pro: Emergency handling: Slow but predictable. **Steering:** Quick and responsive. **Acceleration/torque:** The V8 provides very quick acceleration and smooth power delivery (0–100 km/h: 7.1 seconds), while the V6 performs fairly well with the automatic 4-speed transmission. **Transmission:** The clutch is reasonably smooth and the automatic shifts reasonably well most of the time. **Routine handling:** Models equipped with the sport suspension (which includes larger tires) provide sure and predictable handling on dry roads. The upgraded base suspension also makes the car more stable and controllable on most roads. **Braking:** Disc/disc; excellent braking performance for a car this heavy (100–0 km/h: 37 m).

Con: Emergency handling: The 3.8L V6 is rough and noisy when pushed. The V8s are a bit too powerful for the amount of traction available to the rear wheels, making for lots of wheelspin and instability on slippery surfaces. **Transmission:** The manual transmission is notchy at times and the automatic sometimes hesitates between gears. **Routine handling:** The rear end tends to slip out under hard cornering. The GT rides harshly on rough roads. These cars are a bit clumsy around town because of a wide turning circle.

Comfort/Convenience

Pro: Standard equipment: Even the base Mustangs come well equipped with lots of standard features. **Driving position:** Excellent, with plenty of headroom and legroom for tall drivers. **Controls and displays:** First class. Complete and well laid-out dual-cowl dashboard and controls. **Climate control:** Efficient, easy to adjust, and quiet. **Interior space/comfort F/R:** Comfortable front seats on all models. **Cargo space:** The split folding rear seat frees up much-needed trunk storage space. Easy loading, thanks to the low liftover. **Quietness:** Improved body rigidity has reduced interior noise somewhat.

Con: Rear visibility is obstructed by the roof pillars and high parcel shelf. **Interior space/comfort F/R:** Front seats need at least a few centimetres more travel, and the rear seat is best left to children, especially with the convertible version. **Entry/exit:** The wide doors make for clumsy entry and exit in tight spots. GT ride comfort is below average. **Trunk/liftover:** Very shallow trunk with a small opening. Low liftover.

Mustang

List Price (negotiable)	**Residual Values** (months)			
	24	36	48	60
Mustang: $23,495 (18%)	$17,000	$14,000	$11,000	$9,000
Cvt.: $28,095 (20%)	$21,000	$17,000	$15,000	$13,000
GT: $31,795 (21%)	$23,000	$21,000	$18,000	$16,000
Cvt.: $35,795 (22%)	$25,000	$23,000	$20,000	$18,000

Technical Data

Powertrain (front-drive)
Engines: 3.8L V6 (193 hp)
- 4.6L V8 (260 hp)
- 4.6L V8 (305 hp)
- 4.6L V8 (390 hp)

Transmissions: 5-speed man.
- 6-speed man. (Cobra)
- 4-speed auto.

Dimension/Capacity
Passengers: 2/2
Height/length/width: 53.4/181.5/71.8 in.
Headroom F/R: 38.2/35.9 in.
Legroom F/R: 42.9/31.3 in.
Wheelbase: 101.3 in.
Turning circle: 35.4 ft.
Cargo volume: 10.9 cu. ft.
Tow limit: 1,000 lb.
Fuel tank: 59L/reg.
Weight: 3,300 lb.

Safety Features/Crashworthiness

	Std.	Opt.
Anti-lock brakes	■	■
Seat belt pretensioners F/R	—	—
Side airbags	—	—
Traction control	■	■
Head restraints	*	*
Visibility F/R	*****	**
Crash protection (front) D/P	*****	*****
Crash protection (side) F/R	***	***
Convertible	**	***

TAURUS ★★

Ford's Taurus will soon get the axe.

RATING: Below Average. This is the last model year for the Taurus; its Sable twin was dropped a few years back when Mercury dealers shut down in Canada. Wagons are competent performers, but they're outclassed by imports and most minivans as far as reliability and overall performance are concerned. **Strong points:** Quiet running; good handling and road holding; comfortable ride; better-than-average crash protection; and bargain prices (we know why). **Weak points:** Insufficient storage space; limited rear headroom and access; and a history of serious mechanical and body deficiencies that can drive you straight to the poorhouse. **New for 2004:** Nothing.

OVERVIEW: Despite their past quality deficiencies, the Taurus and Sable are Ford's most popular family cars due to their attractive combination of a reasonable base price, good crashworthiness, impressive handling, and interior and ride comfort. The base 3.0L Vulcan V6 is adequate, though dated, and the high-performing 24-valve V6 provides plenty of power for most driving needs. Other nice standard features include heated outside mirrors, a 60/40 split-fold rear seatback for additional cargo space, a driver's foot rest, and reserve power to operate the power windows and moon roof after the engine is shut off. Wagons get four-wheel disc brakes.

Cost analysis/best alternatives: An identical 2003 version discounted by at least $3,000 is the better buy. Your best bet is to buy an extended bumper-to-bumper warranty and trade in at the 5-year mark—just before the extra warranty expires. Other sedans worth considering are the Honda Accord, Mazda 626, Nissan Altima, and Toyota Camry. Wagons worth considering are the Subaru Outback and Impreza, and Volvo V40. **Options:** Expect to spend about $300 more for the flexible-fuel Taurus (ethanol or methanol), currently sold only to fleets. The automatic climate control system, a rear integrated child safety seat, power seats, and a heavy-duty suspension are wise choices. Don't buy the optional leather seats; they're slippery and not all that durable. The digital instrumentation is another useless option; it's gimmicky and distracting. On the other hand, Ford's optional pedal extensions minimize the danger from airbag deployments and the InstaClear windshield is a boon in Canadian winters. **Rebates:** As the Toyota Camry and Honda Accord eat into Ford's sales and stock continues to pile up in dealers' lots, generous dealer incentives and customer rebates will start kicking in before the year is up. Expect at least $3,000+ rebates on the 2003s and at least $2,000 plus low-interest financing on this year's models. **Delivery/PDI:** $995. **Depreciation:** Higher than average as the Ford Five Hundred replacement nears its debut for 2005. **Insurance cost:** Higher than average. **Parts supply/cost:** Parts are easily found, and they are reasonably priced. **Annual maintenance cost:** Average to higher than average as the vehicle ages. **Warranty:** Bumper-to-bumper 3 years/ 60,000 km; rust perforation 5 years/unlimited km. **Supplementary warranty:** A wise investment, particularly when you replace the AC, automatic transmission, or fuel pump around the fifth year of use. **Highway/city fuel economy:** 7.7–12L/100 km with the 3.0L.

Quality/Reliability/Safety

Pro: Important safety improvements, once found only on luxury vehicles, have worked their way down to Ford's popular mid-sized family sedans. Taurus, for example, has standard "thinking" seat belts and airbags that come in two parts to protect head and shoulders and upper torso separately. They can detect whether a driver has buckled the belt and prevent the airbag from deploying in fender-bender accidents, or cause it to deploy more slowly in high-speed impacts. Furthermore, the car's sides have been reinforced to prevent intrusion into the interior, and the Taurus' armrests have been redesigned to crumple, thereby protecting passengers from possibly severe internal injuries in a side impact. Optional adjustable pedals, needed by short-statured drivers to avoid airbag injury, are also a worthwhile safety feature. Test-drive a vehicle with the extensions, though, before ordering the option.

Con: Reliability: There have been an inordinate number of reliability complaints recorded by government and private consumer protection agencies. **Quality control:** Way below average. As you will see below, owner complaints called into NHTSA still show a disturbing pattern of powertrain, brake, suspension, fuel and electrical system, and body component deficiencies. **Warranty performance:** Erratic, and sometimes perverse. The same company that set up a generous 7 years or 160,000 km after-warranty assistance program for 1994–95 engine failures won't admit that its automatic transmissions are also crap and merit a similar compensation program. And engine problems still dog Ford's latest models, with intake manifold defects topping the list. **Owner-reported problems:** Engine failures, leaks, sudden shutdown; automatic transmission shifts out of First gear too soon, shifts slowly, constantly bangs through the gears, and often chooses the wrong gear; AC, electrical system (lots of blown fuses), steering, fuel system, fuel pumps, front suspension, warped brake rotors, and ignition problems. **Service bulletin problems:** NHTSA reports that there is a service bulletin that confirms the automatic transmission torque converter clutch fails to engage (problem dates back to 1996 models).

Technical Service Bulletins Summary

Make: Ford
Model: Taurus
Year: 2003

Service Bulletin Number: 03123

Component:

Powertrain: automatic transmission: torque converter

Summary Description:

Torque converter clutch not engaging when commanded and/or diagnostic trouble codes P0741 or P1744 stored in memory. Models from 1996 to 2003. *TT

Transmission may not go into Reverse; troubleshooting transmission malfunctions; fuel cap lamp illumination; engine cooling fan-induced body boom, rough engine idle sensation; unusual engine noise at idle; incorrectly installed gear-driven camshaft position sensor synchronizer assemblies may cause engine surge, loss of power, or MIL to light; no-start or hard starts (see below).

No-Start/Hard Start/Rough Idle

Article No.: 03-3-50 **Date:** February 17, 2003

DRIVEABILITY - IDLE AIR CONTROL (IAC) VALVE

2000–03 TAURUS
2002 THUNDERBIRD
2000–03 EXPLORER, RANGER
2001–03 EXPLORER SPORT TRAC, EXPLORER SPORT
2000–02 LS
2000–03 SABLE, MOUNTAINEER

ISSUE: Some vehicles may exhibit driveability conditions.

These may include:

^No-start, difficult to start, stalls
^Low idle
^Rough idle
^High idle
^Hesitation/surge while accelerating or at steady speed

These conditions may be intermittent with no Diagnostic Trouble Codes (DTC) and no Malfunction Indicator Lamp (MIL).

Rattling, clunking front suspension; power window grunting noise. **NHTSA safety complaints/safety:** Complete brake failure; airbags deploy for no reason or fail to deploy; airbag warning light stays lit; rear left wheel and rim flew off due to defective lug nuts; sudden acceleration; vehicle surges and then shuts off when fuel tank is filled; frequent stalling, hard starts, and poor idle caused by chronic fuel pump failures or a contaminated fuel pressure sensor; strong fuel smell comes from the air vents; fuel gauge stuck on Full; fuel tank is easily punctured; engine had to be replaced because block heater was incorrectly mounted on the engine; engine belt tensioner shattered; excessive steering vibration; interior rear-view mirror location obstructs visibility; dash reflects upon the front windshield; seat belts may not reel out or retract; seat belt continually tightened around child and had to be cut off; adjustable brake and accelerator pedals are set too close together and are often too loose; right rear wheel almost fell off due to faulty stabilizer bolt; rear brake lines rub together; high beam lights are too dim; cigarette lighter pops out and falls under passenger seat; Firestone tire blowout.

Road Performance

Pro: Emergency handling: Better than average. **Steering:** The speed-sensitive variable-assist power steering makes the car easier to handle, but not as much as Ford claims. **Acceleration/torque:** Better than average with the 3.0L engine

(0–100 km/h: 8.7 seconds). The base 3.0L V6 is a decade-old engine that offers adequate power for most driving situations, while the twin-cam Duratec version provides exhilarating, quiet acceleration equal to the European-influenced chassis dynamics. **Transmission:** The 4-speed automatic transmission shifts smoothly and responsively. **Routine handling:** The sedan's handling, both around town and on the highway, is better than average due primarily to its solid suspension and stiff body construction.

Con: Ford's 3.0L Vulcan V6 is okay for rentals and city commuting, but it's unable to take full advantage of the car's excellent handling characteristics. Wagons handle poorly in turns and over uneven terrain. **Braking:** Unimpressive braking (100–0 km/h: 41 m). ABS produces strong pedal pulsations.

Comfort/Convenience

Pro: Standard equipment: Nicely styled with plenty of convenience features, but four-wheel disc anti-lock brakes are still optional. The three-way front-seat console is a nice touch. It's a flip-fold affair that's a regular seat, an armrest, or a cup- and coin-carrying console. Armrest switches make it easy to activate the door locks and power windows. Climate control system is efficient and quiet. **Interior space/comfort F/R:** Loads of room both in the front and in the rear. The standard seats are comfortable for most people, but the wagon's fabric-covered seats are more comfortable and supportive than the leather-covered sedan seats. The sedans will seat five in comfort, while the wagons offer an optional third seat. **Cargo space:** There's lots of cargo area in the wagon, and its innovative two-way tailgate allows the entire unit to swing out for large cargo or just the window glass to open for small packages. **Trunk/liftover:** The sedan's huge trunk has a low liftover and split rear seatbacks to facilitate easier loading.

Con: Driving position: Short drivers will have to strain to reach the accelerator and see over the dash and tall drivers will find thigh support lacking. Leather seats offer little lateral support. **Controls and displays:** The massive dash and digital treatment are gimmicky and take getting used to. The radio is set too low and too far forward, and the JBL Premium Sound System's push-button controls require a Ph.D. to understand. Radio reception quality is disappointing. The wagon's liftgate doesn't rise high enough to clear most people's heads. **Climate control:** The rear window defroster is hard to find on the left side of the steering column. Air conditioning push buttons are difficult to understand and manipulate. **Entry/exit:** Difficult third seat access and limited head room for the third seat passenger. **Quietness:** Both the sedan and wagon's noise levels are unusually high, with lots of wind noise and tire drumming from the rear.

Taurus

List Price (negotiable)	**Residual Values** (months)			
	24	**36**	**48**	**60**
Taurus LX: $24,995 (17%)	$14,000	$12,000	$10,000	$7,500
Wagon: $28,355 (18%)	$17,000	$15,000	$13,000	$9,500

Technical Data

Powertrain (front-drive)
Engines: 3.0L V6 (155 hp)
• 3.0L V6 (200 hp)
Transmissions: 5-speed man.
• 4-speed auto.
Dimension/Capacity (LX)
Passengers: 2/3
Height/length/width:
55.1/197.5/73.1 in.
Headroom F/R: 39.2/36.2 in.
Legroom F/R: 42.2/38.9 in.
Wheelbase: 108.5 in.
Turning circle: 35.4 ft.
Cargo volume: 15.8 cu. ft.
Tow limit: 1,750 lb.
Fuel tank: 60L/reg.
Weight: 3,350 lb.

Safety Features/Crashworthiness

	Std.	**Opt.**
Anti-lock brakes	■	■
Seat belt pretensioners F/R	■	—
Side airbags	■	■
Traction control	■	■
Head restraints F/R	***	***
Visibility F/R	*****	*****
Crash protection (front) D/P	*****	*****
Crash protection (side) F/R	***	***
Crash protection (offset)	*****	—

GRAND MARQUIS, TOWN CAR

RATING: Recommended; if you're going to buy a Lincoln, the Town Car can't be beat. These '50s-era behemoths are best suited for highway cruising, vacationing, and trailer towing. **Strong points:** Lots of interior room; quiet running; easy entry/exit; reliable; and excellent resale value. A natural gas V8 engine is available. **Weak points:** Difficult trunk access; terrible fuel economy; and some factory-related powertrain deficiencies that are carried over year after year, contrary to the practice of other automakers who pay suppliers for improved components. **New for 2004:** *Grand Marquis:* Nothing significant. *Town Car:* A higher-torque transmission and satellite radio. The Cartier version is replaced by the Ultimate; Executive models are now for fleet sales only.

OVERVIEW: The industry's lowest-priced six-passenger V8 sedans, the Crown Victoria and Grand Marquis have always been a favourite with police, taxi drivers, farmers, and retirees. With the addition of a 4.6L OHC V8 several years ago, the Grand Marquis GS and LS (a $3,000 upgrade) soldier on as Ford's rear-drive, full-sized sedans.

Although some fleets and police departments have replaced their Crown Victorias and Grand Marquis with 200-hp 3.8L V6-equipped Impalas following reports of rear-end collision fires and to save fuel costs, the rank and file don't like the change. This is even though acceleration tests conducted by the Michigan State Police found that the Ford's additional 15 horses (215-hp 4.6L V8) didn't result in significant performance improvements when compared with the more fuel-efficient and smaller front-drive Impala. Still, the State Police found the rear-drive's pursuit performance hard to resist:

> We're adamant about a rear-drive car...especially for pursuit situations. On snow and ice, you can pull a rear-wheel car around; you can spin it around, but when you accelerate a 4X4 car, the front end will go right out from under you.... It will slide sideways and you will lose control.

Aside from the sheer wastefulness of the design and the high fuel cost of running one of these boats, they're fairly reliable and predictable highway cruisers. Handling isn't very precise, though it is more predictable than with front-drives, and everyone is going to be comfortable inside. The mechanical design is also straightforward and easy to troubleshoot (electronic and emission components excepted).

The base model and LS are joined by the 2004 Marauder, a $46,975 high-performance model equipped with a 302-hp 4.6L V8, limited-slip differential, front side airbags, performance handling suspension, and lots of convenience features. An electronically controlled automatic transmission has been around since early 1992.

The Lincoln Town Car is still, to many people, the epitome of large-car, six-passenger luxury, and it's a popular rear-drive base for Ford's full-sized sedans. Essentially a stretched version of the Crown Victoria/Grand Marquis, the Town Car's air-spring rear suspension provides a smooth ride and prevents tail dragging, even when fully loaded. Its 235-hp 4.6L V8 has fewer horses than the Marauder, but it gives quiet, smooth performance nevertheless.

Cost analysis/best alternatives: Look for a second-series, discounted 2003: You'll get an improved rack-and-pinion steering system, a more rigid frame, and a retuned suspension. Other cars worth considering are the Nissan Altima and Maxima, and the Toyota Avalon. **Options:** Invest in the "Handling and Performance" option to reduce body roll and increase traction, and consider a power seat. All models, except for the GS, come with standard adjustable pedals. **Rebates:** In the late fall expect $3,000+ rebates and a continuation of zero percent financing. **Delivery/PDI:** $1,020; *Town Car:* $1,045.

Depreciation: Slower than average. **Insurance cost:** Higher than average. **Parts supply/cost:** Parts aren't hard to find; parts costs are about average when compared with other American cars in this class, but cheaper than the European equivalent. Asian parts are less expensive, however. **Annual maintenance cost:** A bit higher than average. Although repairs are relatively easy to carry out, these cars have complicated brake, fuel, and electronic systems that are a pain in the butt—and wallet—to troubleshoot. **Warranty:** *Crown Victoria* and *Grand Marquis:* Bumper-to-bumper 3 years/60,000 km; rust perforation 5 years/unlimited km. *Town Car:* Bumper-to-bumper 4 years/80,000 km; rust perforation 5 years/unlimited km. **Supplementary warranty:** Not needed. **Highway/city fuel economy:** 9.2–13.9L/100 km.

Quality/Reliability/Safety

Pro: Quality control: Body components and construction are fairly good. **Reliability:** Overall reliability is above average. **Warranty performance:** Better than average. Ford's luxury car owners are treated like VIPs—at first. Things go downhill from there. These cars have a nice array of safety features, like adjustable pedals (optional on the Crown Victoria, standard with the Grand Marquis LS Premium) and optional ABS, dual-stage airbags, crash-severity and seat-position sensors, and safety belt pretensioners.

Con: Owner-reported problems: Main problem areas are the 4.6L engine oil and coolant leaks (an Owner Notification Program covers engine intake manifolds that are cracked at the coolant crossover duct); constant stalling; automatic transmission glitches; excessive vibration in the driveshaft and differential assembly; prematurely worn brakes, and problems with the fuel and electrical systems. **Service bulletin problems:** Premature wear of the axle shaft or axle bearing; front springs and steel wheels may make for a harsh ride; moisture in the headlamps and erratic headlamp operation; inaccurate fuel gauge; power steering assist calibration; excessive power steering pump noise; front wheel area click or rattle; anti-theft system may lock the steering wheel or cause the transmission to stick in Park; rear parking brake clicking. **NHTSA safety complaints/safety:** Vehicle struck from behind and exploded in flames; tire tread separation; missing upper control arm bolt; brake booster failed; fan belt comes off in rainy weather, causing overheating and loss of power steering, water pump, and other accessories; horn "sweet spot" too small and takes too much effort to sound; sunlight causes a reflection of the defrost vents onto the windshield; poor dash panel illumination.

Road Performance

Pro: Emergency handling: A bit slow, but predictable and sure-footed. Less body lean in corners thanks to the upgraded suspension. **Steering:** Improved steering transmits more road feel than before. **Acceleration/torque:**

Respectable, though not impressive, acceleration with plenty of low-end torque (0–100 km/h: 10.2 seconds). The smooth, quiet-running 4.6L V8 provides more than enough power for a comfortable ride. Towing capacity is 2,250 kg (5,000 lb.) with the Class III Towing or Handling and Performance packages offered with the Grand Marquis. **Transmission:** Flawless, most of the time. **Routine handling:** Fairly good for vehicles this large. Ride isn't overly soft. In effect, the rear end no longer sways when you turn the steering wheel. **Braking:** Better-than-average braking with the four-wheel disc brakes (100–0 km/h: 38 m).

Con: Transmission: Sometimes hesitates between gear changes. Handling still takes a back seat to ride quality.

Comfort/Convenience

Pro: Standard equipment: Lots of standard convenience features. **Driving position:** Very comfortable. **Controls and displays:** Everything is within easy reach and well presented. **Climate control:** Powerful, quiet ventilation system. **Entry/exit:** Large doors make entry/exit a breeze. **Interior space/comfort F/R:** Spacious interior: three passengers are comfortable in the back, where there's lots of head, leg, and shoulder room. The individual front seats are large, comfortable, and supportive (the power seat option is recommended), particularly in the lumbar region. **Cargo space:** Plenty of storage areas. **Trunk/liftover:** Huge trunk. The Town Car's trunk, for example, is an impressive 22 cubic feet, with a low liftover. **Quietness:** Very quiet interior.

Con: Conservative styling. Poor rear visibility. **Controls and displays:** Confusing power seat and controls. Distracting continuous digital readout of your fuel economy. The Town Car's electronic dash is too gimmicky. **Climate control:** Slow to warm up. **Interior space/comfort F/R:** The driver's right knee frequently hits the dashboard/radio housing. Seat belt anchor pokes into driver's right hip. The split-bench front seat in the Crown Victoria isn't very comfortable, due to its insufficient seatback padding and side support. Town Car's moon roof cuts down on rear headroom. Rear seats also lack sufficient seatback padding. **Trunk/liftover:** Deep-dish trunk is not very practical for everyday baggage. You have to do some acrobatics to get at the full-sized spare tire, which is placed far forward in the trunk.

Grand Marquis, Town Car

List Price (negotiable)	**Residual Values** (months)			
	24	**36**	**48**	**60**
Grand Marquis: $36,325 (23%)	$23,000	$17,000	$15,000	$13,000
Town Car Signature: $57,345 (25%)	$39,000	$31,000	$26,000	$22,000

Technical Data

Powertrain (rear-drive)
Engines: 4.6L V8 (224–239 hp)
• 4.6L V8 (302 hp)
Transmission: 4-speed auto.

Dimension/Capacity
Passengers: 3/3
Height/length/width:
56.8/212/78.2 in.
Headroom F/R: 39.4/38 in.
Legroom F/R: 42.5/39.6 in.
Wheelbase: 114.7 in.
Turning circle: 35.4 ft.
Cargo volume: 20. 6 cu. ft.
Tow limit: 2,000 lb.
Fuel tank: 71L/reg.
Weight: 3,800 lb.

Safety Features/Crashworthiness

	Std.	**Opt.**
Anti-lock brakes (4W)	■	—
Seat belt pretensioners F/R	■	—
Side airbags	■	■
Traction control	■	—
Head restraints F/R	*	*
Visibility F/R	*****	**
Crash protection (front) D/P		
Crown Victoria and Gr. Marquis	*****	*****
Town Car	****	*****
Crash protection (side) F/R	****	****
Crash protection (offset) D/P	—	—

LINCOLN LS

RATING: Average. Lincoln handles complaints well, but factory-related deficiencies are simply too numerous to recommend the LS. **Strong points:** Impressive V8 acceleration; comfortable ride; crisp handling; comfortable interior; and good warranty performance. **Weak points:** Erratic automatic transmission shifting; confusing climate controls; poor-quality mechanical, electrical, and body components. **New for 2004**: Reworked automatic transmission to enhance performance; improved suspension to reduce noise, vibration and harshness; additional options.

OVERVIEW: Lincoln's LS rear-drive mid-sized sedan comes with a high-performance 232-hp variant of the Taurus 3.0L V6 mated to a standard automatic gearbox. Also available: a 280-hp 3.9L V8 based on the Jaguar XK8 coupe coupled to a semi-automatic transmission. Both engines are identical but produce fewer horses than the Jag equivalent.

The V6 version is priced in the BMW 3 Series, Lexus ES 300, and Mercedes C-Class territory while delivering standard equipment and interior space that rivals the 5 Series, GS, and E-Class.

Cost analysis/best alternatives: The 2003s got a bit more power, revised styling, and curtain side airbags and are a better buy than this year models, if they're discounted by at least 20 percent. Alternative models include the Acura

TL, Audi A4, BMW's 3 Series, Infiniti G35, and the Lexus ES 300. **Options:** Traction control is a worthwhile option on the V6. Another option worth considering: the sports package, which includes thicker anti-roll bars, upgraded brake pads, re-calibrated variable-ratio steering, an auto-manual shifter (if ordered with an automatic), body-coloured bumper trim, a full-sized spare, and bigger wheels and tires. **Rebates:** Sales have been slow; expect rebates of $3,000+ for a 2003 version and $2,000 plus zero percent financing for a 2004. **Delivery/PDI:** $1,045. **Depreciation:** Faster than average. **Insurance cost:** Higher than average. **Parts supply/cost:** Parts availability hasn't been a problem and parts cost has been reasonable. **Annual maintenance cost:** Predicted to be average. **Warranty:** Bumper-to-bumper 4 years/80,000 km; rust perforation 5 years/unlimited km. **Supplementary warranty:** Not necessary. Lincoln customer relations staffers treat you like a VIP. **Highway/city fuel economy:** *6-cyl.:* 8.9–13.8L/100 km; *8-cyl.:* 9.1–14.2L/100 km.

Quality/Reliability/Safety

Pro: Warranty performance: Better than average.

Con: Reliability: In spite of Ford's good warranty performance noted above, powertrain, brake, and electrical problems still compromise the LS's overall reliability. **Quality control:** Below average. Apparently, Lincoln's return to rear-drive has opened up a Pandora's box of powertrain, AC, and body glitches. **Owner-reported problems:** Poor transmission shifting, excessive drivetrain and body noise and vibrations, inconsistent braking response, electrical system short circuits, and faulty AC performance. **Service bulletin problems:** Engine overheating; erratic transmission shifting; transmission vent fluid leakage; suspension noise when accelerating; battery may go dead after several days of non-use; MIL indicator may stay on; inoperative power window; repeat heater core failures; inadequate front seat cushion padding. **NHTSA safety complaints/safety:** Lower ball joint failure caused accident (vehicle not part of ball joint recall); sudden acceleration in Forward or Reverse gear; airbag deployment for no reason; airbag failed to deploy; lurching, hesitating automatic transmission shifting; chronic stalling; engine head gasket leaks oil on the AC compressor; brakes failed after a cold start; brake pedal becomes hard and resists application, or becomes mushy and goes to the floor without effect; faulty airbag monitor; all kinds of warning lights come on for no reason; defective steering causes violent swerving from side to side; automatic door locks engage by themselves, locking out driver; interior and exterior lights, dash panel lights and gauges go haywire.

Road Performance

Pro: Emergency handling: Traction control works quite well in emergency situations. **Acceleration/torque:** The V8 is the engine of choice for moving the LS (0–100 km/h: 7.3 seconds). It is the only engine available with a manual transmission. **Steering:** Acceptable, but not as responsive as European and Japanese luxury vehicles. **Routine handling:** Pretty good, despite the soft suspension. **Braking:** Quite good.

Con: The V6's performance is seriously compromised by the Lincoln's heft, and turns in 0–100 km/h times that are one to two seconds slower than V6 competitors'. **Transmission:** The 5-speed automatic shifts slowly, not as decisively as competitors' gearboxes, and far less smoothly than Lexus drivetrains. A manual transmission (first one since 1951) is available only with the V6-equipped LS. Suspension provides a soft ride and is not as sporty as the competition.

Comfort/Convenience

Pro: Standard equipment: Very well appointed for the price. **Interior space/comfort F/R:** The LS is roomier than both the Bimmer and Mercedes E-Class, and a bit longer and taller than the Jaguar S-Type. Comfortable seats. **Trunk/liftover:** Easy to load or unload.

Con: Driving position: In spite of the many seat adjustments available, it's hard to find a comfortable driving position. Don't believe the hype that this is a five-seater. **Climate control:** Confusing and not very accessible. **Entry/exit:** Rear access is awkward. **Cargo space:** Small trunk and limited interior storage space. **Quietness:** Tire noise intrudes into the interior.

Lincoln LS

List Price (very negotiable)	**Residual Values** (months)			
	24	**36**	**48**	**60**
Lincoln LS: $39,115 (21%)	$27,000	$23,000	$19,000	$14,000

Technical Data

Powertrain (rear-drive)
Engines: 4.6L V8 (224 hp)
• 4.6L V8 (239 hp)
Transmissions: 5-speed man.
• 5-speed auto.
Dimension/Capacity
Passengers: 2/2
Height/length/width:
56.1/194/73 in.
Headroom F/R: 40.4/37.5 in.
Legroom F/R: 42.6/37.7 in.
Wheelbase: 115. in.
Turning circle: 39 ft.
Cargo volume: 14 cu. ft.
Tow limit: 2,000 lb.
Fuel tank: 55L/prem.
Weight: 3,655 lb.

Safety Features/Crashworthiness

	Std.	**Opt.**
Anti-lock brakes (4W)	■	—
Seat belt pretensioners F/R	■	—
Side airbags	■	—
Traction control	■	—

Head restraints F/R	**	**
Visibility F/R	*****	*****
Crash protection (front) D/P	*****	*****
Crash protection (side) D/P	****	*****
Crash protection (offset)	*****	—

THUNDERBIRD ★

Not your father's Thunderbird.

RATING: Not Recommended. A textbook example of style and nostalgia overriding function and performance. Ford will drop the T-Bird in 2005. **Strong points:** Good steering and handling (some handling is sacrificed for ride comfort) and better-than-average crashworthiness ratings. **Weak points:** Heavy, thirsty, and slow. Low-profile tires make for a firm ride; minimal trunk room even though the car is quite large for a two-seater; outrageously overpriced; IIHS gives head restraints a "Poor" rating. Early reports note that the body construction is astonishingly poor, producing considerable creaks and rattles, and, when passing over bumps, shaking its contents like a soda fountain's milkshake mixer (how's that for '50s nostalgia?). Long-term quality and reliability has yet to be determined. Also keep in mind the car's uncertain future (remember the Merkur XR4Ti?) and small dealer network, which can complicate servicing and warranty support. **New for 2004:** Nothing. Ford says the Thunderbird will be dropped in 2005, explaining that the car was never meant to be around that long and that lousy sales have nothing to do with its demise.

Sure.

OVERVIEW: After a brief hiatus the Thunderbird name returns affixed to a $56,775 retro-styled, two-seat, rear-drive convertible that looks nothing like its 1955–57 namesake, or the $25,095 1997 model it replaces (now worth

$6,500). This 'bird shares variations of the engine and chassis used by the Lincoln LS and Jaguar S-Type, as well as their 5-speed automatic transmission. Power is supplied by a retuned 280-hp 3.9L V8.

Standard features include an independent suspension, four-wheel disc ABS brakes, side airbags, 17-inch alloy wheels, and a CD changer. The Premium package adds traction control (a good idea, considering that rear-drive rear ends tend to fishtail under load) and chrome wheels. The power-folding softtop houses a heated glass rear window; a removable hardtop with the T-Bird's signature side porthole windows is also available.

NHTSA frontal crash tests gave the Thunderbird a four-star rating for driver protection and five stars for the front passenger. Side crash protection was given five stars for both occupants.

Sizzle over substance

The resurrected Thunderbird is mostly a publicity ploy to show Ford's creativity, in much the same way that Honda's first Insight was its statement that it was going "green." Neither vehicle is that exceptional, nor expected to earn back its cost of development, nor be priced so that the average Joe or Jane can afford it. But what a statement, eh? So, while most of the motoring press goes gaga over the new Thunderbird, remember, it's basically a re-styled Linc-Jag propelled by tons of press releases in a desperate attempt to woo buyers by digging deep into Ford's past to show it can make quality cars at affordable prices.

The best Thunderbirds, in my opinion, other than the classic and unaffordable 1954–57 models, were those versions built from 1985–97. They were fairly reliable, relatively inexpensive, sporty, comfortable, and well equipped. Plus, they were fun to drive and easily maintained.

So why didn't the 2002–03 sell?

First off, this Thunderbird sequel is way overpriced. Sure, you get lots of bells and whistles for your $50,000+, but other cars offer just as much for far less money. Secondly, this is one dull-looking luxury roadster with few features that distinguish it from a half dozen imports of the same genre. Other minuses: a tiny, shallow trunk, a cheap-looking, boring instrument panel, limited headroom, an unwieldy folding top cover, and excessive air turbulence when driven with the top down.

Ford wants us to believe that the new Thunderbird iteration is true to the heritage of its classic forbears and represents good value for the money. Unfortunately, this Thunderbird proves just the opposite. Head restraints are rated "Poor" by IIHS, the official launch and delivery was delayed several times due to factory-related problems, the base MSRP is double what a Thunderbird cost just a few years ago, and the car will be discontinued in 2005.

Just the thing to boost our confidence in the product, eh?

Will Thunderbird 2 be a collector's car? Not likely. The only collectors I can imagine will be those Ford Oakville executives who are taking early retirement and can't get a Mini Cooper.

FREESTAR/WINDSTAR ★

RATING: *Freestar:* Not Recommended during its first year on the market. The Freestar is essentially a renamed and reworked Windstar. *Windstar:* One of the most unreliable minivans you'll find. 2003 is its last year on the market. **Strong points:** *Freestar:* A 4.2L V6 engine; new cabin appointments; optional side curtain airbags; easier rear access; flat-folding third-row seats; and larger four-wheel disc brakes. *Windstar:* A five-star crashworthiness rating. **Weak points:** *Freestar:* Crashworthiness, safety, and long-term reliability have yet to be determined. It's likely first year production glitches and parts borrowed from the Windstar will compromise the Freestar's reliability. Larger brakes on the Freestar should improve pad and rotor durability. *Windstar:* Mediocre handling and restricted side and rear visibility. Driver's seat isn't comfortable for big, tall drivers who complain of the lack of legroom, seat contouring, and lower back support. An abundance of clunks, rattles, and wind and road noise. These minivans have maxi-failures that include failure-prone engines, transmissions, brakes, electrical systems, and suspension components. Sure, the Windstar combines an impressive five-star safety rating, plenty of raw power, an exceptional ride, and impressive cargo capacity. But self-destructing automatic transmissions, defective engine head gaskets and bearings, and "do you feel lucky today?" brakes and front coil springs can quickly transform your dream minivan into a nightmare. And, as a counterpoint to Ford's well-earned Windstar crashworthiness boasting, take a look at the summary of safety-related complaints recorded by the U.S. Department of Transportation: coil spring breakage blowing the front tire, sudden acceleration, stalling, steering loss, windows exploding, wheels falling off, horn failures, sliding doors that open and close on their own, and vehicles rolling away while parked. **New for 2004:** Everything.

OVERVIEW: *Freestar:* The Freestar is a re-engineered, re-styled Windstar without any important dimensional changes. It features upgrades in safety, interior design, steering, ride and performance. An optional "safety canopy" side curtain air bag system offers protection in side impact collisions and rollovers for all three rows of seating. There's also better access to the third row seat, which now folds flat into the floor.

All models come with a 201-hp 4.2L V6, derived from the 3.8L. A thriftier 5-speed automatic transmission may replace the mandatory four-speed gearbox by mid-year.

In the States, Mercury will sell an upscale Freestar called Monterey to replace that brand's defunct Nissan-based Villager.

Windstar: The Windstar is a front-drive minivan that looks a bit like a stretched Mercury Villager. It's longer, larger, and lower than most other minivans. It's also one of the few minivans not built on a truck platform (it uses the Taurus platform instead), and as such it has some of the carlike handling characteristics of Chrysler's minivans and some of the engine and automatic transmission problems experienced by Taurus and Sable owners. It's offered in two body styles: a seven-passenger people-hauler and the less expensive basic cargo van. Buyers get a standard 200-hp 3.8L V6 hooked to a 4-speed automatic transmission.

Cost analysis/best alternatives: The 2004 Freestar sells for $27,195. A 2003 Windstar costs $26,595. Buyers should steer away from both the Windstar and Freestar; however, if you must make a choice, get the Freestar after it has been on the market six months or more. Be wary of the upscale Windstar SEL: As with the Chrysler Town & Country, it loses its value more quickly than other versions. The GM Venture and Montana, Honda Odyssey, and Toyota Sienna are recommended alternatives. **Options:** Dual integrated child safety seats for the middle bench seat and adjustable pedals are worthwhile options. Sliding side doors are an overpriced convenience feature and are failure-prone as well. **Rebates:** $3,000 rebates or zero percent financing on 2003 Windstars. Look for $2,000 rebates and low-interest financing on the Freestar by late spring. **Delivery/PDI:** $1,025. **Depreciation:** *Windstar:* Much faster than average. **Insurance cost:** Above average. **Parts supply/cost:** Reasonably priced parts are easy to find, mainly due to the entry of independent suppliers. Digital speedometers are often defective and can cost almost $1000 to repair. **Annual maintenance cost:** Average while under warranty; outrageously higher than average thereafter, due primarily to powertrain breakdowns not covered by warranty or insufficiently covered by parsimonious "goodwill" gestures. **Warranty:** Bumper-to-bumper 3 years/60,000 km; rust perforation 5 years/ unlimited km. **Supplementary warranty:** An extended bumper-to-bumper warranty is a good idea. **Highway/city fuel economy:** *Windstar:* 9.5–13.3L/100 km.

Quality/Reliability/Safety

Pro: Adjustable pedals help protect drivers from airbag injuries. Be careful, though; some drivers have found them set too close together and say they often felt loose. Other nice safety features: airbags that adjust deployment speed according to occupant weight and a sliding door warning light.

Con: Warranty performance: Below Average. Owners say Ford is unusually tight-fisted in giving out refunds for the correction of factory-related deficiencies. **Quality control:** *Windstar:* Below Average, behind General Motors and DaimlerChrysler; way below average when compared with the Japanese competition. The last five years have been a disaster from a quality control standpoint. Stingy "goodwill" payouts, with owners paying over one-third the cost, have been used as a substitute for better quality control. It just doesn't make sense that Ford won't offer a free 7-year retroactive powertrain warranty to protect its customers from engine and tranny breakdowns. **Reliability:** *Windstar:* Serious problems affecting the automatic transmission, which may shift erratically and is historically failure-prone. Ford insiders and service bulletins confirm that this transmission problem is both hardware and software related and affects much of Ford's model lineup—it's a chronic complaint from owners reporting to NHTSA. Electrical system and brake defects have also frequently sidelined the Windstar. **Owner-reported problems:** *Windstar:* Transmission, suspension (coil springs), and power-sliding door malfunctions lead the list of factory-related problems, but there have also been many complaints concerning computer modules, engine oil leaks, timing cover gasket coolant leaks, early engine camshaft replacement, AC failures, premature brake rotor and caliper wear, and excessive brake noise. Poor body seam sealing allows water to leak into cabin; owners report a smattering of paint delamination complaints (see Part Three); and windshield glare from the dash. **Service bulletin problems:** *Windstar:* Front end grinding and popping noise when passing over bumps or making turns; sliding door malfunctions; airbag warning light stays lit. **NHTSA safety complaints/safety:** *Windstar:* Vehicle caught on fire while parked; driver-side airbag failed to deploy; sudden acceleration or complete engine shutdown; loss of braking; hood flew up while under power; lug nuts failed, causing front right wheel to fall off; rear window shattered while vehicle was parked and when defogger engaged; power-sliding door makes noises as if it were loose, won't retract when encounters an object, won't hold in place when opened, won't close flush, and suddenly pops open; back door won't open or close properly; with ignition switched On, transmission can be shifted without pressing the brakes; brake and gas pedal mounted too close together; automatic door lock pinches your arm if it's near the locks; joints not connected under the quarter wheel weld; driver's seat crossbar pops in and out; Check Engine light constantly comes on for no reason; excessive steering vibrations when turning; vehicle sways from side to side when passing over bumps; when interior rear-view mirror is set for night vision, it gives a distorted view, making objects hard to see; tire jack collapsed.

Road Performance (Windstar)

Pro: Emergency handling: High-speed handling is acceptably stable and predictable. **Steering:** Adequate; not as effortless as the Asian and Chrysler/GM competition. **Acceleration/torque:** The 3.8L V6 is usually competent and smooth, with lots of low-end torque (0–100 km/h: 10.7 seconds).

Transmission: The electronic 4-speed automatic responds well and shifts smoothly, when it's working properly (see "Con"). **Routine handling:** Easy to drive. Smooth and supple ride under most driving conditions improves as the load increases.

Con: Engine known to stall out due to fuel pump, computer module, and electrical system glitches. Erratically performing automatic gearbox may slip out of gear, lurch into gear, or simply refuses to engage whichever gear you choose. The Windstar's city manners aren't impressive: excessive body swish and sway takes its toll on tires and makes highway driving less carlike than the Chrysler, Honda, and Toyota minivans. **Steering:** Excessive torque steer tug on the steering wheel during rapid acceleration. It also has a large turning radius. **Quietness:** Lots of wind noise on the highway. **Braking:** Disc/drum; average braking that's a bit difficult to modulate.

Comfort/Convenience (Windstar)

Pro: Standard equipment: Comfortable and well-appointed interior. Superior sound system. One innovative touch that many families will appreciate: a wide-angle mirror housed in the ceiling console lets the driver watch the little darlings in the rear without turning around. **Driving position:** Excellent driving position gives most drivers a commanding view of the road. **Controls and displays:** Well laid-out instrument panel with easy-to-read gauges. **Climate control:** Adequate and easily adjusted. **Entry/exit:** Acceptable. Dual sliding doors. **Cargo space:** Lots of small storage spaces in addition to the large amount of space for larger items. Another nice touch is the optional power lock switch just inside the rear hatch, which saves you from having to walk to the front of the van to lock up. **Trunk/liftover:** Rear hatch is easy to open and shut, and the low floor improves cargo handling.

Con: There's not a lot of choice when it comes to engine, body format, and cabin layout. **Controls and displays:** Headlight switch is partly obscured by the steering wheel. Tiny, flat buttons on the Windstar's radio make it hard to calibrate. Weak AC and heater airflow to the floor makes an auxiliary heater a good idea. **Interior space/comfort F/R:** The Windstar's driver's seat isn't comfortable for big, tall drivers who complain of the lack of legroom, seat contouring, and lower back support. The driver's shoulder belt can ride uncomfortably on a tall driver's collarbone because it's anchored too far back. It's also a long reach to put on. Middle seating behind the driver provides less leg and knee room than Chrysler minivans. Middle and rear seat benches use short cushions and the rear seat is practically inaccessible. Getting them in and out of the vehicle is more of a chore than Ford lets on (the rear bench seat alone weighs almost 50 kg). Restricted side/rear visibility. **Cargo space:** Unlike Chrysler, Ford still hasn't figured out a way to reconfigure the Windstar's storage space so that it will hold bulky items like 4' x 8' sheets of building material. The right sliding door takes a lot of effort to open and close.

Quietness: An abundance of clunks, rattles, and wind and road noise. Engine sounds gruff when accelerating.

2003 Windstar

List Price (very negotiable)	**Residual Values** (months)			
	24	**36**	**48**	**60**
LX: $26,595 (20%)	$17,000	$15,000	$12,000	$10,000
SEL: $37,465 (25%)	$24,000	$20,000	$17,000	$13,000

Technical Data

Powertrain (front-drive)
Engine: 3.8L V6 (200 hp)
Transmission: 4-speed auto.
Dimension/Capacity
Passengers: 2/2/3
Height/length/width: 68/201.2/75.4 in.
Headroom F/R1/R2: 39.3/ 41.1/37.9 in.
Legroom F/R1/R2: 40.7/36.8/35.6 in.
Wheelbase: 120.7 in.
Turning circle: 42 ft.
Cargo volume: 136.4 cu. ft.
GVWR: 5,540 lb.
Payload: 1,800–1,831 lb.
Tow limit: 3,400 lb.
Fuel tank: 98L/reg.
Weight: 3,891 lb.

Safety Features/Crashworthiness (Windstar)

	Std.	Opt.
Anti-lock brakes (4W)	■	—
Seat belt pretensioners F/R	■	—
Side airbags	■	■
Traction control	■	■
Head restraints F/R	****	***/**
Visibility F/R	*****	**
Crash protection (front) D/P	*****	*****
Crash protection (side) D/P	*****	*****
Crash protection (offset)	*****	—

2004 Freestar

List Price (very negotiable)	**Residual Values** (months)			
	24	**36**	**48**	**60**
Wagon: $27,195 (20%)	$21,000	$15,000	$12,000	$10,000
SEL: $37,695 (23%)	$25,000	$20,000	$16,000	$14,000

Technical Data

Powertrain (front-drive)
Engine: 4.2L V6 (201 hp)
Transmission: 4-speed auto.
Dimension/Capacity
Passengers: 2/2/3
Height/length/width: 68.8/201/76.6 in.
Headroom F/R1/R2: 38.8/39.7/37.9 in.
Legroom F/R1/R2: 40.7/38/35.6 in.
Wheelbase: 120.8 in.
Turning circle: 42 ft.
Cargo volume: 136.4.
GVWR: 5,860 lb.
Payload: N/A
Tow limit: 3,501 lb.
Fuel tank: 98L
Weight: 4,275 lb.

General Motors

Will ToyoGM follow DaimlerChrysler?

You bet it will.

GM would already be bankrupt if it weren't for its truck division profits and sweetheart deals with a number of Asian partners. Even at that, the company is in dire straits. Though GM has been giving out huge rebates and other sales incentives for the past decade, shoppers are still turning up their noses at the company's passenger cars. Dealers complain that the automaker is selling the "deal," while Japanese manufacturers are selling brand and quality.

The decline and fall of Detroit's "Big Three"

Manufacturer	Sales	% Change 2001	Market Share 2002	Market Share 2001
GM	827,989	-15.6	23.9	26.6
Ford	485,689	-17.0	14.0	15.9
Chrysler	310,669	-6.5	10.0	9.7

The above figures are just for passenger cars and do not include Swedish Saab, which GM owns; Volvo, Jaguar, and Aston Martin, which Ford owns; or the German Mercedes cars of DaimlerChrysler. However, they do include cars sold to rental fleets and the cars going to workers and suppliers under factory discount plans. If these cars weren't counted, the Detroit companies probably would lose a few more points of the retail business.

In August 2003, Toyota outsold the Chrysler group for the first time in Canadian automotive history, then it beat Ford in worldwide sales for the first six months of 2003. Toyota also posted a 20 percent increase in September car and truck sales while General Motors, Ford, and DaimlerChrysler all saw sales skid. GM's car sales fell 18.4 percent, DaimlerChrysler car sales declined 19.9 percent, and Ford Canada's car sales plummeted 26.8 percent.

General Motors has serious financial liabilities that are draining cash flow. Its biggest mistake has been to divert profits into risky auto companies that have bled the company dry and taken development funds away from new products. The Chevrolet division, for example, has a co-venture agreement for the European market with money-losing Lada; has bought perpetually on-strike Daewoo to market cars that will inevitably undercut its own dealers' small-car sales; has promised to buy near-bankrupt Fiat's ("Fix it again, Tony") auto interests; and has thrown billions of dollars into a black hole called Saab.

As if the foregoing weren't bad enough, GM is also burning a lot of cash propping up its shaky employee pension and healthcare liabilities, paying ever-increasing sales incentives, and fighting off a renewed 2004 Japanese SUV and truck offensive.

The 2004 Nissan Titan seeks to beat GM on its own full-sized truck turf.

New models and new alliances

GM wants to make more attractive, reliable, and profitable products as it shakes its reputation for making look-alike, low-quality vehicles. And, ironically, the company is counting on a fistful of Asian and European competitors to save its hide through new manufacturing and marketing alliances. GM's also making up for stagnant car sales by concentrating on new SUV and pickup offerings, which means we'll see more dual factories like the U.S./Canada plants churning out the Pontiac Vibe and its Toyota Matrix 4X4 twin.

Unlike Ford, GM's dealer network isn't in turmoil and relations are cordial and respectful, despite some bad blood and lawsuits related to the Oldsmobile franchise phase-out. This dealer truce may not last for long, though, if GM's Saturn division doesn't start making real profits soon to make up for all the money GM has blown on the division. Another looming problem is General Motors' review of its Isuzu partnership in Canada. Sixty Saturn-Saab-Isuzu dealers don't want to lose the Isuzu franchise yanked after working with GM since 1986. Finally, GM's Daewoo subterfuge (selling Daewoos as re-badged Suzukis and Chevys) will inevitably cut into GM's own compact car sales, ticking off dealers who have invested heavily in Cavalier and Sunfire sales and servicing.

In Canada, Chevrolet dealers will sell the entry-level Aveo; the Optra, a compact based on the Daewoo Lacetti; and the Epica, a mid-sized vehicle based on Daewoo's Magnos. Suzuki will sell the subcompact Swift, based on the Kalos, and Verona.

Desperate to get new products onto the market, GM says it will reinvent its Cadillac and Saturn divisions over the next five years by converting most Cadillacs back to rear-drive (yippee!) and expanding the Saturn model lineup with a 2005 minivan. GM's new 2004 models are: the Buick Rainier; Cadillac CTS V, Armoured DeVille, XLR, and SRX; Chevrolet Aveo, Malibu Maxx, and redesigned Malibu; GMC Envoy EUV and Canyon; and the Pontiac GTO and redesigned Grand Prix.

Curiously, most of the Cadillacs carry initials rather than names, making buyers even more confused when comparing the EXT and ESV trucks, and CTS, DHS, DTS, STS, STL, SRX, and XLR passenger cars.

Incidentally, the XLR is built alongside the Chevrolet Corvette at GM's assembly plant in Bowling Green, Kentucky. It is the first car other than the Corvette to be built there. The SRX, a premium sport wagon, is built alongside the Cadillac CTS in Lansing, Michigan.

Warranty performance, quality control

The company has put more money into warranty repairs and continues to let its dealers make most post-warranty decisions. So it's not surprising to see that GM warranty complaints have lessened a bit over the years. This being said, GM's quality is still nowhere near the high quality cars and trucks sold by the Japanese automakers.

GM's quality control has improved somewhat as well. Nevertheless, GM engines and automatic transmissions, though not as failure-prone as Ford and Chrysler components, still aren't as reliable or as durable as the Asian competition, who've been embarrassed themselves over the recent disclosure of their own spate of powertrain defects. Furthermore, GM brake and electronic components often fail prematurely and cost owners big bucks to diagnose and repair. The quality of body components and their assembly also remain far below Japanese and European standards.

Interestingly, these problems are shared by all three Detroit-based auto manufacturers and will slowly diminish as Chrysler, Ford, and GM move back to rear-drives and merge operations and divisions with Japanese manufacturers.

CAVALIER, SUNFIRE 

RATING: Average. Despite their attractive prices and styling, these small cars have a history of chronic powertrain and safety-related deficiencies. **Strong points:** Very reasonably priced; good acceleration; plenty of front passenger and cargo room; comfortable riding; and well appointed for an entry-level vehicle. Split rear seatbacks add to storage capacity. **Weak points:** Steering and handling are only average. Limited rear passenger room; cheap interior; mediocre fit and finish; and small trunk opening. Problematic rear entry and exit (coupe). Real world fuel economy is much less than what's advertised. Crash safety and quality control still need improvement. **New for 2004:** Nothing significant. The Daewoo-sourced Aveo will be sold alongside the Cavalier to add some choice this year. The 2005 Cavalier Cobalt will make its debut in mid-2004.

OVERVIEW: These homegrown compact twins have been around for decades. They have a good powertrain set-up, exceptional styling (especially the Sunfire coupe) and lots of interior room. The Sunfire is identical to the Cavalier, except for its more rakish look. The Cavalier LS Sport and 1SC-equipped Sunfire are performance versions of the base compacts.

Cost analysis/best alternatives: A second-series (one made after March) 2003 can be a good buy if it's sufficiently discounted. If not, go for a 2004 early in the new year, when sales incentives are more generous. With competitors like the Honda Civic, Toyota Corolla and Chevrolet's own Aveo, GM will continue to pay shoppers to take these cars off their hands. Other models worth a look: the Hyundai Accent, Elantra or Tiburon, Mazda's Protegé, or the Nissan Sentra. **Options:** There's very little that's worthwhile, except for the side airbags, enhanced suspension, power windows and locks, and remote keyless entry. Forget about the rear spoiler; it looks silly and doesn't add to the car's performance. The LS, LS Sport, and performance packages aren't worth the extra cost. **Rebates:** $2,000 rebates and other incentives on the 2003s and 2004s. **Delivery/PDI:** $850. **Depreciation:** Slower than average. **Insurance cost:** Average. **Parts supply/cost:** Parts are easy to find and they're reasonably priced, with heavy discounting by independent suppliers. **Annual maintenance cost:** Average. **Warranty:** Bumper-to-bumper 3 years/60,000 km; powertrain 5 years/100,000 km; rust perforation 6 years/160,000 km. **Supplementary warranty:** Yes, but only for the powertrain if you'll be keeping the car for more than five years. **Highway/city fuel economy:** *Base engine and automatic 4-speed.:* 6.7–10L/100 km.

Quality/Reliability/Safety

Pro: Reliability: Although not as reliable as the Japanese competition, it's much better than what's offered by Ford and DaimlerChrysler. **Warranty performance:** Better than average. Important safety features include a remote keyless entry with a panic button, top child-seat anchors for all three rear positions, and side airbags that protect the head and upper torso.

Con: Quality control: Hit or miss, often leading to early powertrain and brake failures. Cheap body hardware is fragile and often poorly assembled. **Owner-reported problems:** Historically, engine head gasket failures have been a frequent problem, made worse by GM denials that the problem exists and refusal to admit that a "secret warranty" covers the defect. This dishonesty infuriates customers like this 2003 Cavalier owner:

> I was informed by a customer service rep that my vehicle was covered for head gasket repair under a "Special Policy" recall. At the urging of a customer service rep from Chevrolet, I took my vehicle to the dealer for repair. I was notified by their service department that this "Special Policy" had expired July 2003 and that it would cost over $1,400 to fix.

Many complaints of automatic transmission failures and grinding when shifting. PCM computer modules have also been one of the most common sources of complaints; symptoms include stalling and a shaky idle. Fuel injection and cooling systems are temperamental as well. The power steering may lead or pull, and the steering rack tends to deteriorate quickly, usually requiring replacement sometime shortly after 80,000 km. The front MacPherson struts also wear out rapidly, as do the rear shock absorbers. Many owners complain of rapid front-brake wear and warped brake discs after a year or so. One owner reported the following brake repairs to NHTSA:

> Front brake rotors are warping and had to be turned at 1,600 miles [2,575 km] and 1,800 miles [2,897 km]. They then were replaced at 2,800 miles [4,506 km]. They would cause the vehicle to jump when braking.

More recent owner-reported problems: front vacuum leak causes vehicle to lose power; hard starts; chronic stalling; slipping transmission; grinding noise when shifting gear; rattling noise when shifting from First to Second gear; frequent steering failures and noisy steering; excessive pulsation when braking; airbag warning lamp comes on continuously; dashboard noise; fuel sloshing in fuel tank; window may fall off its track and slide between the door panels. **Service bulletin problems:** No starts, hard starts, and poor idling may be due to clogged fuel injectors; firm shifts and shudder; transmission slips, or no shift (see below).

A/T - Slips/No Shift/No Reverse/Second/OD Gears

Bulletin No.: 03-07-30-012 **Date:** April 2003

Transaxle or transmission slips, won't shift, no Reverse, Second, Third or Overdrive, erratic shifts (Replace driven sprocket support Assembly)

2003 Chevrolet Cavalier, Malibu
2003 Oldsmobile Alero
2003 Pontiac Grand Am, Sunfire
with 4T40E or 4T45E Transmission

Neutral, or RPM flare while in Drive, no 1–2 upshift; momentary loss of steering assist (see below).

Steering - Momentary Loss of Power Assist

Bulletin No.: 03-02-32-001 **Date:** April 2003

Momentary reduction of power steering (P/S) assist on initial start-up with low ambient temperatures (Replace P/S pump)

2002–2003 Chevrolet Cavalier
2002–2003 Oldsmobile Alero
2002–2003 Pontiac Grand Am, Sunfire
with 4-cylinder, 2.2L gasoline engine

Condition: Some customers may comment on a momentary reduction of power steering (P/S) assist on initial start-up with ambient temperatures about -12°C (10°F). The system returns to full assist after a few seconds.

Rattle and knocking; steering or suspension popping, creaking sound when turning; front-end clunks; rattling or thumping sound heard from the passenger side floorboard when vehicle passes over rough roads; grinding or growling in Park; wheel squeaking; dash rattle; instrument panel lens cracks or discolouring; faulty heater; inaccurate fuel gauge tied to faulty fuel module, fuel level sensor or other defects (see Grand Am, Alero section); and hard to change temperature settings. **NHTSA safety complaints/safety:** Tire jack suddenly collapsed; engine fires; Reverse bulbs exploded, causing back light area to catch fire; leaking fuel tank; plastic fuel tank is easily punctured; fuel spews out when refuelling; seatback broke when vehicle was rear-ended; seatback spring assembly suddenly snapped for no reason; right and left wheel axle broke off vehicle; chronic hesitation, stalling, and surging; sudden acceleration; clutch will not disengage, causing sudden acceleration; faulty ABS; brake failure due to leaking master cylinder fluid; ABS locked up, causing vehicle to go into a skid; airbags failed to deploy; seat belt failed to retract; automatic transmission wouldn't go into Reverse; suddenly shifted into Neutral on the highway; failed to engage upon start-up; locks up in Second gear; vehicle rolled away even though parked with parking brake engaged; vehicle rolls backward when stopped on an incline; when vehicle is in Drive with foot on the brake it lurches forward, stalls, and produces a crashing sound; during highway driving the vehicle suddenly accelerated without steering control; rear leaf spring U-bolts broke, causing entire rear end to drop; front right side of the vehicle collapsed due to wheel bolts shearing off, causing the wheel to detach completely; springs are too weak, causing poor stability and control; floor mat impedes clutch pedal travel; sudden brake cable breakage while driving; brake grinding noise; warping of the front and rear brakes; when stopping, vehicle makes a thumping noise; when driving with door locked, door came ajar; hood flew up while driving; misaligned driver's door; windshield water leaks; side window exploded; sunroof shattered; defogger doesn't correct chronic windshield fogging; Check Engine light came on due to loose fuel cap; horn wouldn't work in an emergency; early burnout of the turn signal and brake light bulbs; prematurely worn Goodyear tire blew out; plastic bumper fell apart, causing front wheel damage; and the trunk lid remains open at such a low angle that it's easy to hit your head.

Road Performance

Pro: Acceleration/torque: The fuel-injected 2.2L 4-cylinder engine provides adequate power for all driving needs. The optional suspension package offers better handling at highway speeds and the best ride control on bad roads. **Braking:** Disc/drum; acceptable.

Con: Emergency handling: Excessive lean when cornering under power and standard tires corner poorly. **Steering:** Power steering feels over-assisted, resulting in insufficient road feel. **Transmission:** The 5-speed manual transaxle has an abrupt clutch. **Routine handling:** Base models don't handle as well as do most other vehicles in this class.

Comfort/Convenience

Pro: Standard equipment: Nicely equipped with standard AC and rear defroster, a Getrag 5-speed manual transmission, and a user-friendly instrument panel and centre console. **Driving position:** Generous up-front headroom, although tall drivers may find the driving position a bit confining. **Controls and displays:** Well laid-out dashboard and controls. **Entry/exit:** Lots of foot space and large door openings make for easy access into the front or rear areas. **Interior space/comfort F/R:** Interior room will seat four adults comfortably. **Cargo space:** Fairly generous for a compact. **Trunk/liftover:** Good-sized trunk with a low liftover. **Quietness:** Quiet interior with minimal road/wind noise intrusion.

Con: No Sunfire sedan; the coupes aren't easy to enter or exit. **Climate control:** Air conditioning controls aren't easily accessed. Rear seat room is skimpy.

Cavalier, Sunfire

List Price (negotiable)	**Residual Values** (months)			
	24	**36**	**48**	**60**
Cavalier VL: $15,995 (12%)	$11,000	$8,000	$6,000	$4,500
Sunfire SL: $15,995 (12%)	$11,000	$7,500	$6,500	$5,000

Technical Data

Powertrain (front-drive)
Engines: 2.2L 4-cyl. (140 hp)
Transmissions: 5-speed man. OD
• 4-speed auto.

Dimension/Capacity
Passengers: 2/3
Height/length/width:
54.8/180.3/67.4 in.
Headroom F/R: 38.9/37.2 in.
Legroom F/R: 42.3/34.6 in.
Wheelbase: 104.1 in.
Turning circle: 35.6 ft.
Cargo volume: 13.2 cu. ft.
Tow limit: 1,000 lb.
Fuel tank: 53L/reg.
Weight: 2,677 lb.

Safety Features/Crashworthiness

	Std.	Opt.
Anti-lock brakes (4W)	—	■
Seat belt pretensioners F/R	■	—
Side airbags	—	■
Traction control	—	■
Head restraints F/R	*	*
Visibility F/R	*****	*****
Crash protection (front) D/P		
2d	****	—
Crash protection (side) F/R		
2d	**	—
4d	***	—
Crash protection (offset)	—	—

ALERO, GRAND AM 

RATING: Below Average. Rating downgraded due to large number of serious safety-related complaints. Performance and quality control are no match for the Honda and Toyota competition. **Strong points:** *Grand Am:* Well appointed with many standard features, like traction control and AC. Competent V6; good steering and handling; and average quality control. *Alero:* Impressive V6 acceleration; even though it's the same engine found in the Grand Am, it performs better in the Alero. Logical, user-friendly gauges and controls; fairly spacious interior for cargo and passengers; standard traction control; and quiet running. **Weak points:** *Grand Am:* Manual transmission has been dropped; mediocre ride over rough terrain; excessive 4-cylinder noise; noisy interior; a four-seater with barely enough room in the coupe. Difficult rear seat access (coupe); awkward radio controls; rear visibility obstructed by spoiler; problematic trunk access; annoying body creaks and rattles; and long-term powertrain reliability is still a question mark. *Alero:* Excessive 4-cylinder engine noise and torque steer; steering isn't as crisp as the Grand Am's; difficult rear seat access (coupe); and long-term powertrain reliability is still questionable. **New for 2004:** Alero will be dropped later this year due to the phase-out of GM's Oldsmobile division. Grand Am gets a radio speaker upgrade. The 2005 Grand Am will be renamed the G6 and adopt GM's new "Epsilon" platform used by the 2004 Malibu.

OVERVIEW: Taking its styling cues from GM's redesigned Grand Prix, the Grand Am offers a roomy, comfortable interior in two- and four-door body styles. Sharing its platform and mechanical components with the Oldsmobile Alero, which targets a slightly more upscale and performance-minded clientele, the base Grand Am SE uses a 140-hp 2.2L power plant; other trim levels use a 175-hp 3.4L V6.

The upscale Alero, Oldsmobile's entry-level model, debuted in 1999 as the replacement for the slow-selling Achieva—often referred to as the "under-Achieva." Alero shares the Grand Am's chassis and powertrains, although it is styled quite differently. 2004 is its last model year.

Cost analysis/best alternatives: These are slow sellers that haven't changed much. 2003 Aleros represent the best deal since they are being phased out and come very well equipped. Look for a discount of at least 20 percent. Other cars worth considering are the Honda Accord, Nissan Altima, Subaru Legacy, Toyota Camry, and Toyota Avalon. **Options:** The best engine choice for power, smoothness, and value retention is the 3.4L V6; it gives you much-needed power and is quite fuel efficient. Early reports indicate the new 2.2L engine is quieter, but its lack of power is noticeable, particularly when coupled to an automatic transmission. Stay away from the Computer Command Ride option; true, it allows you to choose your own suspension setting, but the settings aren't quite what they pretend to be. **Rebates:** $3,000+, plus zero percent financing. **Delivery/PDI:** $900. **Depreciation:** Faster than average. **Insurance cost:** Higher than average. **Parts supply/cost:** Easy-to-find parts at reasonable prices. **Annual maintenance cost:** Average. **Warranty:** Bumper-to-bumper

3 years/60,000 km; rust perforation 6 years/160,000 km. **Supplementary warranty:** A good idea. **Highway/city fuel economy:** *2.2L and 4-speed auto.:* 6.7–10L/100 km; *3.4L:* 7.8–11.8L/100 km.

Quality/Reliability/Safety

Pro: Warranty performance: Average.

Con: Quality control: Bad. GM's quality improvements, evident on its trucks, vans, and SUVs, have yet to work their way through to these passenger vehicles. **Reliability:** The automatic transmission and engine have proven to be unreliable in the past. **Owner-reported problems:** Powertrain malfunctions, including sudden transmission failure and poor shifting; vehicle runs hot; premature brake pad and rotor wear; electrical problems; suspension squeaks; and substandard body assembly producing squeaks and rattles, water leaks (especially into the trunk area), and poor paint adhesion. Report that the sunroof may leak water into the electrical panel, causing short circuits. Rear visibility obstructed by spoiler. **Service bulletin problems:** Poor shifting and inaccurate gauges (see below).

Erratic Readings, Shifts, and Cruise Control Operation

Bulletin No.: 02-07-30-044 **Date:** November 2002

Erratic speedometer, erratic shifting, cruise control inoperative, DTCs P0502, P0758, P1860 set (Repair Vehicle Speed Sensor wires and reposition harness)

2003 Oldsmobile Alero
2003 Pontiac Grand Am
with 2.2L engine (VIN F - RPO L61), automatic transmission (RPO MN4) and non-anti-lock brakes (RPO J41)

^Erratic speedometer readings
^Erratic shifting
^Cruise control inoperative

Cause: These conditions may be caused from the extra length of Vehicle Speed Sensor (VSS) harness intermittently contacting the right axle shaft. This contact may cause a rub-through to the wires enclosed in that harness, resulting in the above conditions.

Firm transmission shifts and shudders or transmission slips or fails to shift (see Cavalier section); loss of power steering; noisy front suspension; inaccurate fuel gauge readings (see below).

Inaccurate Fuel Gauge Readings

Bulletin No.: 01-06-04-008D **Date:** April 2003

Inaccurate or erratic fuel gauge reading, fuel pump-related driveability concerns (Install new fuel tank sender)

2000–03 Chevrolet Cavalier, Malibu
2000–03 Oldsmobile Alero
2000–03 Pontiac Grand Am, Sunfire

Condition: Some customers may comment about inaccurate or erratic fuel gauge readings. A typical comment might be that it appears from the gauge reading that there is fuel available, yet the tank is nearly empty.

Cause: This condition may be the result of the corrosive effect of certain fuel blends on the contact surfaces of the fuel tank sender.

Front door window glass comes out of run channel; and there is wind noise from rear of vehicle. **NHTSA safety complaints/safety:** *Grand Am:* Transmission suddenly failed while cruising at speeds over 100 km/h; chronic stalling; vehicle continued accelerating after passing another car on the highway; car constantly surges and hesitates; airbags failed to deploy; inadvertent airbag deployment; inadequate headlight illumination; vehicle tends to wander all over the road; pulls to one side when accelerating; power-steering fluid leakage; steering shudders upon braking; faulty power-steering pump; steering feels too loose; excessive steering vibration when accelerating; ABS light comes on, followed by brake failure; brake pedal set too low; premature wearout of the front and rear brake pads and rotors; front brake pads wear out every 5,000 km; parking brake failure; fuel pump failure; shoulder belts twist in their housing; in the GT two-door, shoulder belts ride abnormally high on the shoulder and neck area due to guide design; power door locks lock on their own; headlights blink on and off; theft alarm sounds for no reason; middle rear lap seat belt is too short to secure a child safety seat and GM says an extension isn't available; loose driver's seat; head restraints and optional rear spoiler blocks rear vision; windshield wipers shut off intermittently; rear windshield shattered when defogger was activated; side door glass shattered behind mirror for no reason; inside door edge is razor sharp; slight front impact causes the battery tray to either break off, or results in the battery sliding off the tray. *Alero:* Engine compartment fire; right front wheel separated because the lug nuts and bolts sheared off; tapped brakes to turn cruise control off and vehicle accelerated; inadvertent airbag deployment; airbags failed to deploy; rear main oil seal leak blew oil onto exhaust pipe; fuel tank leakage; water leakage onto the back of the instrument panel causes the instruments and gauges to malfunction; other electrical shorts cause instrument panel gauges and controls to fail; windshield water leaks cause electrical shorts; sun visor fell on driver's head; hard to find horn "sweet spot" in an emergency; side mirror spring mounts break when passing over unpaved roads; seat latch doesn't lock immediately when moved; dash is reflected onto the windshield.

Road Performance

Pro: Emergency handling: Better than average. **Steering:** Steering is predictably responsive, less so with the Alero. **Acceleration/torque:** Fair acceleration (0–100 km/h: 9.9 seconds) with the larger 4-cylinder engine. V6 acceleration is impressive, delivering power smoothly. **Transmission:** Well suited to the V6. **Routine handling:** Better than average on the highway. Suspension improvements have made for better handling and have smoothed out the ride somewhat. **Braking:** Disc/drum, disc/disc; reasonably good braking (100–0 km/h: 37 m) with some brake fade after repeated stops.

Con: Excessive torque steer (twisting) when accelerating; impact of rough roads isn't completely absorbed by the suspension, causing vehicle instability.

Comfort/Convenience

Pro: Standard equipment: Well appointed. Nice, heavily padded interior. Good interior and exterior styling. **Driving position:** Comfortable cockpit with an easily adjusted, supportive driver's seat and plenty of headroom and legroom for two. Alero's bucket seats are particularly comfortable. **Climate control:** Efficient heating/defrosting and ventilation system. Rear windows roll all the way down. **Interior space/comfort F/R:** Sedan has better-than-average room in front and back. **Trunk/liftover:** Standard split folding rear seatbacks expand cargo space a bit.

Con: Alero's dash vents are rather crude and the seat fabric feels cheap. **Controls and displays:** Not user-friendly; hard-to-see recessed gauges. **Entry/exit:** The doors on two-door models are heavy and awkward to open in tight spaces. The front seat belts interfere with getting in and out. **Interior space/comfort F/R:** The coupe's back seat will only hold two adults comfortably, due to its narrower seats and reduced headroom. Rear visibility is also somewhat limited. **Cargo space:** Limited storage space. **Trunk/liftover:** High sill and small opening makes loading difficult. **Quietness:** Excessive engine, wind, and road noise; interior creaks and rattles, and annoying exhaust drone.

Alero, Grand Am

List Price (negotiable)	**Residual Values** (months)			
	24	**36**	**48**	**60**
Grand Am SE: $21,770 (17%)	$15,000	$12,000	$9,500	$7,500
Alero GX: $22,230 (17%)	$17,500	$14,500	$12,000	$8,500

Technical Data

Powertrain (front-drive)
Engines: 2.2L 4-cyl. (134–140 hp)
• 3.4L V6 (175 hp)
Transmissions: 5-speed man.
• 4-speed auto.
Dimension/Capacity
Passengers: 2/3
Height/length/width:
55.1/186.3/70.4 in.
Headroom F/R: 37.8/36.5 in.
Legroom F/R: 43.1/33.9 in.
Wheelbase: 107/111.2 in.
Turning circle: 40.5 ft.
Cargo volume: 14/15 cu. ft.
Tow limit: 1,000 lb.
Fuel tank: 53L/reg.
Weight: 3,066 lb.

Safety Features/Crashworthiness

	Std.	**Opt.**
Anti-lock brakes (4W)	■	■
Seat belt pretensioners F/R	—	—
Side airbags	—	■
Traction control	■	■
Head restraints F/R	***	***
Alero	**	**
Visibility F/R	*****	***
Coupe	*****	*

Crash protection (front) D/P	****	*****
4d	****	****
Crash protection (side) F/R	*	****
4d	***	***
Crash protection (offset)	*	—

MALIBU, MAXX, CLASSIC ★

RATING: *2004 Malibu and Maxx*: Not Recommended during the cars' first year on the market. **Strong points:** Good V6 powertrain set-up. Well appointed; comfortable though firm ride; responsive, European-style handling; plenty of passenger and luggage space; and fewer squeaks and rattles. **Weak points:** Base 4-cylinder engine and transmission may not be adequate for highway driving with a full load and an automatic transmission. Bland interior. No crashworthiness or reliability data. Expect lots of first-year factory-related glitches and slow parts delivery. Short, 3-year powertrain warranty. *Malibu Classic:* V6 base engine performs quite well. Comfortable ride, easy routine handling, and plenty of interior space. **Weak points:** Mediocre high-speed handling; no traction control; uncomfortable seats; stiff suspension may be too firm for some; excessive interior noise; premature tire failures; and rotten quality control. **New for 2004:** Everything. The 2003 Malibu will be carried over unchanged and renamed the Malibu Classic; it will be sold mostly to fleets and rental-car agencies.

OVERVIEW: *Malibu and Maxx*: The 2004 Malibu is an entirely new front-drive, mid-sized sedan that's reasonably priced and well appointed. It is joined by the Malibu Maxx a five-door hatchback that features a skylight over the rear passengers and a rear seat with 18 cm of forward-and-back travel to provide increased passenger and cargo space.

Both cars make good use of GM's venerable base 2.2L 4-cylinder power plant, found on many of its compact cars. In its "Ecotec" configuration, the engine has five more horses (145); however, buyers will be disappointed that the more powerful, 170-hp 3.1L V6 is no longer a standard feature. Instead, buyers will pay extra for a torquier and smoother 200-hp 3.5L V6, unless they opt for the LS or top-of-the-line LT models. Unfortunately, the only transmission available is a 4-speed automatic, a major disappointment for both performance aficionados and drivers looking for better fuel economy. Rumours have it, though, that the performance crowd will be rewarded when the 2005 Maxx comes equipped with a 235-hp V6.

These cars are the first North American models to use GM's "Epsilon" platform, which is also the platform for the Saab 9-3 and Opel Vectra. GM says the European experience has shown that this stiffer platform enhances handling and ride comfort and reduces body squeaks and rattles. Another benefit is that the cars have a wider stance and more interior room. But that extra room comes at a cost. It's actually 5 cm shorter, with a wheelbase that's also slightly shorter. On the other hand, the 2004 is a centimetre or so wider and almost 3 cm higher. Overall volume is larger by 18 percent but much of that is taken from a 12 percent smaller trunk.

The new Malibu's styling is quite conservative and uncluttered. Interior room is maximized by a fold-flat passenger seat and a 60/40-split rear folding bench seat. Other useful standard amenities: a driver's seat power height adjuster; a telescoping steering column that also tilts; power windows, door locks, and outside mirrors; and power adjustable brake and accelerator pedals (LS and LT).

From a safety perspective, these cars offer almost everything. The LT sedan and Maxx have four-wheel disc brakes, and ABS and traction control are standard on the LS and LT and optional on the base sedan. A side curtain airbag is optional on all models.

Malibu Classic: This is a smaller, mid-sized, carried-over sedan that replaced the long-discontinued Celebrity and Ciera. It's a conventionally styled (boring?) feature-laden car, which offers a standard 170-hp 3.1L V6 coupled to a 4-speed, floor-mounted automatic transmission, ABS, AC, tilt wheel, electric trunk release, power door locks, and an AM/FM/CD stereo.

Cost analysis/best alternatives: Go for the practically identical 2003 Classic, if only for the base V6 and expected sizeable discount and rebates. If you really want the redesigned Malibu or Maxx, don't expect rebates now; wait at least until mid-2004 for prices to stabilize and factory improvements to work their way into the lineup. Be especially patient in waiting for the Maxx; it's not likely to hit dealer showrooms before early 2004. The Honda Accord's got more usable interior space, and quicker and more accurate steering, and Toyota's Camry is plusher, though not as driver-oriented. Both Japanese competitors are also far more reliable. Other cars worth considering: Hyundai Elantra or Tiburon, Mazda6, or the Nissan Sentra. **Options:** The LS package, side curtain airbags, and premium tires. **Rebates:** $3,000+ rebates and zero percent financing on the Classic. **Delivery/PDI:** $900. **Depreciation:** Average. **Insurance cost:** Higher than average. **Parts supply/cost:** Malibu Classic uses GM generic parts that are found everywhere and are reasonably priced. **Annual maintenance cost:** Average. **Warranty:** Bumper-to-bumper 3 years/60,000 km; rust perforation 6 years/160,000 km. **Supplementary warranty:** A wise investment. **Highway/city fuel economy:** *3.1L:* 7.4–11.8L/100 km. Interestingly, the V6 isn't that much thirstier than the 4-cylinder.

Quality/Reliability/Safety

Pro: Quality control: Below-average quality control. **Reliability:** No problems of a serious nature, although chronic brake rotor warpage is both annoying and time-consuming, as the following email from *Lemon-Aid* reader Ron Phillips relates:

> I have a '99 Chevrolet Malibu and I am having nothing but problems with the brakes. The car in general is very good but today is the second time that I have brought my vehicle in for new or machined rotors. The service advisor at the dealership that I took it to said that this is a problem for all '99 Chevy Malibus as the rotors are made from a lighter material and they warp instantly. If I hit the brakes on the highway (at higher speeds) the whole vehicle shakes very erratically and it becomes a safety concern.

In another complaint recently sent to NHTSA, the owner of a 2002 Malibu relates having the same problem with his Malibu's brakes:

> At 12,000 miles [19,312 km] I had to replace the front brake pads and turn the rotors—needed rotors turned again at 16,000 miles [25,749 km].

Warranty performance: Average. **Safety:** The automatic headlight control that turns on at dusk is a useful feature and the optional curtain side airbags are effective in protecting the head and upper torso.

Con: Owner-reported problems: Premature engine piston failure caused "piston slap" noise (pistons on national back order); chronic stalling; erratic transmission shifting; pervasive odour of oil burning on the exhaust; driver's seat rocks back and forth. **Service bulletin problems:** No-start, condition accompanied by Service Engine Soon warning lamp; harsh transmission shifts; harsh 1–3 upshifts; fluid leakage from automatic transaxles; rear creaking or popping noise; inaccurate fuel gauge; brake pedal, underbody, or suspension clunk, rattle noise; rear doors take a lot of effort to open or close; hood cable freezing; primary hood latch doesn't engage. **NHTSA safety complaints:** Airbags deployed inadvertently or failed to deploy; sudden brake loss; premature warpage of the brake rotors; steering loss; sudden steering lock-up; chronic steering shimmy; frequent stalling for unexplained reasons; stalling caused by fuel line leak; hit a small bump in road and vehicle stalled with a complete loss of electrical power; tire blew and cruise control wouldn't disengage; fuel tank filler tube is loosely connected to the frame; excessive vibration at any speed; brake, steering wheel pulsation; engine rattle caused by defective piston; automatic transmission jumped from Park to Reverse; transmission doesn't lock when the key is in the accessory position; very loose steering; faulty high-beam switch; original equipment tires don't have gripping power;

door creaking; outside mirrors are too small; interior rear-view mirror vibrates; inaccurate fuel gauge; suspension squawking noise.

Road Performance

Pro: Emergency handling: Acceptable. Cornering under speed is well controlled, with little front-end plowing or excessive body roll. **Steering:** Acceptable for everyday driving. **Acceleration/torque:** Brisk acceleration and plenty of low-end torque with the V6 engine (0–100 km/h: 8.8 seconds); the V6 delivers power in a smooth, quiet manner. **Transmission:** Smooth and predictable shifting, most of the time. **Routine handling:** Better than average, thanks to an independent suspension that doesn't sacrifice solid handling for passenger comfort.

Con: Hard cornering produces considerable body lean and numb steering. Push-rod V6 is a bit rougher than the overhead-cam V6 used by the competition. Downshifting is a bit slow when passing other vehicles. **Braking:** Disc/drum; antiquated rear drum brakes provide mediocre braking with standard ABS (100–0 km/h: 39 m). Some brake fade after repeated stops.

Comfort/Convenience

Pro: Standard equipment: Well appointed for the price, but the better-equipped LS version has more of what you need. **Driving position:** Seating is adequate, though a bit low. Large side mirrors help provide good all-around visibility. **Controls and displays:** Very well thought-out instrumentation with gauges that are easy to access and read. For example, the ignition switch is located on the instrument panel, a radical departure for GM. **Climate control:** Excellent ventilation system and first-class controls that are easy to use. AC vents are mounted high enough to direct cool air at your face. **Entry/exit:** Easy access to both front and rear interiors. **Interior space/comfort F/R:** More head and legroom than the Chrysler Stratus. Unusually spacious rear seat area can accommodate three adults. **Cargo space:** Lots of interior storage space. **Trunk/liftover:** Plenty of trunk space that can be expanded through the split folding rear seatbacks. Low liftover and practical cargo net.

Con: Bland styling. Seats could use better side bolstering and lumbar support. Some gauges are hard to read in direct sunlight. Outside mirrors are too small. Heating to the floor area is a bit slow. Door checks may not hold the doors open when parked facing uphill. Rear bench seat is hard and flat. Rear seatbacks don't lie flat when folded. Trunk hinges intrude into the trunk area, possibly damaging contents in a stuffed trunk. Parcel shelf compromises rear visibility. The 2.2L engine's noise intrudes into the passenger compartment. **Quietness:** Excessive engine, tire, and suspension noise.

Malibu

List Price (negotiable)	**Residual Values** (months) 24	36	48	60
Malibu Sedan: $22,160 (18%)	$17,500	$15,000	$12,000	$9,000
LS: $24,695 (18%)	$18,500	$16,000	$13,000	$10,000
LT: $29,220 (18%)	$21,000	$19,000	$15,000	$13,000
Classic Sedan: $21,500 (18%)	$15,000	$13,000	$10,000	$7,000

Technical Data (Malibu)

Powertrain (front-drive)
Engines: 2.2L 4-cyl. (145 hp)
• 3.5L V6 (200 hp)
Transmission: 4-speed auto.
Dimension/Capacity
Passengers: 2/3
Height/length/width:
57.5/188.3/69.9 in.
Headroom F/R: 39.9/38.5 in.
Legroom F/R: 41.9/38 in.
Wheelbase: 106.3 in.
Turning circle: 38 ft.
Cargo volume: 15.4 cu. ft.
Tow limit: 1,000 lb.
Fuel tank: 6.2L/reg.
Weight: 3,297 lb.

Safety Features/Crashworthiness

	Std.	Opt.
Anti-lock brakes	■	—
Seat belt pretensioners F/R	■	■
Side airbags	—	—
Traction control	■	■

Technical Data (Classic)

Powertrain (front-drive)
Engine: 3.1L V6 (170 hp)
Transmission: 4-speed auto.
Dimension/Capacity
Passengers: 2/3
Height/length/width:
56.4/190.4/69.4 in.
Headroom F/R: 39.4/37.6 in.
Legroom F/R: 41.9/38 in.
Wheelbase: 107 in.
Turning circle: 40 ft.
Cargo volume: 17.3 cu. ft.
Tow limit: 1,000 lb.
Fuel tank: 54L/reg.
Weight: 3,052 lb.

Safety Features/Crashworthiness

	Std.	Opt.
Anti-lock brakes	■	—
Seat belt pretensioners F/R	■	—
Side airbags	—	—
Traction control	—	—
Head restraints F/R	**	*
Visibility F/R	*****	*****
Crash protection (front) D/P	*****	*****
Crash protection (side) F/R	***	****
Crash protection (offset)	***	—

CENTURY, GRAND PRIX, IMPALA, MONTE CARLO, REGAL ★★★

RATING: *Century, Grand Prix, Impala, Monte Carlo, Regal:* Average. **Strong points:** Nice array of standard features; good choice of powertrains; comfortable ride; easily accessed and roomy interior. **Weak points:** Excessive engine noise at high speeds; rear seating uncomfortable for three; bland styling; and obstructed rear visibility due to high-tail rear end. Most importantly, these cars have been hobbled by a chintzy 3-year powertrain warranty that's clearly insufficient, knowing GM's past engine and transmission deficiencies. **New for 2004:** *Century:* Standard four-wheel disc brakes; last model year. *Grand Prix:* Improved brake components, a re-styled interior, a fold-flat front passenger seat, less body cladding, more sound-proofing, wider-opening doors, and the entry-level SE version is no more. *Impala*: A new Impala SS debuts equipped with a supercharged, 240-hp V6 engine. *Monte Carlo:* Also gets an SS designation after borrowing the Impala's supercharged engine. *Regal:* Nothing significant; awaits a 2005 redesign.

OVERVIEW: These front-drive, mid-sized cars share the same platform and mostly exhibit the same performance characteristics with similar powertrains—except for the supercharged V6 engines.

Century: In its last model year, this sedan is loaded with convenience features and powered by a 3.1L V6. One gets the impression the Century targets young retirees who prefer to be driven, rather than drive. *Grand Prix:* Touted as GM's sport sedan, it comes as a two- or four-door model and is powered by a 3.1L V6 and an optional 3.8L V6—or supercharged variation—borrowed from the Pontiac Bonneville. Larger brakes, upgraded power steering, and a 5-cm longer body, an 8-cm longer wheelbase, and a 5-cm wider track are also featured. *Impala, Monte Carlo:* Impala is a more upscale, nicely appointed sedan that seats six and comes equipped with a base 3.4L and an optional 3.8L engine. Four-wheel ABS, traction control, and tire-inflation monitors are standard on the LS, along with a host of other features. Side airbags are no longer a standard feature. Unlike the other vehicles in this section, the Impala has posted impressive sales over the past several years. A coupe cousin to the

Impala, the Monte Carlo is a mid-sized sport coupe that incorporates more creases in its styling, along with round tail lights, a longer wheelbase, and shorter length. Side airbags are optional on all models. *Regal:* The Regal is designed to appeal to sport-oriented buyers who will appreciate the car's 3.8L V6, a supercharger variant, and other high-performance add-ons. This hasn't happened, so next year the 2005 model gets a complete redesign and absorbs the Century in its lineup. In the meantime, Buick will continue to use special edition versions to attract younger buyers. ABS, traction control, a driver's side airbag and tire-pressure monitors, are found only with uplevel models, so both performance and safety will cost you more this year.

Cost analysis/best alternatives: Get discounted 2003 models, since this year's few extra features are overpriced, unproven, or simply unnecessary. Keep in mind that there are plenty of more reliable, better-performing vehicles available from the competition. Consider the Acura Integra, Honda Accord, Mazda6, Nissan Altima and Maxima, and Toyota Camry and Avalon. Those wanting a bit more performance should consider the BMW 3 Series. **Options:** The rear-mounted child safety seat and upper torso-protecting side curtain airbags. Stay away from the Impala's rear spoiler; it obstructs rear visibility and is of doubtful utility. **Rebates:** This is where GM's fight for market share (instead of profits) becomes hand-to-hand combat. Its cars are in a marketing segment that's steadily losing ground to Japanese entries and GM is throwing discounts, rebates, low-financing plans, extended 5-year+ payment plans, balloon payments, and overnight test-drives into the mix. The longer you wait, the cheaper these cars will become. Expect juicy financing deals and $3,000+ rebates on all 2003 models and more incentives on the 2004s. **Delivery/PDI:** $1,000. **Depreciation:** Average. **Insurance cost:** Higher than average. **Parts supply/cost:** Moderately priced parts that aren't hard to find. **Annual maintenance cost:** Higher than average. Independent garages can perform most non-emissions servicing. **Warranty:** Bumper-to-bumper 3 years/60,000 km; rust perforation 6 years/160,000 km. **Supplementary warranty:** Essential, especially after the third year of ownership. **Highway/city fuel economy:** *Base sedan with 3.1L engine:* 7.4–11.8L/100 km; *3.4L:* 7.2–12L/100 km; *3.5L:* 8–12.7L/100 km; *3.8L:* 7.6–12.3L/100 km.

Quality/Reliability/Safety

Pro: Warranty performance: Average. **Safety:** Head-protecting side curtain airbags are a worthwhile option.

Con: Tire-inflation monitor is unproven. **Quality control:** Below average for powertrain and body construction, which puts it on par with Chrysler and a bit ahead of Ford. **Reliability:** The 3.1L, 3.4L, and 3.8L V6 engines have had more than their share of fuel system, intake manifold, and computer module problems that keep cars off the road. **Owner-reported problems:** Premature engine head gasket failures; "piston slap" noise; early transmission breakdowns. Electrical system problems are common on cars loaded with power accessories. Front and rear brakes rust easily, wear out early, and the discs warp far too often. Shock absorbers and MacPherson struts wear out or leak prematurely. The power rack-and-pinion steering system degenerates quickly after three years and is characterized by chronic leaking. Poor body fit, particularly around the doors, leads to excessive wind noise and water leaks into the interior. Door locks freeze up easily. General tires are often out of round. **Service bulletin problems:** *All models:* Here's the "smoking gun" GM internal service bulletin admitting to poor-quality engine intake manifolds. Note that this defect has existed over eight model years and imagine how many owners have paid to fix this defect, which GM admits here, *is clearly its own fault* (see below).

Engine Oil or Coolant Leak

Bulletin No.: 03-06-01-010A **Date:** April 2003

Engine oil or coolant leak (Install new intake manifold gasket)

2000–03 Buick Century
2002–03 Buick Rendezvous
1996 Chevrolet Lumina APV
1997–2003 Chevrolet Venture
1999–2001 Chevrolet Lumina
1999–2003 Chevrolet Malibu, Monte Carlo
2000–2003 Chevrolet Impala
1996–2003 Oldsmobile Silhouette
1999 Oldsmobile Cutlass
1999–2003 Oldsmobile Alero
1996–99 Pontiac Trans Sport
1999–2003 Pontiac Grand Am
2000–03 Pontiac Grand Prix, Montana
2001–03 Pontiac Aztek
with 3.1L or 3.4L V6 engine (VINs J, E - RPOs LGB, LA1)

Condition: Some owners may comment on an apparent oil or coolant leak. Additionally, the comments may range from spots on the driveway to having to add fluids.

Cause: Intake manifold may be leaking, allowing coolant, oil, or both to leak from the engine.

Correction: Install a new-design intake manifold gasket. The material used in the gasket has been changed in order to improve the sealing qualities of the gasket. When replacing the gasket, the intake manifold bolts must also be replaced and torqued to a revised specification. The new bolts will come with a pre-applied threadlocker on them.

This engine defect affects 16 models, including the Chevrolet Monte Carlo (above), since 1996.

Firm shift and shuddering (see Cavalier section); grinding, growling while in Park; inoperative power seat adjuster; and poor radio reception. **NHTSA safety complaints:** (Keep in mind that the safety problems noted below may apply to any one of the following practically identical cars.) *Impala:* Airbags failed to deploy; frequent complaints of dash area and engine compartment fire; front harness wires overheat; excessive current load from fuel pump may burn the ignition block wire terminal; inhalation injuries caused by the melting of the wiring harness plastic; electrically heated seat burned the driver's back; the connection that goes to the brake pedal piston collapsed, causing total brake failure; chronic stalling; engine sputters, hesitates; Service Engine and battery lights come on (dealer unable to correct problem); when traction control is activated, wheel slip computer is also activated and security system kills the engine and prevents it from being restarted; driver's side wheel fell off; vehicle jerks when passing over rough pavement; car rolls back at a stop; excessive steering wheel vibration; excessive front-end vibration; fuel sloshes in tank when accelerating or stopping; brake rotors had to be replaced at 7,500 km; aluminum subframe "flexes," causing a clicking sound when turning; popping sound upon stopping, starting, and turning; transmission line broke and poured fluid all over the road; transmission centre support bearing was put in backwards, causing the transmission to fail; left and right control arm, lower control arm, balljoint, and steering failure; AC R-134A refrigerant leaks into car interior; driver's seat adjuster failed, causing seat to suddenly move backward, causing loss of vehicle control; the rubber seal on the windows, which sometimes acts as a squeegee when lowering and raising the window, has been removed with a new design, which allows road salt to enter and short-circuit the window mechanism; front driver's side windshield wiper doesn't clean the windshield completely; poor design allows dirty windshield washer fluid to be deflected off the windshield and cut the view out of the side windows; a hazy film collects on the inside of all the windows (dealer says it's normal). *Monte Carlo:* Engine cradle mounting welds came apart from the steering gear, and driver's and passenger's seat belts tighten up progressively to the point where they are extremely uncomfortable. *Century:* Sudden brake failure; car rolls backward when stopped in gear on an incline; constant reflection of curved dashboard in windshield with or without sunlight; passenger-side windshield wiper channels water directly in the line of vision on the upstroke, temporarily blocking driver's vision; front-seat head restraints won't stay in raised position; vent behind shifter handle becomes very hot when heater is on; gear shift lever continually sticks; water leaks onto the interior carpet; air dam deflector on the front of the vehicle is mounted too low and hits the road on dips; defective radio volume control; horn is hard to find. *Grand Prix:* Chronic stalling; false airbag deployment; airbag failed to deploy; seatbacks designed with an inertia lock that only locks when braking aggressively, allowing unoccupied seatback to flop around and distract driver; ABS failure; prematurely worn rear brake pads create excessive metal-to-metal grinding noise; wheel lug nuts and bolts sheared off, causing wheel to fall off; windshield wipers malfunction; windshield wiper system freezes up in cold weather; lap/shoulder seat belts become twisted when reaching up and pulling down from the guide loop; cruise control suddenly engaged on its own and wouldn't release; sometimes cruise control causes the vehicle to suddenly accelerate and then slow down; engine coolant won't siphon back into radiator; right door speaker and dash rattling. *Regal:* Vehicle was on a medium incline, with the ignition on and the shift indicator in the Drive position, when driver took foot off the brake pedal and the vehicle rolled backward; excessive shaking on smooth roads; audio speaker failure; loose power steering; passenger-side airbag cover came loose.

Road Performance

Pro: Emergency handling: Above average, thanks to standard traction control. **Steering:** Precise and predictable. **Acceleration/torque:** The 3.1L and 3.4L V6 engines produce sufficient power for smooth acceleration and work well with the 4-speed automatic transmission (0–100 km/h: 10.5 seconds), but could use more high-speed torque. For extended highway use, you'll find the 3.8L power plant better suited to your needs. **Transmission:** The automatic transmission is quiet and smooth under most conditions. **Routine handling:** Handling and ride are better than average due to recent suspension and steering refinements. **Braking:** Disc/drum, disc/disc; Above average braking.

Con: There's excessive engine noise intrusion into the interior. The automatic 4-speed transmission is sometimes slow to downshift.

Comfort/Convenience

Pro: Standard equipment: Reasonably well appointed, with such innovative features as automatic headlights that turn on at dusk and a system that prevents the doors from locking automatically if the key is in the ignition. Large side mirrors help to overcome the obstructed rear view. **Driving position:** The cockpit area is much more user-friendly this year; lots of room; good all-around visibility; optional power driver's seat is a boon for short drivers. A tilt steering wheel is standard. **Controls and displays:** Improved instrumentation and easy-to-read gauges and controls. **Climate control:** Excellent heating-defrosting-ventilation system that even includes a pollen filter. **Entry/exit:** Not difficult. **Interior space/comfort F/R:** Front bench seats accommodate two with plenty of head and legroom. Rear seats have enough space for three adults. Grand Prix seating is quite firm, but not uncomfortable. **Trunk/liftover:** Large trunk and a low liftover. **Quietness:** Fairly quiet, except for some engine noise.

Con: Luxury models use tacky imitation wood and cheap cloth covers. Flimsy cupholders. Climate controls are set too low, interfering with middle passenger's knees, and low dash vents can chill a driver's hands. The Grand Prix sport sedan's headroom has been sacrificed to give the car a sleeker appearance.

Seats are too soft and lack support on all models, with the exception of the Grand Prix. Insufficient rear legroom. **Cargo space:** Absence of truly functional interior storage areas. There are lots of little storage spaces, but they're generally small and narrow. **Trunk/liftover:** *Century:* Quite a stretch to get over the wide, bumper-level shelf. *Grand Prix:* High trunk sill and narrow opening make for difficult loading and unloading, and large deck-lid hinges reduce usable trunk space. Rear seatbacks don't fold down to increase trunk space.

Century, Grand Prix, Impala, Monte Carlo, Regal

List Price (negotiable)	**Residual Values** (months)			
	24	**36**	**48**	**60**
Century: $26,100 (22%)	$18,500	$13,500	$11,500	$8,000
Grand Prix GT1: $27,995 (22%)	$19,500	$14,500	$12,500	$9,000
Impala: $26,675 (21%)	$19,500	$14,500	$12,500	$8,500
Monte Carlo LS: $28,000 (22%)	$18,500	$14,500	$11,500	$8,500
Regal: $30,450 (23%)	$21,500	$18,500	$15,500	$13,500

Technical Data

Powertrain (front-drive)
Engines: 3.1L V6 (175 hp)
• 3.4L V6 (180 hp)
• 3.8L V6 (200 hp)
• 3.8L V6 (240 hp)
• 3.5L V6 (215 hp)
• 3.8L V6 SC (260 hp)
Transmissions: 5-speed man.
• 4-speed auto.
Dimension/Capacity (Grand Prix)
Passengers: 2/3
Height/length/width: 54.7/196.5/72.7 in.
Headroom F/R: 38.3/36.7 in.
Legroom F/R: 42.4/35.8 in.
Dimension/Capacity (Century)
Height/length/width: 56.6/194.6/72.7 in.
Headroom F/R: 39.3/37.4 in.
Legroom F/R: 42.4/36.9 in.
Dimension/Capacity (Regal)
Passengers: 2/3
Height/length/width: 56.6/196.2/72.7 in.
Headroom F/R: 39.3/37.4 in.
Legroom F/R: 42.4/36.9 in.
Wheelbase: Grand Prix: 110.5 in.; Regal: 109 in.
Turning circle: 40 ft.
Cargo volume: 16/17 cu. ft.
Passengers: 2/3
Tow limit: 1,000 lb.
Fuel tank: 64L/reg./prem.
Weight: approx. 3,400 lb.

Safety Features/Crashworthiness

	Std.	**Opt.**
Anti-lock brakes (4W)	■	—
Seat belt pretensioners F/R	■	—
Side airbags	■	■
Traction control	■	—
Head restraints F/R	*	*
Grand Prix	***	*
Visibility F/R	*****	***
Crash protection (front) D/P		

All excl. Grand Prix and Impala	****	***
Grand Prix	****	****
Impala	*****	*****
Crash protection (side) F/R		
Century, Monte Carlo, and Regal	***	***
Grand Prix	**	***
Impala	****	****
Crash protection (offset)	****	—

BONNEVILLE, LESABRE, PARK AVENUE, ULTRA ★★★

RATING: Average. **Strong points:** *All models:* Great powertrain performance; comfortable ride; and plush interior. *Bonneville:* Terrific styling; a number of powertrain choices; rear seat to trunk pass-through; and fairly good fuel economy. *LeSabre:* Nice array of useful standard equipment at an affordable base price, but costs go up quickly as the options grow. *Park Avenue:* Quiet running; plenty of passenger and cargo room. Rear seating is a bit better than in the LeSabre. Low trunk liftover. **Weak points:** *All models:* Fuel-thirsty; poor-quality mechanical and body components. The suspension is on the soft side; the cars are a bit jittery on rough roads and handling isn't all that responsive. *Bonneville:* Some torque-steer on the SSEi; braking sometimes hard to modulate; lots of wind and road noise; cramped seating for five; uncomfortable front seating for the middle passenger; insufficient rear headroom for tall passengers; rear bench seat lacks support; and the tacky-looking dash isn't well laid-out. *LeSabre:* Climate controls aren't particularly user-friendly and not all gauges can be easily read. Rear seating is squeezed for three. Cushion is both too low and soft to provide confident, supportive seating. Trunk hinges intrude into storage area. *Park Avenue:* Ponderous handling caused partly by a mediocre suspension and over-assisted steering with the base model; obstructed rear visibility; hard braking accompanied by severe nosedive; and interior gauges and controls aren't easily deciphered or accessed. Front bench seat is too narrow. Ultra requires premium fuel. **New for 2004:** Nothing significant on any of the above models.

OVERVIEW: Full-sized luxury sedan aficionados love the flush glass, wraparound windshield and bumpers, and clean body lines that make for an aerodynamic, pleasing appearance. But these cars are more than a pretty package; they provide lots of room, luxury, style, and—dare I say—performance. Plenty of power is available with the 205-hp 3.8L V6 engine and the 240-hp supercharged version of the same power plant. It does a 0–100 km/h time in under nine seconds, impressive considering the heft of these vehicles, and improves low- and mid-range throttle response. Power is transmitted to the front wheels through an electronically controlled transmission that features "free-wheeling" clutches designed to eliminate abrupt gear changes. Although the Park Avenue is a bit larger and more expensive that the Bonneville and LeSabre, its mechanicals, performance, and overall reliability are quite similar to its smaller brethren. The Ultra version comes with a supercharged engine, StabiliTrak, larger 17-inch wheels, lower-profile tires, and a plusher interior.

Cost analysis/best alternatives: Get a discounted 2003 model. Other vehicles worth considering: the Grand Marquis, Infiniti I30, Lexus ES 300, or Nissan Maxima. **Options:** Go for the side airbags and traction control. **Rebates:** $3,000+ by early spring. **Delivery/PDI:** $950. **Depreciation:** Faster than average. **Insurance cost:** Higher than average. **Parts supply/cost:** Parts aren't hard to find, but they can be pricey (particularly the supercharged engine components). **Annual maintenance cost:** Higher than average. **Warranty:** Bumper-to-bumper 3 years/60,000 km; rust perforation 5 years/160,000 km. **Supplementary warranty:** It'll come in handy going into the fourth year. **Highway/city fuel economy:** 7.8–12.4L/100 km; 7.7–13.2L/100 km with the High Output engine. Supercharged V6 sips fuel, unlike the V8-equipped competition.

Quality/Reliability/Safety

Pro: Quality control: Average. **Warranty performance:** Better than average. **Safety:** A personalized vehicle security system disables the starting and fuel systems if a non-matching key is used. Rear shoulder belts have a strap to pull the belt away from the neck of small passengers and children. Right-side mirror tilts down when Reverse is engaged. Safety belts mount to the outboard front seats for better neck protection.

Con: Reliability: Mediocre. **Owner-reported problems:** Engine, transmission, brake, and electronic module malfunctions. This 2003 LeSabre owner found out that GM's intake manifold fix was no fix at all:

> I recently received a letter from GM stating that there may be a problem with coolant leaks around gaskets at the upper intake

manifold or at the lower intake manifold which might "cause high engine temperatures." The letter says it is a "voluntary customer satisfaction program." The suggested fix is to take the vehicle in to the local dealer and have them change some of the fasteners and then "add cooling system sealant" to the radiator. It seems to me that putting cooling system sealant in a brand new car (and thus reducing the life of the radiator) is an unacceptable fix for a possible gasket problem. I would think that the company needs to replace the intake manifold gaskets instead, if they are the faulty part. However, when I called the number that GM supplied they said that they would not replace the gasket and that if we did not have the repair done as stated in their letter and the gasket subsequently failed after the warranty period was over, we would have to pay the full cost of the repairs.

Service bulletin problems: *Bonneville* and *LeSabre:* Firm shift, shudder; grinding, growling while in Park; and rear seatback trim loose or sagging (LeSabre), and inoperative front heated seat. **NHTSA safety complaints:** *Bonneville*: Defogger was activated and fire ignited; fire erupted in the rear deck speaker; and the sun washes out gauge readings (see the service bulletins found in the Impala section). *Lesabre*: Engine intake manifold failure; engine surging while driving on the highway; brake pedal goes almost to the floor without stopping the vehicle; trunk lid fell and struck driver's head; improper headlight illumination; and horn is difficult to activate due to the hand pressure required. *Park Avenue*: Front outer tie-rod end failure (yet, they are said to be lubricated for life).

Road Performance

Pro: Acceleration/torque: The 3.8L V6 engine is competent, quiet, and smooth running, with lots of low-end torque. Impressively fast with the supercharged engine (0–100 km/h: 8 seconds). **Transmission:** The F31 electronic 4-speed transmission works imperceptibly. Cruise control is much smoother without all those annoying downshifts we've learned to hate in GM cars. **Routine handling:** Acceptably predictable, though quite slow with variable-assist power steering and shocks that are a bit firmer than usual. Body roll has been reduced thanks to the retuned suspension. With its stiffer suspension the Ultra performs well on winding roads, while the softly sprung Park Avenue is best for city use. **Emergency handling:** Fairly quick and sure-footed. **Braking:** Good braking with or without ABS (100–0 km/h: 41 m).

Con: Acceleration/torque: Acceleration isn't breathtaking with the base 3.8L engine; at higher revs torque falls off quickly. **Steering:** Power steering is a bit vague at highway speeds. Panic braking causes considerable nose-diving that compromises handling.

Comfort/Convenience

Pro: The entire body has a more solid feel than what one normally finds with GM. **Standard equipment:** Very well equipped, with many performance and convenience features included as standard equipment. **Controls and displays:** Complete and easy-to-read instruments and gauges, with only a few exceptions. **Climate control:** Much improved interior ventilation and AC performance. **Entry/exit:** No problem. **Interior space/comfort F/R:** Comfortable front seating for two. **Cargo space:** Plenty of storage areas that are easily accessible. **Trunk/liftover:** Huge trunk has a low liftover.

Con: Driving position: Driver seat lacks lateral support. Wide rear pillars block visibility, but large side mirrors compensate. Rear seating lacks adequate legroom. **Quietness:** Excessive wind, tire, and road noise.

Bonnevile, LeSabre, Park Avenue, Ultra

List Price (very negotiable)	**Residual Values** (months)			
	24	**36**	**48**	**60**
Bonneville SE: $33,920 (22%)	$24,000	$18,000	$14,000	$12,000
LeSabre Custom: $33,510 (22%)	$25,000	$19,000	$15,000	$13,000
Park Avenue: $47,050 (23%)	$34,000	$27,000	$23,000	$18,000
Ultra: $52,605 (23%)	$37,000	$31,000	$26,000	$22,000

Technical Data

Powertrain (front-drive)
Engines: 3.8L V6 (205 hp)
• 3.8L V6 (240 hp)
Transmission: 4-speed auto.
Dimension/Capacity (LeSabre)
Passengers: 2/3
Height/length/width:
57/200/73.5 in.
Headroom F/R: 38.8/37.8 in.
Legroom F/R: 42.4/39.9 in.
Wheelbase: 112.2. in.
Turning circle: 40.7 ft.
Cargo volume: 18 cu. ft.
Tow limit: 3,000 lb.
Fuel tank: 70L/reg.
Weight: 3,567 lb.

Safety Features/Crashworthiness

	Std.	**Opt.**
Anti-lock brakes	■	—
Seat belt pretensioners F/R	■	—
Side airbags	■	■
Traction control	■	■
Head restraints	*****	*****
Park Avenue	*	*
Visibility F/R	*****	***
Crash protection (front) D/P	****	*****
Crash protection (side) F/R	****	****
Crash protection (offset)	*****	—

AURORA (2003) ★★

RATING: Below Average. **Strong points:** V8 provides impressive acceleration; plenty of passenger room; comfortable ride on smooth terrain; good handling; and well-appointed interior. **Weak points:** Stiff, jostling suspension over rough pavement; numb steering; uncomfortable rear seating; obstructed rear visibility; complicated radio and climate controls; poorly assembled body components; and an unusually large number of automatic transmission problems. **New for 2004:** Aurora has been dropped due to the Oldsmobile division shutdown (they should have closed Saab and Saturn, instead). V6 was dropped on the 2003 model.

OVERVIEW: The Aurora is a five-passenger, front-drive sports sedan that targets the Lexus ES 300, Infiniti I30, Mazda Millenia, and Toyota Avalon. It uses the same platform as the Cadillac DeVille, Pontiac Bonneville, and Buick LeSabre and comes equipped with a 250-hp 4.0L V8 Northstar engine—a smaller version of the Cadillac power plant. The only transmission offered is a floor-mounted, 4-speed automatic—putting the Aurora at a disadvantage with its competitors, who mostly offer 5-speeds. Standard features include dual airbags, four-wheel ABS, traction control, adjustable shoulder height manual seat belts, cruise control, keyless remote entry and security system, and power-assisted everything.

Cost analysis/best alternatives: The 2003 Aurora's V8 power comes with a huge fuel penalty. Other cars worth considering are the Acura TL, Ford Crown Victoria and Mercury Marquis, Lexus ES 300, Lincoln Town Car, Nissan Maxima, and Toyota Avalon. **Options:** The traction control and anti-skid system. Bucket seats provide more comfortable support. **Rebates:** Discounts, $3,500+ rebates, and zero percent financing on all leftover 2003 models. **Delivery/PDI:** $1,050. **Depreciation:** Faster than average. **Insurance cost:** Higher than average. **Parts supply/cost:** Moderately priced parts that aren't hard to find. **Annual maintenance cost:** Above average. **Warranty:** Bumper-to-bumper 3 years/60,000 km; rust perforation 5 years/160,000 km. **Supplementary warranty:** A good buy, considering the car's powertrain deficiencies and the closing down of the Oldsmobile franchise. **Highway/city fuel economy:** 8.4–13.6L/100 km.

Quality/Reliability/Safety

Pro: Warranty performance: Average. The shutting down of the Oldsmobile division will mean there will be fewer dealers to support your claim for goodwill warranty assistance.

Con: Quality control: Terrible. One gets the impression that GM's thinking is "out of sight, out of mind." GM has given up on this "has been's" factory-related defects. **Reliability:** Below average. Owners can't rely on Auroras that carry failure-prone transmissions. **Owner-reported problems:** Chronic engine and automatic transmission malfunctions and breakdowns; poor braking performance; and subpar body construction. Interior and exterior fit and finish just doesn't measure up to the competition; creaks, squeaks, and rattles are common. **Service bulletin problems:** Intermittent no-start, no-crank condition; engine overheating in cold weather; sudden engine shutdown; poor transmission and engine performance may be caused by debris in the transaxle valve body and case oil passages, says TSB #02-07-30-013; incorrect First gear ratio; delayed Reverse engagement; harsh shifting upon start-up; transmission whining noise and cooling line leaks; leakage from the quick-connect fitting at the case cover; no Fourth gear, or slipping in Fourth gear; oil leakage from the oil level sensor; intermediate shaft clunk; water contamination of the ABS sensor; excessive vibration on smooth roads; broken sunroof deflectors; faulty windshield wipers; horn blows on its own, or refuses to blow; loose, sagging attachment arms/straps in the rear compartment seatback trim panel. **NHTSA safety complaints/safety:** Vehicle suddenly accelerated when started up and put into Reverse; sudden loss of steering; when turning, feels like tire is rubbing underbody; high headrest obstructs rear visibility; interior and exterior lights go out periodically; rear tail lights dim intermittently; ABS and traction control warning lights come on for no reason; will not go into First gear with cupholder extended; defrost grate reflects on to the windshield.

Road Performance

Pro: Acceleration/torque: Impressive acceleration with both the base V6 and optional V8 engines, although the larger engine makes for easier passing and merging with traffic. **Transmission:** Smooth and quiet shifting. **Routine handling:** Better-than-average handling and ride quality. **Braking:** Excellent braking, although pedal effort is a bit high.

Con: Steering: GM's Magnasteer system lacks road feel. **Emergency handling:** Some excessive body lean and jostling when pushed; not as nimble as the competition. Firm suspension may be uncomfortable for some. Brake pedal may be hard too modulate due to its firmness.

Comfort/Convenience

Pro: Standard equipment: Very well appointed. Impressive styling combines with a roomy interior. Well thought-out interior ergonomics. Plush interior includes comfortable front seats and generous amounts of headroom and legroom. **Driving position:** Excellent. **Controls and displays:** User-friendly, except for radio controls. **Climate control:** Efficient and quiet. **Interior space/comfort F/R:** Roomy interior provides generous room for five adults. Front seating is exceptionally comfortable, with just the right amount of thigh and lumbar support. **Cargo space:** Lots of storage areas that are easily accessed. **Trunk/liftover:** Large trunk offers a large and low opening. **Quietness:** Excellent soundproofing keeps out engine, wind, and tire noise.

Con: Rear visibility is obstructed by the high tail and the centre-mounted brake light. Audio system and climate controls aren't easily adjusted. **Entry/exit:** Five-passenger seating can't compete with Avalon's six-passenger capacity, and the middle occupant won't be comfortable for long. Rear seat cushion is too small and not sufficiently padded.

Aurora

List Price (very negotiable)	**Residual Values** (months)			
	24	**36**	**48**	**60**
Aurora: $46,590 (26%)	$31,000	$25,000	$22,000	$15,000

Technical Data

Powertrain (front-drive)
Engine: 4.0L V8 (250 hp)
Transmission: 4-speed auto.
Dimension/Capacity
Passengers: 2/3
Height/length/width: 56.7/199.3/72.9 in.
Headroom F/R: 38.6/37.7 in.
Legroom F/R: 42.5/38 in.
Wheelbase: 112.2 in.
Turning circle: 43 ft.
Cargo volume: 14.9 cu. ft.
Tow limit: 3,000 lb.
Fuel tank: 76L/prem.
Weight: 3,627 lb.

Safety Features/Crashworthiness

	Std.	Opt.
Anti-lock brakes (4W)	■	—
Seat belt pretensioners F/R	■	—
Side airbags	■	—
Traction control	■	—
Head restraints F/R	****	****
Visibility F/R	*****	*****
Crash protection (front) D/P	****	****
Crash protection (side) F/R	***	****
Crash protection (offset)	*****	—

CORVETTE 

RATING: *Base model:* Average—a brawny, bulky sport coupe that's slowly evolving into a more refined machine. *Z06:* Below Average—better suited to the racetrack than the 401 or Burrard Street bridge. Uncomfortably harsh suspension and more road/engine noise than you'll find with the base Corvette. **Strong points:** Powerful powertrain; easy handling; user-friendly instruments and controls; and lots of standard convenience features. **Weak points:** Poorly performing "skip shift" manual gearbox; hard ride (Z51 or Z06 suspension); limited rear visibility; inadequate storage space; no spare tire (you get a can of tire sealant); poor-quality fit and finish; cabin amenities and materials not up to the competition; large number of complaints of sudden steering wheel lock-up (1997–2001 models). **New for 2004:** Nothing. GM is putting all its money into a more powerful and better-performing 2005 C6 model due to arrive in mid-2004. Spy photos show a sleeker, cleaner body, non-retractable headlights, and unusually large side-view mirrors.

OVERVIEW: Introduced in 1953 as a futuristic show car, and the first American car to use fuel injection (in 1957), Chevrolet has sold more than a million Corvettes with an estimated 600,000 still on the roads. Its peak year was 1984, when over 84,000 coupes and convertibles were sold in North America for about a third of today's price.

Corvette returns with its base 350-hp 5.7L LS1 power plant, harnessed to either a 4-speed automatic or a 6-speed manual transmission. The engine is made from aluminum, and most transmission components have been shifted to the rear for better handling and increased passenger room. There are three body styles available: a convertible with manual fabric top and glass rear window; a hatchback coupe with removable roof panel; and a fixed-roof Z06 hardtop.

The high-performance Z06 returns along with additional options for the coupe and convertible, including a head-up display that projects instrumentation readouts onto the windshield; a power telescoping steering column; and an active handling system. The Z06's 405-hp V8 only comes with a 6-speed manual transmission.

Cost analysis/best alternatives: Get the 2003 base model, unless you really want the Z06's extra horses and kidney-pounding suspension. Keep in mind that premium fuel and astronomical insurance rates will further drive up your operating costs. Other sporty models worth considering: the Dodge Viper, or a Porsche 911 or Boxster. **Options:** Remember, performance options rarely increase performance to the degree promised by the seller; the more performance options you buy, the less comfort you'll have, more things can go wrong, and simple repairs increase in complexity. Stay with the 6-speed manual transmission and base FE1 suspension. Although you may find the Z51 set-up a bit harsh, you will definitively find the Z06 suspension set-up uncomfortably firm. Last year's ineffective, driver-adjustable Selective Real Time Damping option isn't worth the extra bucks. Run-flat tires from Goodyear are an excellent investment. Forget about the head-up instrument display that projects speed and other data onto the windshield; it's annoyingly distracting and is usually turned off. **Rebates:** The Corvette is so popular that GM normally doesn't have to offer rebates to boost sales, but there will be lots of discounting early in 2004 as buyers wait for the 2005's arrival. **Delivery/PDI:** $850. **Depreciation:** A bit slower than average, but nowhere near the resale prices often quoted by salespeople. **Insurance cost:** Astronomical. **Parts supply/cost:** Good availability, but parts are pricey. **Annual maintenance cost:** Higher than average. **Warranty:** Bumper-to-bumper 3 years/60,000 km; rust perforation 6 years/160,000 km. **Supplementary warranty:** A smart idea to protect you from the steering lock-up and oil-burning, erratic tranny shifts. **Highway/city fuel economy:** 8.8–14.1L/100 km with the base engine and an automatic transmission.

Quality/Reliability/Safety

Pro: Warranty performance: Above average. Key-controlled lockout feature discourages joy riding by cutting engine power in half. Both Corvette versions are equipped with an impressively effective PASS KEY theft-deterrent system that uses a resistor pellet in the ignition to disable the starter and fuel system when the key code doesn't match the ignition lock. **Safety:** The tires have built-in low-pressure sensors that warn the driver by way of a light on the centre console. Run-flat tires eliminate the need for a spare tire.

Con: Quality control: Below average, especially body construction. **Reliability:** Spotty. The Corvette's sophisticated electronic and powertrain components have low tolerance for real-world conditions. Expect lots of visits to the dealer's repair bays. Steering wheel lock-ups will also increase your familiarity with dealer service personnel. **Owner-reported problems:** Excessive oil consumption; harsh transmission shifts; 2–4 band and 3–4 clutch damage; rear differential and transmission pump leaks; noisy transmission engagement; light brake drag; brake light remains lit; B-pillar wind noise. Electronically controlled suspension systems have been glitch-plagued over the past several years. Squeaks and rattles due to the car's structural deficiencies. The car was built as a convertible, and therefore has too much body flex. Servicing the different sophisticated fuel-injection systems isn't easy, even (especially) for GM mechanics. **Service bulletin problems:** Engine knocking or lifter noise (see below).

Engine Knocking or Lifter Noise

Bulletin No.: 02-06-01-038 **Date:** December 2002

Engine knock or lifter noise (Replace o-ring)

2001–02 Chevrolet Camaro
2001–03 Chevrolet Corvette
2001–02 Pontiac Firebird
2002–03 Cadillac Escalade, Escalade EXT
2000–03 Chevrolet Suburban, Tahoe
2001–03 Chevrolet Silverado
2002–03 Chevrolet Avalanche
2000–03 GMC Yukon, Yukon XL
2001–03 GMC Sierra
with 4.8L, 5.3L, 5.7L, or 6.0L V8 engine (VINs V, T, Z, G, S, N, U - RPOs LR4, LM7, L59, LS1, LS6, LQ9, LQ4)

Condition: Some customers may comment on an engine tick noise. The distinguishing characteristic of this condition is that it likely will have been present since new, and is typically noticed within the first 161–322 km (100-200 mi.). The noise may often be diagnosed as a collapsed lifter. Additionally, the noise may be present at cold start and appear to diminish and then return as the engine warms to operating temperature. This noise is different from other noises that may begin to occur at 3219–4828 km (2000–3000 mi.).

Cause: The o-ring seal between the oil pump screen and the oil pump may be cut, causing aeration of the oil.

Cracked or broken transmission case; inoperative AC control; deck lid not flush with left-hand quarter panel; exhaust system noise (see below).

Exhaust System "Jingling" Noise

Bulletin No.: 03-06-05-005 **Date:** April 2003

Subject: "jingle"-type noise coming from exhaust system (fix not yet available)

2002–03 Chevrolet Corvette

Condition: Some customers may comment that a "jingle"-type noise can be heard coming from the exhaust system. The noise is similar to the sound made by shaking a key ring. This noise occurs most often on a deceleration.

Correction: Engineering is currently developing a permanent fix. Once the fix is available, this bulletin will be revised.

NHTSA safety complaints/safety: Government officials have opened an investigation into hundreds of owner reports of sudden steering wheel lock-up while driving on the highway or upon start-up (tow required). This safety hazard has been around for years and proves the point that GM has learned little since its Corvair fiasco.

Catalytic converter caught fire; when fuel tank is full, full leaks from the top of the vent; steering wheel locked up upon start-up; steering wheel locked up in Reverse gear and was corrected only when transmission was put into Park and key re-inserted into the ignition; "Service Column Lock" indication has been a common problem; chronic stalling and engine surging; fuel-injector failures cause vehicle to shudder and stall; engine dies while driving in the rain and brakes don't work; brakes drag and lock up; brake pedal doesn't spring back; overheated rotors; car is nearly uncontrollable at time of brake lock-up; early failure of the engine serpentine belt and tensioner; intermittent electrical problems degrade computer operation; if one wheel loses traction, the throttle closes, starving the engine; excessive cabin heat, even with AC set to Max; driver's seat moves while driving; warped trunk door; seat belts twist easily and tend to pull down uncomfortably against the shoulder; passenger seat belt jams and won't extend or retract; smelly fumes enter the cabin, causing watery eyes and dizziness; dash reflects onto the windshield; the glass rear-view window limits rear vision; front and rear wheel weights fly off the wheels.

Road Performance

Pro: Acceleration/torque: Gobs of torque and horsepower with a top speed of 277 km/h (172 mph). Base engine is fast (0–100 km/h: 5 seconds), but the Z06 is faster (0–100 km/h: 3.9 seconds). **Transmission:** The 6-speed gearbox performs well in all gear ranges and makes shifting smooth, with short throws and easy entry into all gears. **Emergency handling:** Very good, especially with enhanced side-slip angle control to prevent skidding and provide better traction control. No oversteer, wheel spinning, breakaway rear ends (are you listening, Ford?), or nasty surprises. **Steering:** Predictable, with good road feel.

Routine handling: Above average. The car is so low that its front air dam scrapes over the smallest rise in the road, but still gives no-surprise handling and responds quickly to the throttle. The Bilstein FX-3 Selective Ride Control suspension can be pre-set to Touring, Sport, or Performance. Under acceleration, an electronic module automatically varies the suspension as the speed increases. **Braking:** Disc/disc; better than average. The ABS-vented disc brakes are easy to modulate and fade-free.

Con: Optional Z51 suspension's ride is too firm for some and the Z06 suspension is too firm for many. This is one large and heavy sports car, whose weight compromises its fuel economy. True, the 6-speed manual does provide quicker acceleration and better fuel economy, but it's not particularly user-friendly.

Comfort/Convenience

Pro: Standard equipment: Attractively styled and full of power-assisted accessories. Defrosted rear glass window. Rust-resistant fibreglass body. Simple-to-operate convertible top. Innovative passive keyless entry system. **Driving position:** Excellent, with everything within easy reach, good front and rear visibility, and a roomy cockpit for average-statured adults. **Controls and displays:** Complete instrumentation that's easy to decipher, and user-friendly controls. **Climate control:** Easy to adjust and quiet. **Interior space/comfort F/R:** Snug, but not uncomfortable for two adults.

Con: Steering wheel hides some controls. Obtrusive roof pillars cut visibility to all corners. No room in the engine compartment for a sophisticated climate control system. Dual-zone climate control system takes time to warm up. Lowering the top takes patience: you release two latches, lift the plastic panel, and manually fold the top into its storage compartment. **Entry/exit:** Crouch and crawl; like getting into a space capsule. Tall drivers will find the interior confining. **Cargo space:** Insufficient storage space, which is no surprise to sports car enthusiasts. The lowered top cuts trunk space in half. **Trunk/liftover:** Skimpy storage space; high sill makes loading baggage difficult. **Quietness:** In addition to an irritating engine boom, there's excessive tire noise, wind whistling through the car's A- and C-pillars, and the all-too-familiar fibreglass body squeaks caused by excessive body flexing on rough surfaces.

Corvette

List Price (negotiable)	**Residual Values** (months)			
	24	**36**	**48**	**60**
Corvette: $69,940 (25%)	$47,000	$38,000	$31,000	$26,000
Z06: $77,440 (25%)	$53,000	$45,000	$37,000	$32,000

Technical Data

Powertrain (rear-drive)
Engines: 5.7L V8 (350 hp)
• 5.7L V8 (405 hp)
Transmissions: 6-speed man.
• 4-speed auto.
Dimension/Capacity
Passengers: 2/0
Height/length/width:
47.7/179.7/73.6 in.
Headroom: 37.8 in.
Legroom: 42.7 in.
Wheelbase: 104.5 in.
Turning circle: 42 ft.
Cargo volume: 13.3 cu. ft.
Tow limit: N/A
Fuel tank: 72L/prem.
Weight: 3,215 lb.

Safety Features/Crashworthiness

	Std.	Opt.
Anti-lock brakes	■	—
Seat belt pretensioners F/R	■	—
Side airbags	—	—
Traction control	■	—
Head restraints F/R	***	***
Visibility F/R	***	**

DEVILLE ★★★

RATING: Average. As Cadillac reinvents itself in a futile attempt to lure younger buyers, its cars are looking less and less like traditional "caddies" and more like Lego toys. One gets the impression that Cadillac designers aren't sure of their market, so they're trying the angular, slab panel styling just to be different. They are also loading up the cars with every conceivable gadget imaginable to project an image of daring innovation, instead of desperate experimentation. **Strong points:** Brisk acceleration, with the large V8, and handling is fairly good for a car of this size. **Weak points:** Excessive torque steer (twisting to the point that the steering wheel is jerked to one side) when accelerating is remedied via the traction control system by reducing engine power and selectively applying the brakes (feeling confident?). Body and mechanical components aren't top quality and servicing can be complicated. **New for 2004:** Heated and cooled front seats, a heated steering wheel, a CD stereo system, and a manual parking brake release. The adaptive seat option has been dropped.

OVERVIEW: Cadillac's largest car (no, the Escalade is an SUV), the DeVille comes in three model configurations: the Base and DHS models, and the sporty DTS. A powerful 300-hp V8 engine coupled to a 4-speed automatic transmission propels the DTS, while a 275-hp variant powers the other models. Other standard features include: traction control, four-wheel anti-lock disc brakes, front side airbags, and an anti-skid system (DTS).

DeVille owners have reported the following deficiencies: a persistent "rotten-egg" odour in the cabin, electrical failures causing hard starts, poor idling, and constant stalling, right side rear wheel fell off due to a broken axle, tire jack collapse, side door airbag deployed while driving, airbag light stays lit, brake failures, faulty tire-pressure sensors, frequent tire blowouts, turn signals don't self-cancel, distorted windshields, front seat shoulder strap crosses at the neck, and too-short seat belts.

Deville

List Price (very negotiable)	**Residual Values** (months)			
	24	**36**	**48**	**60**
Base: $55,685 (24%)	$40,000	$33,000	$27,000	$20,000

Technical Data

Powertrain (front-drive)
Engines: 4.6L V8 (275 hp)
• 4.6L V8 (300 hp)
Transmission: 4-speed auto.
Dimension/Capacity
Passengers: 3/3
Height/length/width: 56.7/207/74.4 in.
Headroom F/R: 39.1/38.3 in.
Legroom F/R1/R2: 42.4/43.2 in.
Wheelbase: 115.3 in.
Turning circle: 40.2 ft.
Cargo volume: 19.1 cu. ft.
Tow limit: 2,000 lb.
Fuel tank: 70L
Weight: 3,984 lb.

ASTRO, SAFARI ★★★★

Some buyers are turned off by the boxy styling.

RATING: Above Average. These vehicles are more mini-truck than minivan. Believe it or not, these run-of-the-mill minivans are beginning to look quite good when compared to the problem-plagued Chrysler and Ford minivans and the more-expensive Asian competition (VW? Not even in the running). They have fewer safety-related problems reported to the government, are easy to repair, and cost little to acquire. Stay away from the unreliable AWD models; they're expensive to repair and not very durable. **Strong points:** Brisk acceleration; trailer-towing capability; lots of passenger room and cargo space; well laid-out instrument panel with easy-to-read gauges. Very low ground clearance enhances this minivan's handling, but precludes most off-roading. Still, handles better than GM's front-drive extended-length minivans. Good brakes. Reasonable price, and improved reliability and quality control. **Weak points:** Driving position is awkward for most drivers; tall drivers will find the pedals too close; obtrusive engine makes for very narrow front footwells that give little room for the driver's left foot to rest. Difficult entry/exit due to the high step-up and the intruding wheelwell. Harsh ride, limited front seat room, interior noise levels rise sharply at highway speeds, and excessive fuel consumption made worse by the AWD option. **New for 2004:** Nothing significant; Astro was slated to be dropped this year, but may continue on until the 2005 model year.

OVERVIEW: More a utility truck than a comfortable minivan, these boxy, rear-drive minivans are built on a reworked S-10 pickup chassis. As such, they offer uninspiring handling, trouble-prone mechanical and body components, and relatively high fuel consumption. Both Astro and Safari come in a choice of either cargo or passenger van. The cargo van is used either commercially or as an inexpensive starting point for a fully customized vehicle. The Safari is identical to the Astro, except for a slightly higher base price.

The engine is a 4.3L 190-hp V6. Dual airbags are standard. Also offered are an optional rear door and rear bench seats that can be adjusted fore and aft. Carried-over standard features are a 4-speed automatic transmission with Overdrive, power steering, a front stabilizer bar, and anti-lock four-wheel disc brakes. All-wheel drive with a single-speed transfer case with viscous-controlled differential (no switches to throw) is available on the regular length and extended models. The Sport package includes louvred rear-quarter body panels, two-tone paint, foglights, and a front air dam. With the right options, the Astro and Safari have the advantage of being versatile cargo haulers when equipped with a heavy-duty suspension. In fact, Astro's 2,495 kg (5,500 lb.) trailer-towing capability is 907 kg (2,000 lb.) more than that of the front-drive Venture.

Cost analysis/best alternatives: Get a discounted 2003 model with the disc brake upgrade. Front-drive minivans made by Nissan and Toyota have better handling, and are more reliable and economical people carriers; unfortunately, they lack the Astro's considerable grunt, essential for cargo hauling and trailer towing. Also, consider getting a GM Montana or Venture if hauling isn't a high priority. **Options:** Integrated child safety seats, rear AC, and rear Dutch doors. Be wary of the AWD option; it exacts a high fuel penalty. **Rebates:** About $2,500 on all models. **Delivery/PDI:** $1,000. **Depreciation:** Average. **Insurance cost:** Slightly higher than average. **Annual maintenance cost:** Any garage can repair these rear-drive minivans. **Parts supply/cost:** Good supply of cheap parts. A large contingent of independent parts suppliers keeps repair costs down. Parts are less expensive than for other vehicles in this class. **Warranty:** Bumper-to-bumper 3 years/60,000 km; rust perforation 6 years/160,000 km. **Supplementary warranty:** A toss-up. Individual repairs won't cost a lot, but those nickels and dimes can add up. A powertrain-only warranty is a wise choice. **Highway/city fuel economy:** 10.6–14.6L/100 km for two-wheel drive and 11.3–15.3L/100 km with AWD.

Quality/Reliability/Safety

Pro: Quality control: Average. **Reliability:** Average; most of the Astro and Safari's defects are easy to diagnose, leading to a minimum of downtime in the repair bay. **Warranty performance:** Average. Four-wheel disc brakes are an important improvement, considering this minivan's poor braking performance in the past.

Con: Owner-reported problems: Some problems with the automatic transmission, steering, electrical system, heating and defrosting system, and suspension components. **Service bulletin problems:** Hard starts, rough idle, and intermittent misfiring (see below).

Engine Runs Rough

Bulletin No.: 02-06-04-059 **Date:** December 2002

Engine runs rough, Service Engine Soon light on, DTC's P0300 or P0335 set (shim crankshaft sensor)

2002–03 Chevrolet Astro, Blazer, Express, Silverado, and S-10 truck models 2002–03 GMC Jimmy, Safari, Savana, Sierra, and Sonoma truck models with 4.3L V-6 engine (VINs W, X - RPOs L35, LU3)

Condition: Some owners may comment on a rough running condition or a Service Engine Soon (SES) light being illuminated. Upon investigation, the technician may find a DTC code P0300 or P0335.

Cause: The crankshaft sensor may be contacting the reluctor wheel.

A transfer-case shudder can be fixed inexpensively (see below).

Transfer-Case Shudder

Bulletin No.: 99-04-21-005B **Date:** September 2002

Transfer-case shudder (flush and replace transfer-case fluid)

1999–2003 Chevrolet Astro (AWD)
1999–2003 GMC Safari (AWD)
1999–2001 Oldsmobile Bravada
with all-wheel drive (NV136) transfer case (RPO NP4)

Condition: Some customers may comment about a vibration (shudder) during tight turning manoeuvres on dry pavement at vehicle speeds less than 8 km/h (5 mph).

Cause: This condition may be caused by a slip-stick of the clutch plates in the transfer case clutch pack. On 1999 and newer models, the preload that is designed into the clutch pack may contribute to the condition. In addition, contamination of the fluid may interfere with the friction modifiers.

Right rear door handle breakage. GM says that a chronic driveline clunk can't be silenced and is a normal characteristic of its vehicles (yikes!). **NHTSA safety complaints/safety:** Sudden acceleration; chronic stalling; hard starts; cranks, but would not start due to broken electrical connection; sudden total electrical failure, especially when going into Reverse; airbag failed to deploy; harsh, delayed shifting; on a slight incline sliding door will unlatch and slam shut; driver's cargo door latch and hinge slipped off and door flew open; loss of power steering; sudden steering and brake loss when power-steering pump shaft snapped; when applying brakes, pedal stiffens and then vehicle surges ahead; extended stopping distance with ABS, especially in rainy weather; brake pedal set too high; differential in transfer case locked up while driving; defective axle seals; vehicle rolls backward on an incline while in Drive (dealer adjusted transfer case to no avail); front passenger seat belt locked up; fuel gauge failure; faulty AC vents; intermittent windshield wiper failure; and water can be trapped inside the wheels and freeze, causing the wheels to be out of balance.

Road Performance

Pro: Acceleration/torque: The V6 engine is more than adequate for most driving chores and has plenty of reserve power for trailer towing and heavy hauling (0–100 km/h: 11.8 seconds). **Braking:** Disc/disc; excellent braking.

Con: The V6 is thirsty in city driving. **Emergency handling:** Ponderous, but still fairly predictable. **Steering:** Very light power steering, but still handles and manoeuvres like a large truck. **Transmission:** Clunky automatic transmission. **Routine handling:** Handling isn't very precise; overall behaviour is competent but sloppy. A heavy-duty suspension will improve both ride and handling. Busy, harsh ride caused by stiff springs.

Comfort/Convenience

Pro: Standard equipment: Adequate. **Controls and displays:** Well laid-out instrument panel with easy-to-read gauges. **Climate control:** Very good climate control system. **Interior space/comfort F/R:** Spacious interior provides plenty of room for passengers. All seats are comfortable, although the front two are the best. The rear seat can be removed by a single person. **Cargo space:** An incredible amount of cargo space to handle all kinds of bulky items. **Trunk/liftover:** No problem. In fact, the optional Dutch doors provide better rear-view visibility and includes a convenient lift-open glass and defroster.

Con: Driving position: The driving position is awkward for most drivers. Tall drivers will find the pedals too close. Obtrusive engine makes for very narrow front footwells that give little room for the driver's left foot to rest. Some dashboard controls are hard to reach. **Entry/exit:** Difficult due to the high step-up and the intruding wheelwell. **Quietness:** Interior noise levels rise sharply at highway speeds.

Astro, Safari

List Price (very negotiable)

Base: $29,125 (21%)

Residual Values (months)

24	36	48	60
$19,000	$16,000	$13,000	$11,000

Technical Data

Powertrain (rear-drive)
Engine: 4.3L V6 (190 hp)
Transmissions: 4-speed auto.
• AWD

Dimension/Capacity
Passengers: 2/3/3
Height/length/width: 75/189.8/77.5 in.
Headroom F/R1/R2: 39.2/37.9/38.7 in.
Legroom F/R1/R2: 41.6/36.5/38.5 in.
Wheelbase: 111.2 in.
Turning circle: 40.5 ft.
Cargo volume: 98 cu. ft.
GVWR: 5,600 lb.
Payload: 1,647 lb.
Tow limit: 5,700 lb.
Fuel tank: 95L/reg.
Weight: 3,964 lb.

Safety Features/Crashworthiness

	Std.	Opt.
Anti-lock brakes (4W)	■	—
Seat belt pretensioners F/R	■	—
Side airbags	—	—
Traction control (AWD)	—	■
Head restraints F/R	*	*
Visibility F/R	*****	**
Crash protection (front) D/P	***	*
Crash protection (offset)	*	—

MONTANA, SILHOUETTE, VENTURE

RATING: Above Average, thanks to GM's extra warranty protection for the powertrain. **Strong points:** A comfortable ride, easy handling, dual sliding doors, plenty of comfort and convenience features, flexible seating arrangements, good visibility fore and aft, lots of storage bins and compartments, and good crash scores. One of the quietest minivans in its class. **Weak points:** Average acceleration with a light load; unproven AWD; less effective rear drum brakes and excessive brake fading; tall drivers will find insufficient headroom, and short drivers may find it hard to see where the front ends; low rear seats force passengers into an uncomfortable knees-up position; narrow cabin makes front-to-rear access a bit difficult; seat cushions on the centre and rear bench seats are hard, flat, and too short, and the seatbacks lack sufficient lower back support; cargo may not slide out easily, due to the rear sill sticking up a few inches; a high number of safety-related failures and engine exhaust manifold defects; and disappointing fuel economy. **New for 2004:** An upgraded audio system.

OVERVIEW: These practically identical front-drive, mid-sized minivans are slightly longer, though narrower, than a Dodge Caravan and come in regular and extended wheelbase versions. The extended wheelbase models get more bells and whistles, including optional power slide-open passenger- and driver-side doors. The Venture and Silhouette seat seven, while the Montana manages to squeeze in one more passenger. Longer versions have dual sliding doors with optional power assist.

Cost analysis/best alternatives: Get a 2003 leftover discounted 15–20 percent. Other vehicles worth considering: Honda Odyssey, Mazda MPV, Nissan Quest, and Toyota Sienna. **Options:** GM's 2004 extended-length minivans now offer $5,000 (U.S.) "Sit-N-Lift" electrically powered second seats for the physically challenged. Installed by the dealer, the seat pivots, extends, and lowers to the side of the vehicle, allowing the passenger to access the vehicle while sitting down. The seats will support passengers weighing up to 135 kg (300 lb.); can be used in any extended-length GM minivan going back to the 1997; don't require new holes for installation; can be removed or replaced with a standard seat prior to trade-in; can be used on leased vehicles; and qualify for GM warranties and special programs. Be wary of the AWD option. Since GM put it on Astro vans, premature failure complaints have soared. Many of these vans are equipped with Fire-stone tires—have them removed by the dealer, or ask for an additional $400 rebate. Integrated child safety seats are generally a good idea. A power-assisted passenger-side sliding door is both convenient and dangerous—despite its override circuit that prevents the door from closing on a hand, a number of injuries have been reported. Furthermore, with most sliding doors, mechanical and electronic glitches allow the doors to open when they shouldn't, and they can be difficult to close. Overall, this $1,200 convenience feature is overpriced, failure-prone, and a bit slow in operation. Consider the $235 load-levelling feature—a must-have for front-drive minivans. It keeps the weight on the front wheels, giving you better steering, traction, and braking. Other options that are worth buying: traction control, power side windows, and rear air conditioner, defroster, and heater. The Montana Package offers a firmer suspension and self-sealing tires. **Rebates:** $3,000 and zero percent financing on all models. **Delivery/PDI:** $1,000. **Depreciation:** A bit slower than average. **Insurance cost:** High, but about average for a minivan. **Annual maintenance cost:** Average during the warranty period. Engine, transmission, ABS, and electrical malfunctions will likely cause maintenance costs to rise after the third year of ownership. **Parts supply/cost:** Parts are generic to GM's other models, so they should

be reasonably priced and not hard to find. Body parts are likely to be more problematic and costly. **Warranty:** Bumper-to-bumper 3 years/60,000 km; powertrain 5 years/ 100,000 km; rust perforation 6 years/160,000 km. **Supplementary warranty:** Not needed. **Highway/city fuel economy:** 9.3–13.5L/100 km.

Quality/Reliability/Safety

Pro: Quality control: Average. 1996–2003 models have chronic powertrain problems highlighted by engine manifold and head gasket failures along with frequent automatic transmission breakdowns and clunky shifting. Body components are less rattle-prone. **Reliability:** Average. **Warranty performance:** Average.

Con: Fit and finish is still not up to the Asian competition. Chassis could still use reinforcing to mute the rattles heard when passing over rough roads. **Owner-reported problems:** Some engine head gasket and intake manifold defects, EGR valve failures, electrical glitches, excessive front brake noise and frequent repairs (rotors and pads), blurry front windshield, and assorted body deficiencies. **Service bulletin problems:** Engine intake manifold leaks (see Buick Century); poor engine and automatic transmission operation (see below):

> I just wanted to let you know that after contacting you back in January regarding our 2001 Chevy Venture head gasket problem, I have just received my judgment through the Canadian Arbitration Program.
>
> I used the sample complaint letter as well as the judgment you have posted in the *Ford Canada v. Dufour* court case. This, combined with an avalanche of similar Chevy Venture complaints that are posted on the Internet, helped us to win a $1,700 reimbursement of the $2,200 we were looking for.
>
> The reason for our not receiving the full amount is that the arbitrator stated that GM Canada would have only replaced one head gasket instead of replacing both as we had done, and that a dealer would have supplied us with a car free of charge and therefore did not allow us the car rental expense we incurred.
>
> We are still extremely happy with the results and thank you for your books and website; you have a fan for life.
>
> D.W.

Transmission shudder, chuggle, hard shifting, and won't downshift (see Cavalier section); delayed gear engagement; defective catalytic converters that cause a rotten-egg smell in the interior may be replaced free of charge under the emissions warranty; roof rusting, leaking, or rust perforation may require a new roof under GM's "Special Policy," (see following).

Front/Rear Roof Rust Perforation

Bulletin No.: 02-08-67-006B **Date:** March 2003

Roof perforation (replace roof)

1997–2003 Chevrolet Venture
1997–2003 Oldsmobile Silhouette
1997–2003 Pontiac TranSport/Montana

Important: Implementation of this service bulletin by "GM of Canada" dealers requires prior District Service Manager approval.

Condition: Some customers may comment that there is rust forming around the front or rear portion of the roof.

Cause: During production, the E-coating (ELPO primer) may have been missed in concealed areas of the front or rear portions of the outer roof panel.

Water in jack compartment; troubleshooting assorted squeaks and rattles; static noise from radio speakers; exhaust system rattle or buzz; grinding or growling when in Park on an incline; and repair tips for a faulty AC compressor. **NHTSA safety complaints/safety:** Fire ignited under driver's seat; airbag deployed when door was slammed; airbags failed to deploy; windshield suddenly exploded outward while driving with wipers activated; Firestone tire blowout; brakes activate on their own, making it appear as if van is pulling a load; steering idler arm fell off due to missing bolt; loose fuel tank due to loose bolts/bracket; fuel tank cracked when passing over a tree branch; plastic tube within heating system fell off and wedged behind the accelerator pedal; bracket weld pin that secures the rear split seat sheared off; faulty fuel pump causes chronic stalling; no-starts; surging; sudden acceleration; vehicle suddenly lost power while going uphill, slid back, and stalled; van will roll back while in gear on an incline; Service Engine light comes on constantly; Service Engine light remains on due to transmission bearing assembled backwards in the transmission box; premature transmission failures; power-sliding door has a gap at the base large enough to trap a child; when reversing, the door will not stop; sliding door slams shut on an incline—person's wrist broken when power-sliding door opened; broken sliding door track trim panel; centre rear lap seat belt isn't long enough to secure a rear-facing child safety seat; middle-row passenger-side seat belts jam in the retracted position; front passenger shoulder belt cuts into passenger's neck; children can slide out of the integrated child safety seat; rear seat belts failed to release; rear hatch handle broke, cutting driver's hand; electrical harness failures result in complete electrical shutdown; headlights, interior lights, gauges, and instruments fail intermittently (electrical cluster module is the prime suspect); excess padding around horn makes it difficult to depress horn button in an emergency; weak-sounding horn; loud noise emanating from underneath the vehicle; frequent windshield wiper motor failures; heater doesn't warm up vehicle sufficiently; antifreeze smell intrudes into the interior; and the weld that holds the lift wheel pin is not adequate to support the weight of the trailer.

Road Performance

Pro: Emergency handling: Better than average, due to the responsive powertrain, steering, and ABS. **Steering:** The steering takes some effort, but it's responsive and fairly predictable. Longer, loaded versions have a smoother ride than the base versions, which tend to be choppier. **Acceleration/torque:** Nice powertrain set-up with smooth acceleration for most light-duty work. The 3.4L engine is smooth and responsive (0–100 km/h: 10.7 seconds). These minivans use the same quiet-running V6 power plant that harnesses a few more horses than the Chrysler minivans' top-line 3.8L 6-cylinder, providing good mid-range and top-end power. Chrysler's engines do have a bit more torque, however. **Transmission:** The electronically controlled 4-speed automatic transmission shifts smoothly and quietly—another advantage over Chrysler and Ford. **Routine handling:** Pretty good. Sedan-like handling is better than most of the competition, particularly when equipped with the load-levelling option. Some body roll in hard turns. **Braking:** Unimpressive for the Venture, much better with the Montana and Silhouette.

Con: Some body roll in hard turns. Powertrain isn't suitable for heavy towing or carrying a full passenger load. Automatic transmission doesn't hold well on hills. **Acceleration/torque:** The GM engine is hampered by less torque, making for less grunt when accelerating and frequently downshifting out of Overdrive when climbing moderate grades (there's no Overdrive on/off switch). This is the main reason why owners feel that the GM minivans are underpowered when compared with Chrysler's. **Braking:** Disc/drum; watch out for excessive brake fading; the brakes progressively lose their effectiveness after repeated application.

Comfort/Convenience

Pro: Standard equipment: Fairly well equipped. **Driving position:** Drivers are treated to a carlike driving position. Comfortable left footrest. Concealed windshield wipers, a large windshield, and larger side mirrors enhance visibility fore and aft. **Controls and displays:** Everything is easy to read and within reach. **Climate control:** Efficient and easy to use. AC has a pollen control feature. **Entry/exit:** Easy, especially with the low interior step-in, side interior door handles, power remote sliding door on the curb side and a manual sliding door on the street side of extended-wheelbase models. **Interior space/comfort F/R:** Plenty of room in the first two rows, while the third row is more problematic. Most seats are quite comfortable, particularly on the Venture. There are two reclining bucket seats in the front row, and the second and third rows can accommodate modular or bench seats. Easy-to-fold seats drop down to increase storage space; the second seat flips forward for access to the third seat. **Cargo space:** Lots of storage bins and compartments. **Trunk/liftover:** A breeze. Dual sliding doors mean that you no longer have to walk around to the passenger side to load or unload cargo. Rear seats can be flipped down to carry a 4' x 8' sheet of plywood. **Quietness:** One of the quietest minivans in its class.

Con: Tacky plastic interior. **Driving position:** Tall drivers will find insufficient headroom and short drivers may find it hard to see where the front ends. **Controls and displays:** Steering column stalk controls are confusing. **Climate control:** These minivans take a while to warm up the interior. **Interior space/comfort F/R:** Uncomfortable centre and rear bench seats; cushions are hard, flat, and too short; and the seatbacks lack sufficient lower back support. Low rear seats force passengers into an uncomfortable knees-up position. Narrow cabin makes front-to-rear access a bit difficult. **Cargo space:** One convenience item that can easily become an inconvenience: The two cargo nets found between the front seats and behind the rear seat are easily entangled and complicate the removal of the rear seats. Also, eight of the 12 cupholders are only usable if the seats are folded down. **Trunk/liftover:** Cargo may not slide out easily due to the rear sill sticking up a few inches. **Quietness:** Some wind noise.

Montana, Silhouette, Venture

List Price (negotiable)	**Residual Values** (months)			
	24	**36**	**48**	**60**
Venture: $26,465 (20%)	$18,000	$15,000	$13,000	$10,000
Montana: $29,165 (20%)	$20,000	$17,000	$15,000	$13,000
Silhouette GL: $36,075 (23%)	$26,000	$21,000	$17,000	$15,000

Technical Data

Powertrain (front-drive)
Engine: 3.4L V6 (185 hp)
Transmissions: 4-speed auto.
• AWD
Dimension/Capacity (Silhouette; Montana and Venture)
Passengers: 2/2/3; 2/3/3
Height/length/width: 67.4/187.3/72 in.
Headroom F/R1/R2: 39.9/38.9/38.8 in.
Legroom F/R1/R2: 39.9/36.9/34 in.
Wheelbase: 112 in.
Turning circle: 37.4 ft.
Cargo volume: 119.8 cu. ft.
GVWR: 5,357 lb.
Payload: 1,415–1,577 lb.
Tow limit: 3,500 lb.
Fuel tank: 76L/95L
Weight: 3,803 lb./3,942 lb.

Safety Features/Crashworthiness

	Std.	Opt.
Anti-lock brakes (4W)	■	—
Seat belt pretensioners F/R	■	—
Side airbags	■	—
Traction control	—	■
Head restraints F/R	***	***
Visibility F/R	*****	*****
Crash protection (front) D/P	****	****

Crash protection (side) F/R	*****	****
Crash protection (offset)	*	—

General Motors/Daewoo

You've gotta be kidding!

General Motors doesn't intend to keep these cars around long. They are just handy filler while GM realigns its small-car lineup. The automaker bought Daewoo's bankrupt car division early last year, froze out the existing Daewoo dealer network, and is now sneaking in Daewoos under the Chevy and Suzuki banner.

GM's Aveo: Here today, gone tomorrow.

Daewoo quality control was a joke several several years ago, and it's a joke now. But Daewoo owners aren't laughing. These are low-quality vehicles, not supported by any indigenous dealer structure, and likely to be gone in the near future as GM makes better use of its Toyota and other offshore connections.

GM has never marketed another company's cars and made money doing so (remember the 70s Firenza?). The GM/Daewoo subterfuge of selling Daewoos as re-badged Suzukis and Chevys will inevitably cut into GM's own compact car sales, ticking off dealers who have invested heavily in Cavalier and Sunfire sales and servicing.

In Canada, Chevrolet dealers will sell the entry-level Aveo; the Optra, a compact based on the Daewoo Lacetti; and the Epica, a mid-sized vehicle based on Daewoo's Magnos. Suzuki will sell the subcompact Swift, based on the Kalos, and Verona.

Recommendation? Don't go near these cars during their first year on the market. Sure, they are relatively well equipped and attractively styled, but do you want to be to be saddled with what could be Daewoo's version of an 80s Hyundai Pony, Stellar, or Excel?

Aveo

List Price (negotiable)	**Residual Values** (months)			
	24	**36**	**48**	**60**
Sedan: $13,485 (11%)	$9,000	$7,000	$5,500	$4,500

Technical Data

Powertrain (front-drive)
Engine: 1.6L 4-cyl. (103 hp)
Transmissions: 4-speed auto.
Dimension/Capacity (Silhouette; Montana and Venture)
Passengers: 2/3
Height/length/width: 67.4/187.3/72 in.
Headroom F/R: 39.3/37.6 in.
Legroom F/R: 41.3/35. in.
Wheelbase: 97.6 in.
Turning circle: 16 ft.
Cargo volume: 11.7 cu. ft.
Tow limit: N/A
Fuel tank: 45L
Weight: 2,381 lb.

Safety Features/Crashworthiness

	Std.	**Opt.**
Anti-lock brakes	—	—
Seat belt pretensioners F/R	■	—

Optra

List Price (negotiable)	**Residual Values** (months)			
	24	**36**	**48**	**60**
Base: $16,190 (12%)	$12,000	$10,000	$8,000	$6,000

Technical Data

Powertrain (front-drive)
Engine: 2.0L 4-cyl. (119 hp)
Transmissions: 4-speed auto.
Dimension/Capacity
Passengers: 2/3
Height/length/width: 67.4/187.3/72 in.
Headroom F/R: 39 /37.8 in.
Legroom F/R: 42/36.7 in.
Wheelbase: 102.4 in.
Turning circle: 5.5 ft.
Cargo volume: 12.4 cu. ft.
Tow limit: 1,000 lb.
Fuel tank: 55L/reg.
Weight: 2,756 lb.

Safety Features/Crashworthiness

	Std.	**Opt.**
Anti-lock brakes (4W)	—	■
Seat belt pretensioners F/R	■	—

General Motors/Saab

Another Saab story

Beleaguered with an image that straddles the quirky and bland, Saab sales has been stuck in Neutral for the past decade and may actually have shifted into Reverse during the last year.

Nevertheless, the company continually makes unrealistic sales forecasts it never meets, posting important sales declines during a decade when prestige luxury cars showed unprecedented growth. So, after posting losses of about $2 billion since 1989, GM/Saab has decided to trim its worldwide dealer network from 1,700 to 900 dealers by the end of 2003. Additionally, the automaker will directly compete against its own dealers by setting up company-owned sales sites in several prime locations. Both of these moves are sure to anger the dealer body and further weaken Saab's marketing position and servicing capability.

Saabs are still original with their unique cockpit, console-mounted ignition and window switches, and innovative "night panel" that darkens all but the speedometer. Their unusual, aerodynamic styling is (to put it mildly) distinctive; passenger comfort is unbeatable; and they handle very well, with precise steering and lots of road feel. Unfortunately, lots of less expensive cars offer just as much road performance and are more reliable and easier to repair.

Saab's growth has been undeniably slowed by increased competition from European and Japanese luxury sedans that are more smartly styled, have greater market penetration, and are marketed through attractive and innovative leasing plans. In response to that competition, Saab has taken its products further upscale, dropping its bottom-end models and possibly bringing out its own sport-utility in the near future.

Saab's reliability and servicing problems have lessened through better assembly-line quality control and more experienced mechanics in the GM network. Nevertheless, electrical system and computer module malfunctions are common and hard to troubleshoot. This means your Saab can lose power or stall at any time, despite repeat trips for servicing.

9-3 ★★★

RATING: Average. The 9-3 is a driver's car, with impressive braking performance, sporty handling, and sure-footed cornering. **Strong points:** Many standard safety features; comfortable seating; and lots of storage capability. Generally good build quality and exterior finish. **Weak points:** No hatchback; some turbo lag and torque steer. Convertible has minimal rear seat room, less comfortable seating, and shakes and rattles. The servicing network is rather limited. Engine computer defects cut engine power and make the car dangerous to drive. **New for 2004:** A convertible version debuts this year and will be joined by a wagon in mid-2004.

OVERVIEW: Selling for $34,900 (plus a $1,050 freight and pre-delivery inspection fee), the 9-3 Linear Sport Sedan was totally redesigned last year. It is now set on a new platform and offers many new safety and convenience features. Rear passenger and trunk space, width, and wheelbase were also increased, although the hatchback was dropped in favour of traditional sedan styling.

The car's aerodynamic look is created by the 9-3's steeply raked front windshield and swept-back styling, two factors that limit forward visibility considerably. There are two 4-cylinder turbocharged engines available, a base 175-hp 2.0L, and a 210-hp variant. They are hooked to either a 5- or 6-speed manual transmission, with an optional automatic 5-speed with manual shifting capability (Sentronic) available for either power plant. The 9-3 has independent rear suspension and front MacPherson struts to give the car better-than-average handling and ride comfort. Safety has been enhanced through the use of disc brakes, electronic brake force distribution, cornering brake control, electronic stability, traction control, active head restraints, Seat belt pretensioners F/R, load limiting for all the outer seating positions, and dual-stage front airbags, seat-mounted side airbags, and side roof-rail airbags.

The 9-3 convertible is sold as either an entry level Arc or high-performance Aero. Both models carry a 210-hp 2.0L, turbocharged engine hooked to a 5-speed manual (standard on the Arc); a 6-speed manual (standard on the Aero); or an optional 5-speed Sentronic automatic.

Cost analysis/best alternatives: Inasmuch as the 2004s return relatively unchanged, a leftover 2003 9-3 or 9-5 is your best bet if discounted by about 20 percent. The Acura TL, Audi A4, Infiniti I30, and Volvo 40 series are attractive alternatives to the 9-3 hatchbacks; Mercedes' CLK320, the BMW 3 Series, and the 5-cylinder Volvo C70 are credible alternatives to the $54,000 9-3 Arc convertible ($56,500 for the Aero). Keep in mind that Saabs may be cheaper than their Audi and Volvo counterparts, but they depreciate far more quickly. **Options:** Integrated child safety seats. The GPS navigation aids can be useful, but don't waste your money on the power sunroof, leather upholstery, and heated seats. **Rebates:** About $3,000 before year's end. **Delivery/PDI:** $1,050. **Depreciation:** Faster than average. If you plan to keep yours for only a few years and don't want to lose much money through depreciation, choose the convertible; it keeps its value longer. **Insurance cost:** Slightly higher than average. **Annual maintenance cost:** Average. **Parts supply/cost:** Not easily found; costly at times. **Insurance cost:** Higher than average. **Parts supply/cost:** Parts are moderately expensive and are sometimes hard to find. **Annual maintenance cost:** Higher than average. **Warranty:** Bumper-to-bumper 4 years/80,000 km; rust perforation 6 years/unlimited km. **Supplementary warranty:** A very good idea, considering that these cars are complicated to diagnose and expensive to repair. **Highway/city fuel economy:** *2.0L and auto.:* 8.5–11.9L/100 km.

Quality/Reliability/Safety

Pro: Warranty performance: Above average. GM has treated Saab owners more generously than its other divisions, with the exception of Saturn and Cadillac.

Con: Quality control: Average. **Reliability:** Worse than average, due mainly to the cars' glitch-prone electronics. In fact, most of the 9-3's defects are electrical in nature, tending to take lots of diagnostic time. **Owner-reported**

problems: Major engine malfunctions caused by defective power control modules; electrical and fuel system malfunctions. **Service bulletin problems:** Inoperative starter motor and new synthetic oil to improve manual shift quality. Starter won't crank the engine and tips on improving the AC's temperature regulating mechanism. **NHTSA safety complaints/safety:** Chronic stalling and loss of power caused by faulty computer hardware and software; side airbag failed to deploy; airbag warning light remains lit; aluminum hubs may not support the stress of everyday driving; emergency brake failed to hold; when passing over grooves in the road, vehicle suddenly jerks to the right or left; driver's door lock won't unlock from the inside; Pirelli tire blowout; and OnStar radio failures.

Safety Features/Crashworthiness

	Std.	Opt.
Anti-lock brakes	■	—
Seat belt pretensioners F/R	■	—
Side airbags	■	—
Traction control	—	—
Head restraints F/R	*****	*****
Visibility F/R	*****	*
Crash protection (front) D/P	*****	*****
Crash protection (side) F/R	*****	****
Crash protection (offset)	*****	—

Note: Safety features and crash data also apply to the 9-5.

9-5 ★★★

RATING: Average. **Strong points:** Good acceleration, handling road holding, and braking; "smart" head restraints; comfortable seating (lots of room for six-footers plus); lots of storage capability; improved build quality and fit and finish; good fuel economy; traction control (V6 only). **Weak points:** Turbo lag; the steering feels a bit over-assisted; body lean when cornering; climate controls aren't very user-friendly; servicing network is rather limited. **New for 2004:** The V6 engine has been replaced by a 220-hp 2.3L 4-cylinder. The Linear entry-level sedan will be axed by mid-2004. Saab will get an SUV in 2005 based on the Chevrolet TrailBlazer and GMC Envoy. Go figure.

OVERVIEW: The base 9-5 Arc ($43,000) and Aero ($53,000) use suspension and brake parts taken from GM's Opel Vectra, with a stiffened chassis and four-wheel independent suspension. They hold the road fairly well, though there's still a lot of body lean. Two turbocharged engines are available: the base 2.3L 220-hp 4-cylinder, found on the Arc sedan, and a 250-hp high-performance version of the same engine offered on the Aero.

There's plenty of cargo space and room for five adults in a cabin that's one of the quietest you can find (except for tire noise). In addition to providing a spacious pass-through trunk, Saab pays lots of attention to detail. For example:

work gloves and a plastic bag are provided so you don't soil your hands when changing the tire; the "smart" head restraint is specially designed to prevent neck injuries; the glove box has an air conditioning duct that keeps drinks cold; there are air vents in the back of the centre console for added comfort; rear seats fold flat for added cargo room; the cargo floor slides out for easier loading; and, for added safety, a passenger-side rear-view mirror automatically tilts down so you can get a better view when you engage Reverse.

Some things you won't like about the 9-5: power steering is over-assisted and doesn't transmit sufficient road feel; only two adults can sit comfortably in the rear; suspension feels overly firm; low-profile tires on the Aero have poor wet-weather traction; wagons don't offer third-row seating; confusing dash buttons; the fan-cooled seats are more gimmicky than practical; climate controls are set too low to be easily adjusted and have to be re-calibrated after each start-up; the ignition is still floor mounted; manual transmission-equipped cars must be shifted into Reverse before ignition key can be removed; rear visibility is obstructed; Aero models require premium fuel; and servicing isn't widely available. **Highway/city fuel economy:** *2.3L and manual:* 7.5–11.7L/100 km.

Cost analysis/best alternatives: Alternatives to the 9-5: Acura RL or Mercedes E-Class.

Quality/Reliability/Safety

Pro: Quality control: Above Average. Saab 9-5 owners report fewer factory-related defects than owners of the entry-level 9-3. **Reliability:** Average. **Warranty performance:** Average.

Con: Service bulletin problems: Starter won't crank the engine and tips on improving the AC's temperature regulating mechanism. **NHTSA safety complaints/safety:** Fuel sloshes around in fuel tank when accelerating or stopping; loss of power steering; sulfur fumes invade the interior. **Owner-reported problems:** Electrical, brake, and fuel system glitches; some minor fit and finish complaints.

General Motors/Saturn

SATURN ION, L-SERIES 

RATING: *Ion:* Below Average. There have been some improvements, but Saturn's poor-quality powertrains and uneven servicing give its Asian competitors an important advantage, and even make GM's ubiquitous Cavalier (Cobalt) and Sunfire look good. *L-series:* Below Average. **Strong points:** *Ion:* Plenty of front seat room; Quad Coupe's four doors give good rear seat access;

excellent steering response. *L-series:* Well appointed; comfortable driving position; a fairly large interior with a full range of convenience and safety features, instruments, and controls; standard four-wheel disc brakes, ABS; side curtain airbags that protect the head and upper torso; excellent powertrain matchup with the V6; firm European ride; standard traction control adds to impressive high-speed stability; excellent braking; effective soundproofing and lots of storage areas, including a large accessible trunk. **Weak points:** *Ion:* Limited rear passenger room and rearward visibility (coupe); seats aren't very comfortable; suspension may be too firm for some; poor wet traction; excessive engine and road noise; real-world fuel economy is lower than advertised. *L-series:* Cheap-looking, bland interior isn't as refined as what you would find in a Toyota or Honda and provides a bit less passenger room than the Asian competitors; limited rear visibility and front vision is reduced by the high cowl; steering feels heavy at low speeds and too light at higher speeds; wagons give a jarring ride over bumps; small audio controls and power window switches are found on the centre console; a history of serious engine and transmission breakdowns. **New for 2004:** *Ion:* Interior trim improvements, additional interior soundproofing, and the debut of a Red Line high-performance model. *L-series:* Only an L300 is offered, the manual transmission has been dropped, the base engine gets five more horses, and standard equipment now includes ABS, traction control, and head-protecting side airbags.

OVERVIEW: Remember the hype? Saturn would be GM's *different* car company. Quality would match or exceed anything offered by the Japanese. Customers would be coddled and dealers would offer fair, "no haggle" prices. Sales would soar.

Unfortunately, Saturn never did meet these expectations and the division has never made a profit. Instead, the company has produced low-quality, overpriced cars that have sold poorly now that their novelty has worn off.

Shoppers now know the Emperor has no clothes. Customers who didn't walk out of showrooms when faced with a "take it or leave it" attitude in past years now find the inflated "no haggle" prices can be whittled down by hard bargaining. And Saturn's response to complaints over its poor-quality engines? An extended warranty and a rush order for new engines from...Honda!

Ion

A larger, more comfortable, and more powerful vehicle than its S-series predecessor, the Ion is powered by a 140-hp 2.2L 4-cylinder engine, and gives buyers the choice of either a four-door coupe or sedan. Other features include power steering, a 5-speed manual transmission, an optional 5-speed automatic, and an optional continuously variable automatic transmission. Other standard features: speed-sensitive windshield wipers, split folding rear seatbacks, and plastic body side panels.

This year's arrival of the Red Line high-performance Ion and Vue adds a bit of colour to an otherwise bland model lineup. Carrying a supercharged 200-hp 4-banger hooked to a 5-speed manual transmission, this sporty variant will use its sturdier drivetrain and 17-inch alloy wheels to ratchet up the Ion's performance and handling.

L-series

In an attempt to save money by adapting a European car to the American market, Saturn's mid-sized LS sedan and LW wagon are products of a whole series of cost-cutting compromises engendered by their being spun off GM's Opel Vectra. Some major differences, however, include a lengthened body to make these Saturns more crashworthy, a standard ignition theft-deterrent system, a re-engineered chassis to give a more comfortable ride, the continued use of Saturn dent-resistant polymer body panels, the substitution of a Saturn space frame, and the use of a homegrown 140-hp 2.2L 4-banger constructed with aluminum components. Other components lifted directly from the European parts bin are Opel's 182-hp 3.0L V6 engine, a manual transmission from Saab, and German-made braking systems.

Cost analysis/best alternatives: GM has pledged not to raise 2004 model prices by more than a few hundred dollars—at first. This would make the 2004 L-series a better buy than the Ion, which doesn't have as many upgrades this year. Some Ion alternatives: Honda Civic, Hyundai Accent and Tiburon, and Toyota's Corolla; they perform better and offer higher quality vehicles. Alternatives to the L-series: GM Cavalier (Cobalt) or Sunfire, Honda Accord, Hyundai Elantra or Sonata, Mazda 626, or Toyota Camry. **Options:** Slippery leather seats aren't worth the extra money. **Rebates:** Zero percent financing and $2,500 rebates by year's end. **Delivery/PDI:** $975. **Depreciation:** Average. **Insurance cost:** Average. **Parts supply/cost:** Parts are easily found and reasonably priced. **Annual maintenance cost:** Average. Estimated to increase moderately after the third year of ownership. **Warranty:** Bumper-to-bumper 3 years/60,000 km; Powertrain: 5 years/100,000 km; rust perforation 6 years/160,000 km. **Supplementary warranty:** A good idea in view of Saturn's poor quality control and hard-nosed attitude in treating post-warranty repair claims. **Highway/city fuel economy:** *Ion: 2.2L and auto.: 6.7–10* L/100 km*; L-series: 2.2L and auto.:* 7.2–10.5L/100 km; *3.0L and auto.:* 8.2–11.8L/100 km.

Quality/Reliability/Safety

Pro: The installation of Honda V6 engines in this year's Vue means we'll soon see a similar upgrade with the L300. Unfortunately, this will be too late for owners already hit hard by huge bills for premature engine failures.

Con: Reliability: Below average reliability, particularly after the third year of ownership—two years earlier than when most vehicles start needing major repairs. More problems have been reported with the newer Ion than with the L-series. The new CVT automatic transmission has had production/quality problems at the factory. This usually means a less-than-stellar performance when these transmissions hit the streets. A helpful website to find out what you can expect from your new or used Saturn is 6th Planet (yep, that's Saturn) Used Auto Parts, found at *www.6thplanetusedparts.com.* Barred by Saturn from using its trademarked name, Joe Cutrone sells used Saturn parts. He's an expert mechanic who knows where every Saturn skeleton is buried, which problems are factory-induced, and how much you should pay for parts and most repairs. **Quality control:** Not up to Japanese car standards. Saturn has always been afflicted by serious quality control problems. **Warranty performance:** Great—as long as the warranty is in force, you're greeted with a smile and a handshake. Afterwards, customers report feeling abandoned. **Owner-reported problems:** *Ion:* Engine, transmission, power steering, body, and electrical system glitches. *L-series:* Engine, brake rotors, pads, and booster failures, electrical shorts, and body defects. Here's what Joe Cutrone has to say about Saturn L-series timing belts:

> I have a concern about the new L-series V6 and that's what I would like to discuss in this editorial. I wanted to give you some background information and hopefully provide a little insight from a mechanic's point of view. One thing I have a lot of experience with is European and Japanese timing-belt-equipped engines: their tolerances, valve clearances, and what can happen when a timing belt fails. For those of you that don't know, your L-series engine is designed by Opel, which is owned by GM of Europe. Saturn is telling new L-series owners that the timing belt should be changed at 160,000 km (100,000 mi.). Personally, this is a huge marketing mistake by falsely claiming the lifespan of a belt to reduce cost of ownership. It is also a mistake to use this 160,000-km (100,000-mi.) belt as a selling point.
>
> In response to their claim I would like to warn people how risky it is to try to go 160,000 km (100,000 mi.) on a piece of rubber. To begin with, no other manufacturer has claimed this yet. Has Saturn invented a magic belt? Should they sell this wonder belt to the rest of the automotive world so the other 99 percent of automakers don't have to tell their customers to replace them every 96,000–112,000 km (60,000–70,000 mi.) (which is the accepted norm)? I don't think so.

Service bulletin problems: *Ion:* Engine coolant leak; no-start and hard starts; defective clutch; excessive fuel pump noise; sunroof rattling and flexing; inoperative HVAC blower motor; noisy, troubleshooting radio malfunctions; cracked rear outer door panel. L-series: Engine misfires; no-starts; excessive steering wheel shake; power steering noise and leaks (see below).

Power Steering - Noise/Leak Diagnosis

BULLETIN NO.: 02-T-46 **ISSUE DATE:** June 2002

2000–03	Saturn L-series vehicles
1991–2002	Saturn S-series vehicles equipped with power steering

The purpose of this bulletin is to provide general diagnostic information about power steering noise and/or leak diagnosis on Saturn L-series and S-series vehicles equipped with hydraulic power steering.

Tips on improving AC performance; inoperative power driver's seat; loose or torn seat covers; defroster grille popping noise; insufficient heat and Service Engine light comes on; tires may rub the front inner fender; and radio malfunctions and speaker noise from electrical interference. **NHTSA safety complaints/safety:** *Ion:* Cracked engine head; sudden engine shutdown while driving; hard starts or no-start; automatic transmission shifts harshly from Second to Third gear, or simply fails to go into gear; many complaints of transmission slippage (see the following typical complaint logged by NHTSA):

> The transmission on my 2003 Ion slips constantly. A tech from Spring Hill says this is normal for this type of transmission. Trying to turn left in front of oncoming traffic is a nightmare. This car is in the shop as we speak. The rear quarter panel is cracked and being replaced. The rear bumper is being painted because of scratches from the trunk rubbing. The front bumper is being painted because of runs in the paint. None of the body panels seem to match up. They're attempting to fix a popping noise in the front end. They're replacing the driver's side glass because it is severely scratched. It sometimes runs rough at first crank.

Horn is hard to activate; steering knuckle sheared off; airbags failed to deploy; seat belt retractor fell apart; frequent tire blowouts; steering wheel is hard to turn; turn signal lights often burn out; vehicle loses power and shakes when AC is activated; door seals improperly installed; and the key sticks in ignition. *L-series:* Interestingly, there aren't proportionally as many complaints with the L-series as with the Ion. Airbags failed to deploy; vehicle rolled backward while parked; brakes wear out prematurely; sudden electrical failure; power door locks short out; electrical short causes all lights and gauges to suddenly come on; lights dim when AC is engaged; faulty evaporator surge solenoids; loose driver's seat; seat belt won't lock up at sudden stops; and wipers ice up easily.

Road Performance

Pro: Emergency handling: Excellent with the Ion; fairly good with the L-series. **Steering:** Very little torque steer, owing to the use of equal-length driveshafts. **Acceleration/torque:** Good acceleration with both the 2.2L 4-cylinder and V6. **Routine handling:** Better than average for both vehicles. The L300's European-type handling emphasizes driver control and its four-wheel independent suspension with four-link rear suspension makes for crisp, predictable handling. Steering has been calibrated to enhance road feel at highway speeds and effectively blocks out road harshness. **Braking:** Fairly good.

Con: Ion handling is compromised by wet pavement. **Transmission:** Excessive automatic gearbox shudder when the kickdown is engaged while passing.

Comfort/Convenience

Pro: Standard equipment: Base Saturns come with a wide range of standard equipment. Large glove box and convenient door pockets. **Driving position:** Very good. **Controls and displays:** Most everything's within easy reach. **Climate control:** Efficient, quiet, and easy to adjust. **Interior space/comfort F/R:** Comfortable front seats, with a fair amount of headroom and legroom. **Cargo space:** Plenty of trunk space and storage compartments, particularly with the L-series. The sedan features one of the largest trunks in its class at 17.5 cubic feet; the wagon's easy-to-load cargo area has 79 cubic feet of room with the rear seats folded down. **Trunk/liftover:** Huge trunk has a large opening and low liftover height.

Con: Interior isn't as refined as that found on the Accord or Camry. Steering wheel is too close for comfort and the radio controls are tiny and distracting. Centre-mounted instrument cluster is distracting. L-series visibility hampered by the high rear deck. Rear fold-down armrest isn't very useful. Coupe's third door window doesn't roll down. **Entry/exit:** The rear is not the place to be—doorsills are high, seat cushions are short, low, and too soft. **Quietness:** Lots of rattles and road and wind noise. Excessive engine noise at cruising speed.

Saturn Ion, L-series

List Price (negotiable)	**Residual Values** (months)			
	24	**36**	**48**	**60**
Ion: $14,625 (11%)	$11,000	$8,000	$6,500	$5,500
L300: $27,790 (16%)	$14,000	$11,000	$8,000	$7,000

Technical Data (Ion)

Powertrain (front-drive)
Engine: 2.2L 4-cyl. (140 hp)
Transmissions: 5-speed man.
• CVT
• 5-speed auto.
Dimension/Capacity
Passengers: 2/3
Height/length/width:
56/185/67.9 in.
Headroom F/R: 38.9/36.5 in.
Legroom F/R: 42.2/32.7 in.
Wheelbase: 103.2 in.
Turning circle: 35.4 ft.
Cargo volume: 14.2 cu. ft.
Tow limit: 1,000 lb.
Fuel tank: 50L/reg.
Weight: 2,751 lb.

Technical Data (L300)

Powertrain (front-drive)
Engine: 3.0L V6 (182 hp)
Transmissions:
• CVT
• 5-speed auto.
Dimension/Capacity (L300)
Passengers: 2/3
Height/length/width:
56.4/190.4/68.5 in.
Headroom F/R: 39.3/38 in.
Legroom F/R: 42.3/35.4 in.
Wheelbase: 106.5 in.
Turning circle: 36.6 ft.
Cargo volume: 17.5 cu. ft.
Tow limit: 1,000 lb.
Fuel tank: 59L/reg.
Weight: 2,965 lb.

Safety Features/Crashworthiness (Ion)

	Std.	Opt.
Anti-lock brakes (4W)	■	■
Seat belt pretensioners F/R	■	—
Side airbags	—	■
Traction control	—	■
Head restraints F/R	****	***
Visibility F/R	*****	***
Crash protection (front) D/P	*****	*****
Crash protection (side) F/R	—	***
Crash protection (offset)	***	—

Safety Features/Crashworthiness (L300)

	Std.	Opt.
Anti-lock brakes (4W)	■	■
Seat belt pretensioners F/R	■	—
Side airbags	■	—
Traction control	■	—
Head restraints F/R	*	*
Visibility F/R	***	**
Crash protection (front) D/P	****	*****
Crash protection (side) F/R	***	*****
Crash protection (offset)	***	—

General Motors/Toyota

VIBE/MATRIX ★★★★

RATING: Above Average. Aimed at the youth market, these cars don't provide enough horsepower to justify their "sporty" pretensions. **Strong points:** Brutish, edgy styling; small Matrix steering wheel makes for easier cornering; refined rear suspension gives these cars impressive road manners; plenty of passenger and storage space and handy tie-downs; flexible seating; front passenger and rear seats fold completely flat to carry long cargo; 115-volt power outlet up front; quality workmanship; well-placed interior controls and instruments; a four-star rollover resistance rating; and excellent workmanship. **Weak points:** Weak, noisy, vibrating engines at all speeds; not much low-end power for the base and XR models and even less 4X4 grunt; Vibe GT must be constantly shifted to keep accelerating; firm suspension lets you feel every bump in the road; and dash instrumentation is hard to read with sunglasses. **New for 2004:** Nothing important.

OVERVIEW: Officially classed as subcompact, front-drive, or all-wheel-drive cars, these sporty wagons are a cross between a small SUV and a station wagon, packaged like a small minivan. Vibe is built in Fremont, California, at GM's

NUMMI factory, while the nearly identical Toyota Matrix is manufactured in Toyota's Cambridge, Ontario, plant, alongside the Corolla, whose platform it shares, though the Vibe provides a larger interior volume.

Matrix joins the Subaru Impreza in professing to provide a blend of sports-car performance, SUV versatility, and compact-car affordability. It's offered in front-wheel drive or Toyota's V-Flex 4X4 system.

The front-drive Matrix and Vibe are equipped with a 130-hp engine, 5-speed manual overdrive transmission, and lots of standard features, including an AM/FM CD stereo system, full instrumentation, and flip-up rear hatch glass.

The Matrix XRS and Vibe GT are the top-of-the-line performance leaders, with a 180-hp 4-cylinder engine and a high-performance 6-speed manual gearbox, plus ABS, premium six-speaker stereo, anti-theft system, 17-inch alloy wheels, and unique exterior cladding.

Cost analysis/best alternatives: GM and Toyota co-ventures have produced some cheap, reliable, and durable cars in the past, like the entry-level Corolla/Prizm compacts. The three Pontiac Vibe models are called base, AWD, and GT. Toyota labels them the base, 4X4, and XRS. But, don't get fooled by the name game. These cars are identical, except for exterior styling options packages and prices. Surprisingly, the Toyota entry-level Matrix is several thousand dollars less expensive than its Vibe equivalent. These vehicles hail from factories that have garnered high-quality ratings and also come with similar warranties. If you really need additional horsepower, get either the XRS or GT. But keep in mind that there are better high-performance choices out there, like the Honda Civic Si, Mazda Protegé 5, or a base Acura RSX. Other front-drives worth considering are the Chrysler PT Cruiser, Hyundai Elantra or Tiburon, Honda Civic, Nissan Sentra, or the Toyota Corolla. Of course, the Subaru Impreza or Forester would be other good choices for the 4X4 variant. **Options:** Be wary of options. For example, the XR option only adds side skirts, colour-keyed door handles, a leather-wrapped steering wheel, power windows, and a height-adjustable driver's seat. Of that list, only the windows and seat are worth the extra loonies. Plus, make sure all features listed in the sales brochure are on your car. **Rebates:** Not likely. **Delivery/PDI:** *Vibe:* $950; *Matrix:* $1,020. **Depreciation:** Predicted to be slower than average. **Insurance cost:** Higher than average. **Parts supply/cost:** Good availability, since many parts come from the Corolla bin. **Annual maintenance cost:** Predicted to be much lower than average. **Warranty:** Bumper-to-bumper 3 years/60,000 km; powertrain 5 years/100,000 km; rust perforation 6 years/160,000 km. **Supplementary warranty:** Not needed. **Highway/city fuel economy:** 7.1L/100 km on the highway; 9.3 L/100 km in the city.

Quality/Reliability/Safety

Pro: Warranty performance: Another reason for choosing the Matrix over GM's Vibe. Although Toyota initially dropped the ball in handling engine oil sludge claims on other models, its warranty performance is much better than

GM's and you generally have much more leverage in pressing your claim with Toyota. **Quality control:** Excellent. **Reliability:** Above average. Stainless steel exhaust system. **Safety:** Excellent visibility fore and aft, although the lower rear-window wiper housing obstructs the window area a bit. The rear wiper is particularly effective, even though the washer dribbles rather than sprays.

Con: Owner-reported problems: Owners report that engine surges to 2000 rpm when vehicle is first started. Dealers say this is normal and just as prevalent with the 2003 Corolla. Other complaints: excessive steering wander; many reports of transmission clutch failures (see *www.matrixowners.com*); with transmission in Park, vehicle can roll away; defective crankshaft key; power steering pump failures; a clicking sound emanates from the driver-side dashboard; a rotten-egg smell comes from the exhaust or through the vents; the CD player makes a grinding noise when braking; excessive vibration felt in the interior during acceleration after start-up; excessive dashboard vibration; headlight condensation, driver's seat fabric tears easily; hood popped up while car was underway; black tape covering the window frame tends to bubble; paint blisters quite easily; and gas mileage may not be up to expectations. **Service bulletin problems:** *Matrix:* Loose or deformed front or rear glass door run; AC doesn't put out sufficient cool air; headlights come on when vehicle is turned off; sulfur odour in the interior; front suspension tapping noise; discoloured wheel house moulding. *Vibe:* Engine lacks performance after 7000 rpm; transmission shifts too early when accelerating at full throttle when the engine is cold; harsh shifting; grinding, growling while in Park on an incline; water leak from the A-pillar or headliner area; and low voltage display or dim lights. **NHTSA safety complaints/safety:** Engine surges when braking with AC engaged; excessive steering wander; malfunction light may remain lit; instrument panel lights are dimmed by automatic sensor to a point where they can't be read in twilight hours; and the automatic headlights come on and go off for no apparent reason.

Road Performance

Pro: Acceptable acceleration from start-up in First gear, but not confidence-inspiring for high-speed merging unless you have the 180-hp power plant. **Transmission:** Smooth and quiet shifting with the automatic gearbox. **Emergency handling:** Very good. The double wishbone rear suspension makes for crisp, predictable handling. **Routine handling:** Above average. **Braking:** Disc/drum; disc/disc with the XRS. Braking is good with the base set-up and better than average with the four-wheel discs.

Con: Acceleration/torque: Weak, buzzy base engine can be felt throughout the car; 4X4 models are about 10 percent heavier and get seven fewer horses (123) than the already power-challenged 130-hp front-drive. Says *Forbes* magazine:

> [B]oth all-wheel-drive cars are saddled with a really wretched 4-speed automatic that almost has to be shifted manually to get the car moving. To put it bluntly, the AWD Vibe and Matrix are so pokey they feel like they're towing Winnebagos. To boot, the 1.8L engine doesn't hit its paltry torque peak ... until a screaming 4200 rpm, at which point the vibration—did somebody say Vibe?—in the cabin is worse than a little off-putting.

The manual shift lever's upward and forward position is counterintuitive and feels a bit ragged. Steering feels a bit vague, with too much play. Some torque steer (twisting) evident, especially on wet roads.

Comfort/Convenience

Pro: Standard equipment: Full of power-assisted accessories. The front seats have excellent thigh and torso support. **Controls and displays:** Complete instrumentation that's easy to decipher, and user-friendly controls. **Climate control:** Easy to adjust and quiet. **Interior space/comfort F/R:** Taller and wider than most of its competitors. The Matrix's extra height translates into more front and rear headroom than its competitors. **Entry/exit:** Excellent. Thanks to the tall roof and four large doors, passengers can easily access the interior. **Cargo space:** Surprisingly large and versatile storage areas. **Trunk/liftover:** Fairly spacious storage area with a relatively low liftover.

Con: Driving position: Upright seating position makes it tiring to access the pedals and difficult to comfortably rest your arms. Rear seats are a bit too narrow for three adults. No rear folding centre armrest. **Quietness:** Forget it. Engine noise and vibrations are omnipresent.

Vibe/Matrix

List Price (firm)	**Residual Values** (months)			
	24	**36**	**48**	**60**
Vibe FWD: $20,985 (16%)	$16,000	$12,000	$10,000	$8,500
Vibe AWD: $27,000 (18%)	$21,000	$18,000	$13,000	$11,500
Vibe GT: $26,975 (18%)	$22,000	$19,000	$14,000	$12,500
Matrix FWD: $17,745 (14%)	$14,500	$11,000	$9,000	$7,500
Matrix AWD XR: $22,220 (16%)	$18,500	$14,000	$12,000	$10,000
Matrix XRS: $25,690 (17%)	$21,500	$19,000	$16,000	$13,500

Technical Data

Powertrain (rear-drive)
Engines: 1.8L 4-cyl. (130 hp)
• 1.8L 4-cyl. (180 hp)
Transmissions: 5-speed man.
• 4-speed auto.
• 6-speed man.

Headroom: F/R:40.6/39.8 in.
Legroom: F/R: 41.8/36.3 in.
Wheelbase: 102.4 in.
Turning circle: 35.4 ft.
Cargo volume: 54.1 cu. ft.
Tow limit: 1,500 lb.

Dimension/Capacity

Passengers: 2/3
Height/length/width: 61/171.3.7/69.9 in.
Fuel tank: 50L/reg.
Weight: 2,745 lb.

Safety Features/Crashworthiness

	Std.	Opt.
Anti-lock brakes	■	■
Seat belt pretensioners F/R	■	—
Side airbags	—	■
Traction control	—	—
Head restraints F/R	*****	*****
Visibility F/R	*****	*****
Crash protection (front) D/P	*****	*****
Crash protection (side) F/R		
AWD	*****	****

ASIAN VEHICLES

With just two exceptions (Kia and Daewoo), Asian automakers have a lock on reliable cars, minivans, sport-utilities, and pickups. Whether the vehicles are built in Japan, Canada, Mexico, or the United States, you usually get much more for your money than if you were to buy the equivalent vehicle made by DaimlerChrysler, Ford, or General Motors, or most European automakers for that matter. You can also count on Japanese vehicles to be easy to repair and slow to depreciate. Buyers can get exceptionally good deals by purchasing Japanese vehicles that are re-badged as American models or built as co-ventures in Japanese, American, and Canadian factories (like the Toyota Matrix and Pontiac Vibe). These offerings are generally more reasonably priced and are better supported by an indigenous dealer and servicing network.

On the downside, Japanese vehicles usually sell at a 10–20 percent premium over their Detroit equivalent, though higher resale values wipe out the difference. In the past, Asian automakers have tried to keep prices down through "content-cutting" and offering fewer standard features. But this has produced disastrous results during the past several years. Less content has led to less quality, and manufacturers like Honda and Toyota have had to extend their powertrain warranties up to eight years to cover catastrophic engine and automatic transmission failures.

One gets the impression that Japanese automakers have become complacent after winning so many quality awards from CAA and other groups and are now coasting on their reputation. This would explain why we've seen such an upswing in safety- and performance-related defects over the past few years with such reputable companies as Nissan, Honda, and Toyota. It certainly has nothing to do with American versus Japanese manufacturing plants—many of these companies' poor-quality components have come from factories located on both sides of the Pacific.

Hyundai's Accent is a South Korean bargain.

Fortunately, low interest rates, cheaper offshore plants, and higher volume this year will permit Japanese and Korean automakers to cut prices instead of content. Honda, Hyundai, Nissan, and Toyota have warned Detroit that they will use aggressive discounting and rebates to undercut the competition in almost every vehicle category (only full-sized vans remain unscathed), with particular emphasis on promoting their redesigned Honda Accord, Mazda6, Nissan Quest, and Toyota Sienna.

Nissan's revamped Quest represents a real threat to Toyota's re-engineered Sienna and to the Detriot Big Three's minivans.

Competition is also heating up among the Asian automakers. South Korean vehicles—once the laughing stock of car columnists and consumer advocates—are catching up to the Japanese competition in quality and sales. Hyundai is turning its attention to its Kia subsidiary as it continues to post impressive sales gains, due in large part to low prices, better quality, and longer base warranties. GM's Daewoo acquisition will also inject serious price-cutting into the mix of entry-level cars as Chevrolet and GM's Suzuki partner unload thousands of discounted and rebate-laden small cars on the Canadian market.

Hmmm, wouldn't want to be a General Motors dealer stuck with unsold Cavaliers and Sunfires going into 2004.

Acura

Acura doesn't have the cachet of other luxury Asian brands, like Lexus and Infiniti, but it does distinguish itself by offering performance-oriented luxury-class vehicles in all price ranges. Seven models are sold under the Acura nameplate: the Canada-exclusive, 1.7L EL entry-level compact; the RSX

Integra replacement; a 3.2 TL; a 3.5 RL; an NSX sports coupe; and the MDX sport-utility. A new TSX model joins the lineup this year, but the 3.2 CL coupe has been dropped.

Acura products are mainly fully loaded Hondas with a few additional features. Despite the fact that dealers try to enforce a "no haggle" policy, these cars are generally good buys because maintenance costs are low, depreciation is much slower than average, and reliability is outstanding. What few defects they have are usually related to accessories like the navigation and sound system, as well as creaks and squeaks.

TSX ★

Will the TSX's lack of V6 power and smaller interior drive buyers to the Accord?

RATING: Not Recommended during its first year on the market. Choose a gussied-up V6-powered Accord instead.

OVERVIEW: Essentially a more luxurious European version of the Accord, the TSX only comes with a 200-hp 2.4L 4-cylinder engine that must compete in a luxury sedan niche where V6 power is commonplace. It fits between the RSX and the 3.2TL in Acura's lineup and is hyped as being sportier than the TL, yet not as harsh as the high-performance RSX.

Without a doubt, this is a cleanly styled car, both inside and out, with just a bit of a European and sporty flair in the car's interior layout. More significant, though, is the car's limited interior space. Buyers comparing the TSX with the Accord will quickly discover that the TSX has less head and shoulder room up front, and a more-cramped rear-seat area.

A deal breaker? Perhaps.

Nevertheless, Acura is throwing in a cornucopia of standard safety, performance, and convenience features to make this $34,800 luxury sedan attractive to shoppers who don't feel size and V6 power are everything. In fact, the only extra-cost option is a voice-activated navigation system.

Still, when you consider the $34,800 TSX's price is about $10,000 more than a 4-cylinder Honda Accord, you have to wonder whether all that sizzle will sell the steak.

1.7 EL ★★★★

RATING: Above Average. **Strong points:** High level of performance and comfort; lots of standard safety and convenience features; excellent fuel economy for a luxury subcompact; slow depreciation; and high-quality construction. **Weak points:** Little discounting; insufficient low-end torque; narrow interior with rear seating for only two adults; obstructed rear visibility; and head restraints earned a "Poor" rating by IIHS. **New for 2004:** Nothing important; most of the upgrades were put in last year, including a retuned suspension and steering system for better handling and enhanced ride comfort, reduced engine vibration through improved engine mounts, and upgraded brakes. Additionally, the 2003s were given an upgraded instrument cluster, centre console armrest, and front seats; adjustable head restraints; and an optional "Aero" accessory package.

OVERVIEW: With a $22,200 base price and $24,200 for the Premium, the redesigned 1.7 EL is Acura's best-selling car, for what is essentially a re-styled, more powerful, luxury version of the Honda Civic sedan. In fact, both models are built in Honda's Alliston, Ontario plant.

The VTEC 4-cylinder puts out 127 hp for good all-around acceleration, but maximum torque still requires high revving in the 5000 rpm range. Going uphill with a full load of passengers and baggage overwhelms the 4-banger. Honda's base 5-speed manual gearbox and optional 5-speed automatic set the standard for smooth, quiet shifts, and you get impressive fuel economy to boot (8L/100 km in the city; 6L/100 km on the highway).

Performance and handling are enhanced by a fully independent suspension, tight and responsive steering, and excellent braking, with little brake fade after successive stops.

There's good front and side visibility thanks to the car's low hoodline and large side windows. Rear visibility is hampered a bit by the high rear deck. The interior is quieter and much more luxurious than the Civic LX sedan, plus all gauges and controls are easily seen and accessed. Both the floor-mounted trunk release and gas cap are lockable, and split folding rear seats (also lockable) allow for pass-through storage of long objects. Standard safety features include side airbags, de-powered airbags, and Seat belt pretensioners F/R.

Some of the 1.7 EL's weak points: a chintzy space-saver spare tire and relatively high price tag for what is essentially a better-equipped Civic (still, a low depreciation rate returns a greater portion of your initial investment).

Cost analysis/best alternatives: Get the 2003 model for the safety and performance upgrades and a larger discount. Unfortunately, this is easier said than done, since these cars are quite in demand. Large urban centres are your best hunting grounds. Other vehicles worth a look: the Honda Civic Si, Mazda Protegé 5, Nissan Altima, and Toyota front-drive Matrix. **Options:** A block heater. **Rebates:** $1,500 on the 2004 models. **Delivery/PDI:** $900. **Depreciation:** Much slower than average. **Insurance cost:** Higher than average. **Parts supply/cost:** Moderately priced parts can be found at Acura or Honda dealers. **Annual maintenance cost:** Lower than average. **Warranty:** Bumper-to-bumper 3 years/60,000 km; powertrain 5 years/100,000 km; rust perforation 6 years/unlimited km. **Supplementary warranty:** Not needed.

CL (2003) ★★★★

RATING: Above Average. The Accord grows up. **Strong points:** Good powertrain set-up, steering, and handling; well built; stylish; plenty of standard features; comfortable ride; quality construction and mechanical components. **Weak points:** As with most coupes, the CL has problematic rear-seat access and limited rear headroom. The TL-inspired navigation system is overly complicated; head restraints rated "Poor" by IIHS; serious transmission failures (see *www.acura-cl.com* and do a search for "transmission failure"). **New for 2004:** Nothing; the model has been dropped.

OVERVIEW: Priced at $37,800–$41,800, this four-passenger entry-level luxury coupe shares the TL's platform and base 225-hp 3.2L V6. A high-performance, 260-hp variant powers the Type S.

Sure, we all know that the coupe's mechanical components and platform aren't that different from what's found on the Honda Accord, but when you add up all of its standard bells and whistles, you get a fully loaded small car that costs thousands of dollars less than competing luxury coupes. Base models are well appointed with standard four-wheel ABS, a "smart" passenger-side airbag, traction control, heated front seats, leather upholstery, Xenon headlights, power sunroof, remote keyless entry, and a dash-mounted six-disc CD changer. Type S standard features include all of the above, plus an anti-skid system, a stiffer suspension, 17-inch wheels, and upgraded tires.

The CL gets plenty of power from its smooth-running and quiet 3.2L V6. Handling is better than average, thanks to upgraded suspension, variable-assisted steering, and 16-inch wheels. The ride is comfortable and well controlled. Braking is first-class (100–0 km/h: 35 m).

Front and rear bucket seats are supportive and easily adjusted. There's plenty of room up front, controls and most gauges are user-friendly (the navigation system and tachometer placement are the only exceptions), and the climate controls are efficient and within easy reach. A large trunk with a low liftover and a locking ski pass-through enhances the CL's cargo space.

On the minus side, this is not a car for passengers in the rear. Adults will likely find their heads pressed against the top of the backlight glass, and legroom and footroom is at a premium. Rear access is a crouch-and-crawl affair. Trunk lid hinges intrude into the trunk area and risk damaging cargo when the trunk is closed. Sudden unintended acceleration and hesitation complaints are also thought to be transmission-related.

Cost analysis/best alternatives: Other vehicles worth considering: the BMW 328Ci, Honda Accord, Lexus IS300, Mercedes-Benz CLK320, Nissan Maxima, Toyota Solara, and Volvo C70. **Options:** Don't get the GPS navigation system unless you're technically inclined and patient. **Rebates:** Leftover 2003s will likely carry hefty discounts in the $3,500+ range. **Delivery/PDI:** $950. **Depreciation:** Much slower than average. **Insurance cost:** Higher than average. **Parts supply/cost:** It's not hard to find moderately priced parts at Acura or Honda dealers. **Annual maintenance cost:** Average. **Warranty:** Bumper-to-bumper 3 years/60,000 km; powertrain 5 years/100,000 km; rust perforation 6 years/unlimited km. **Supplementary warranty:** Not needed. **Highway/city fuel economy:** 7.4–12.2L/100 km with the base 3.2L.

Quality/Reliability/Safety

Con: Service bulletin problems: Airbag light may remain lit; windshield wiper smearing and streaking; faulty Delphi Freedom Group batteries will be taken back; moon roof squeaks; seatback panel may loosen; missing speed sensor plug; and troubleshooting engine oil leaks. **NHTSA safety complaints/safety:** Short-statured drivers may be seriously injured by the front airbag's deployment. Transmission slips out of gear, suddenly downshifts, and engine surges. Sudden, unintended acceleration, accompanied by loss of braking ability; and window regulator allows window to slide down into door.

TL ★★★★

RATING: Above Average. **Strong points:** Impressive acceleration; handles well; rides comfortably; well constructed, with quality mechanical and body components; and impressive crashworthiness scores compiled by the NHTSA and IIHS. **Weak points:** Suspension may be too firm for some; uncomfortable rear seating; excessive road noise; problematic navigation system controls; head restraints rated "Poor" by IIHS; and the continuation of manual and automatic transmission breakdowns. **New for 2004:** Nothing significant.

OVERVIEW: Retailing for about $40,800, the TL combines luxury and performance in a nicely styled front-drive five-passenger sedan that uses the same chassis as the Accord and CL coupe. The only engine available, a 270-hp 3.2L V6 mated to a 5-speed automatic transmission or a 6-speed manual coupled to a limited-slip differential, provides impressive acceleration (0–100 km/h in just over 8 seconds) in a smooth and quiet manner. Handling is exceptional with the firm suspension, and responsive, precise steering makes it easy to toss the TL around turns without losing control. Bumps can be a bit jarring, but this is a small price to pay for the car's high-speed stability.

Interior accommodations are better than average up front, but rear occupants may discover that legroom is a bit tight and the seat cushions lack sufficient thigh support. The cockpit layout is very user-friendly, due in part to the easy-to-read gauges and accessible controls (far-away climate controls are the only exception). Visibility fore and aft is unobstructed; however, the optional navigation system is tough to read, hard to calibrate, and subject to malfunction. Invest in maps instead.

Standard safety features include ABS, traction control, front seat belt pretensioners, childproof door locks, three-point seat belts, and a transmission/brake interlock. Crash tests give four stars for driver and passenger crash protection in a frontal collision and four and five stars for side-impact protection. Offset crash ratings are also five-star. On the other hand, head restraints are given a "Poor" rating by IIHS.

Cost analysis/best alternatives: Get the 2003 model if it's sufficiently discounted, and if you don't mind a 4-speed automatic gearbox and 45 fewer horses. Also consider the Audi A4, BMW's redesigned 3 Series; Infiniti's redesigned I30; and Lexus' ES 300. **Options:** Don't waste your money on the satellite navigation system; it's confusing to calibrate and hard to see. **Rebates:** $3,500 rebates now, and some discounting can be expected early next year. **Delivery/PDI:** $950. **Depreciation:** Much slower than average. **Insurance cost:** Higher than average. **Parts supply/cost:** Easily found and moderately priced, especially most mechanical and electronic components, but with the exception of some body parts. **Annual maintenance cost:** Less than average. **Warranty:** Bumper-to-bumper 3 years/60,000 km; powertrain 5 years/100,000 km; rust perforation 6 years/unlimited km. **Supplementary warranty:** Not needed. **Highway/city fuel economy:** 7.4–12.2L/100 km.

Quality/Reliability/Safety

Pro: Quality control: Above average. **Reliability:** Above average. **Warranty performance:** Average.

Con: Owner-reported problems: Manual and automatic transmission defects, faulty brake pads and rotors cause steering wheel "shimmy" and jerking to the right or left when braking, poor body fits, and malfunctioning accessories. **Service bulletin problems:** Troubleshooting a rough running engine; TL and CL automatic transmission warranty extension; correcting excessive vibration; faulty remote audio controls; airbag light comes on by error; power seat malfunctions; moon roof squeaking. **NHTSA safety complaints/safety:** Chronic transmission failures (downshifts without warning, etc.); fire in engine compartment; defective timing belt tensioner; failure of IAC valve led to brake failure; early wheel bearing replacement; dealer says a cell phone left in the passenger seat could cause the airbag light to activate; automatic transmission slipped from Park to Drive and suddenly accelerated on its own; premature brake rotor wear covered by a goodwill warranty, but considered a safety hazard; four tires were replaced in three months due to sidewall bubbling; headlights don't illuminate properly.

RL ★★★★

RATING: Above Average. Basically, a fully loaded, longer, wider, and heavier TL, equipped with a larger engine that produces less horsepower than its smaller brother. **Strong points:** Good acceleration that's smooth and quiet in all gear ranges; exceptional steering and handling; comfortable ride; loaded with goodies; top-quality body and mechanical components, except for the transmissions; and a four-star front impact rating. **Weak points:** Numb steering; head restraints rated "Poor" by IIHS and only "Average" for offset occupant protection; 6-speed manual transmission joins Acura's tranny "Hall of Shame"; problematic navigation system controls. **New for 2004:** No important changes.

OVERVIEW: Retailing for about $55,800, the RL is Honda's—oh, I mean, Acura's—flagship sedan. It's loaded with innovative high-tech safety and convenience features one would expect to find in a luxury car. These include heated front seats, front and rear climate controls, rear-seat trunk pass-through, xenon headlights (get used to oncoming drivers flashing their headlights at you), "smart" side airbags, front seat belt pretensioners, ABS, traction control, and an anti-skid system. No other engine but the 3.5L is available, and the only option offered is Acura's ubiquitous GPS navigation system (see TL comments).

The 225-hp 3.5L V6 mated to a 4-speed automatic transmission provides good acceleration that's a bit slower and more fuel-thirsty than the TL, due partly to the RL's extra pounds. Power is nevertheless delivered in a smooth and quiet manner. The car handles nicely with a less firm ride than the TL, although steering response doesn't feel as crisp. Interior accommodations for four occupants are excellent up front and in the rear, due to the RL's use of a larger platform than the TL. All seats are well cushioned and give plenty of thigh support. Cockpit controls and instruments are easily accessed and the climate control system is efficient and easy to adjust, both fore and aft. Good all-around visibility; however, the optional navigation is annoyingly distracting and not easily mastered.

Cost analysis/best alternatives: Get the practically identical, cheaper 2003 model, if it's sufficiently discounted; also consider BMW's 5 Series, Infiniti's redesigned I30, and the Lexus GS 300/400. You may want to take a look at the TL sedan. **Options:** Forget the satellite navigation system; listen to your wife, or buy maps instead. **Rebates:** Expect $3,000 rebates on the 2004s. **Delivery/PDI:** $950. **Depreciation:** Much slower than average. **Insurance cost:** Higher than average. **Parts supply/cost:** Most mechanical and electronic components are easily found and moderately priced. Body parts may be hard to come by and can be expensive. **Annual maintenance cost:** Less than average. **Warranty:** Bumper-to-bumper 3 years/60,000 km; powertrain 5 years/ 100,000 km; rust perforation 6 years/unlimited km. **Supplementary warranty:** Not needed. **Highway/city fuel economy:** 9.2–13.4L/100 km.

Quality/Reliability/Safety

Pro: Quality control: Above average. **Reliability:** Average for all components except the manual and automatic transmissions, which are quite problematic on Acura's entire vehicle lineup. **Warranty performance:** Above average.

Con: Owner-reported problems: Chronic stalling and premature brake wear. The 6-speed manual transmission often mis-shifts to Second gear when upshifting from Third to Fourth, thereby causing extensive engine damage. Second gear is almost impossible to access in cold weather. **Service bulletin problems:** Airbag light comes on for no reason and troubleshooting automatic transmission malfunctions. **NHTSA safety complaints/safety:** Vehicle suddenly stalled while exiting a freeway; dashboard display is unreadable in daylight; in a crash, seat belt did not restrain driver.

RATING: Average. This is Acura's entry-level performance hatchback that targets drivers who want to shift fast and often and don't mind a firm ride. **Strong points:** Well appointed; good acceleration, steering, and handling above 3000 rpm; user-friendly gauges and controls; excellent braking, especially with the Type S; updated styling; good fuel economy; first tune-up at 160,000 km; low depreciation; a five-star frontal crash rating for both the driver and front-seat passenger and four-star front side impact rating; head restraints rated "Good" up front and "Average" in the rear by IIHS; and high-quality construction, except for transmission components. **Weak points:** So-so acceleration in lower gears; overall acceleration compromised by the automatic transmission; limited front and rear headroom; cramped rear seating, with barely adequate knee and foot space; no sedan available; difficult rear access; high rear hatch liftover complicates loading; excessive road and engine noise; rearward visibility obstructed by small side windows, thick roof pillars, and a tall rear deck; and many manual and automatic transmission complaints (see *forums.clubrsx.com/forumdisplay*). **New for 2004:** Carried over relatively unchanged.

OVERVIEW: This is a better car than the Integra it replaced for the following five reasons: It's more attractively styled, provides a higher level of driving performance, has more legroom, has a 6 percent larger cabin and a 33 percent bigger trunk, and offers more standard features. On the other hand, it's relatively pricey when compared with the Integra (about $5,000 more) and only a two-door hatchback is available.

The $24,400 RSX is built on the Civic platform and is powered by a base 160-hp 2.0L twin-cam iVTEC 4-banger or a torquier 200-hp variant used by the high-performance Type S. Three transmissions are available: a 5-speed manual and 5-speed automatic offered with base models, and a 6-speed manual found on the Type S. Both engines are torquier than the power plants they replaced; however, the base engine has more useful commuting low-end torque than the so-called sportier Type S engine.

Handling and steering performance are enhanced through the use of an upgraded steering and independent suspension system, and more effective four-wheel disc brakes. Standard safety features include side airbags, Seat belt pretensioners F/R, and side-impact protection pads.

Cost analysis/best alternatives: Since there's little that is new this year, get a discounted 2003 version. Other cars worth considering are the Mazda Miata or Toyota Celica GT and GT-S. **Options:** None needed. **Rebates:** $2,000 rebates and zero percent financing on the 2004s. **Delivery/PDI:** $900. **Depreciation:** Slower than average. **Insurance cost:** Higher than average. **Parts supply/cost:** No problem getting moderately priced parts from Acura or Honda dealers. **Warranty:** Bumper-to-bumper 3 years/60,000 km; powertrain 5 years/100,000 km; rust perforation 6 years/unlimited km. **Supplementary warranty:** Acura quality glitches aren't likely to warrant supplementary protection. **Highway/city fuel economy:** *Base engine with manual transmission:*

8.6–7.1 L/100 km; *same engine with automatic:* 9.8–7.1 L/100 km; *Type S:* 9.8–7.6 L/100 km.

Quality/Reliability/Safety

Pro: Quality control: Above average. **Reliability:** Above average for all components except the manual and automatic transmissions, which are quite problematic on Acura's entire vehicle lineup. **Warranty performance:** Above average.

Con: Owner-reported problems: Rough shifting; rotten-egg smell intrudes into the cabin; premature brake and rear strut wear. Airbag light remains lit. Rear bumper cover may fall off, be not aligned or flush, come loose, and chip paint. **Service bulletin problems:** Airbag light comes on for no apparent reason. Electrical shorts cause the headlights to go off and the warning chime and ceiling light to operate erratically. Catalytic converter exhaust leak, noise. **NHTSA safety complaints/safety:** N/A.

RSX

List Price (firm)	**Residual Values** (months) 24	36	48	60
Base: $24,300 (18%)	$19,500	$16,000	$12,000	$10,000
Premium: $27,300 (18%)	$21,500	$18,000	$14,000	$12,000
Type S: $31,300 (19%)	$23,500	$20,000	$16,000	$13,500

Technical Data

Powertrain (front-drive)
Engines: 2.0L iVTEC 4-cyl. (170 hp)
• 2.0L iVTEC 4-cyl. (200 hp)
Transmissions: 5-speed man.
• 4-speed auto.
• 6-speed man.

Dimension/Capacity
Passengers: 2/3
Height/length/width: 55.1/172.2/67.9 in.
Headroom F/R: 37.8/30.1 in.
Legroom F/R: 43.1/29.2 in.
Wheelbase: 101.2 in.
Turning circle: 38.1 ft.
Cargo volume: 17.8 cu. ft.
Tow limit: N/A
Fuel tank: 50L/reg.
Weight: 2775 lb.

Safety Features/Crashworthiness

	Std.	Opt.
Anti-lock brakes	■	—
Seat belt pretensioners F/R	■	—
Side airbags	■	—
Traction control	—	—
Head restraints F/R	****	***
Visibility F/R	*****	**
Crash protection (front) D/P	*****	*****
Crash protection (side) F/R	****	—

NSX ★★★★

RATING: Above Average. **Strong points:** Beautifully styled; loaded with expensive performance and convenience features; handles well; rides comfortably; and is well constructed with quality mechanical and body components. **Weak points:** $140,000? Let's see, one Miata goes into 140 five times...hmmm. Horsepower-challenged by other cars in this price range. Aluminum construction means fender-benders quickly become wallet busters. **New for 2004:** Nothing important. This is its last year on the market.

OVERVIEW: Retailing for about $142,000 the NSX is a luxury sports car that makes no compromises. Its rear mid-mounted, "rev-happy" 290-hp 3.2L VTEC V6 is built for hard driving and taking up permanent residence at the 8000 rpm redline. There's plenty of torque throughout the gear range and the 6-speed manual shifts flawlessly. Handling is optimized through quick and accurate steering, best-in-class brakes, and a no-surprise suspension system.

Last year's improvements consisted of slightly re-styled front and rear ends, a firmer suspension, and larger, 17-inch front wheels.

Cost analysis/best alternatives: Get the 2003 NSX for the suspension, wheels, and styling improvements and a slightly lower price. Also consider the BMW Z8, Chevrolet Corvette, or Porsche 911. **Options:** Don't waste your money on the satellite navigation system; it's confusing to calibrate and hard to see. **Rebates:** Discounting in the 10 percent range can be expected early next year. **Delivery/PDI:** $950. **Depreciation:** Much slower than average. **Insurance cost:** Quite high. **Parts supply/cost:** Easily found and moderately priced, especially most mechanical and electronic components, but with the exception of some body parts. **Annual maintenance cost:** Less than average. **Warranty:** Bumper-to-bumper 3 years/60,000 km; powertrain 5 years/100,000 km; rust perforation 6 years/unlimited km. **Supplementary warranty:** Not needed. **Highway/city fuel economy:** 7.4–12.2L/100 km.

Honda

Like Toyota, Honda continues to post record-breaking sales and profits with every one of its cars, SUVs, and minivans. Furthermore, the company says it will hold the line on 2004 prices, making the popular Odyssey minivan, for example, a veritable bargain, particularly when you see how much more expensive these vehicles are in the States. A U.S.-sold Odyssey LX, for example, would fetch $25,000 (U.S.); the same vehicle sells in Toronto for $32,200 (CDN), plus a disgraceful $1,240 freight/PDI fee.

Before we leave the subject of costs, there's one disquieting note: as with Toyota, Honda/Acura powertrain reliability has deteriorated considerably over the past several years, though not at the same rate as Ford and Chrysler. As you have seen with Acura's lineup, automatic transmissions (not the CVT) tend to perform erratically and break down prematurely. Consequently, Honda has taken a good first step in extending its warranty to eight years to cover 1999–2003 Honda and Acura failures (as suggested in last year's *Lemon-Aid New Cars and Minivans*).

Still, the risk of sudden automatic transmission failure on the highway represents a clear public danger. Honda should recall the entire lineup.

CIVIC HYBRID ★★★

RATING: Average. The Civic Hybrid comes in second to the Toyota Prius, a much more refined hybrid. Remember, though, that servicing can be problematic because there aren't many Hybrids on Canadian roads due to the cars' limited production runs and a lack of government subsidies for "green" cars. A good alternative is a Honda Civic HX or base Toyota Echo—two cars that cost about $12,000 less, get great gas mileage, and don't depreciate as much—savings will buy lots of gas. **New for 2004**: Nothing.

OVERVIEW: Carried over unchanged this year, the Hybrid looks and feels just like a Civic, both inside and out. So you can expect the same high levels of reliability and durability. Plus, its powertrain uses gasoline-electric technology that lets you travel up to 1,047 km (650 miles) on a single tank of gas (at 50+ mpg) and the battery recharges itself as you drive.

The Hybrid has a $28,500 base price (its U.S. price is $10,000 less), plus $1,010 PDI and freight, and you can count on a 3-year/60,000 km base warranty, a 5-year/100,000 km powertrain warranty, an 8-year/160,000 km battery pack warranty, and emissions-related equipment that's covered by a more extensive warranty.

This five-passenger sedan is equipped with Honda's second-generation Integrated Motor Assist (IMA) system, comprising a 93-hp 1.3L gasoline engine combined with a DC brushless motor and high-tech powertrain management that gives 4.6–4.7L/100 km highway and city fuel economy.

On the downside: owners report that similar fuel economy can be achieved with cheaper, conventional Honda's without complex new technology. Furthermore, the car's unique dual power plants can make for risky driving, as this Hybrid owner warns:

> [With a] 2003 Honda Civic Hybrid on a snowy road, coming over a small rise while going around a moderate curve under 40 mph [64 km/h], the battery charging function, activated by driver taking foot off the gas before cresting the hill, produced progressively stronger engine braking effect on the front wheels, equivalent to an unwanted downshift and causing fishtailing and poor response to corrective steering, so that the car slid across the road and into a snow bank and concrete abutment. $6,000 damage. If oncoming traffic, could have been serious injuries or fatality. Due to the strong and variable [hard to predict or control] braking action of the hybrid's charging system, this car should have a skid sensor activating some kind of traction control, including interruption of the charging function as soon as a skid is detected. I think this should be ordered to be retrofitted to all the Hybrids.

Hybrid

List Price (negotiable)	**Residual Values** (months)			
	24	**36**	**48**	**60**
Base: $28,500 (13%)	$22,500	$19,000	$15,000	$12,000

Technical Data

Powertrain (front-drive)
Engines: 1.3L 3-cyl. (93 hp)
• Electric motor
Transmissions: 5-speed man.
• CVT

Dimension/Capacity
Passengers: 2/3
Height/length/width: 56.3/175/66.7 in.
Headroom F/R: 39.8/37.2 in.
Legroom F/R: 42.2/36 in.
Wheelbase: 103.2 in.
Turning circle: 37 ft.
Cargo volume: 10.1 cu. ft.
Tow limit: N/A
Fuel tank: 50L/reg.
Weight: 2,661 lb.

Safety Features/Crashworthiness

	Std.	**Opt.**
Anti-lock brakes	■	—
Seat belt pretensioners F/R	■	—
Side airbags F/R	■	—
Head restraints F/R	**	**
Visibility F/R	***	**

S2000 ★★★★

RATING: Above Average, steps into the void created by the Prelude's departure. **New for 2004:** Nothing significant.

OVERVIEW: This world-class sport roadster is lightweight, with a peppy 2.2L 4-cylinder 240-hp engine. **Highway/city fuel economy:** 8.2–11.6L/100 km.

Performance comes at a high price, though: $48,600, plus $1,240 PDI/transport. Furthermore, only 8,000 a year are earmarked for North America (expect less than 10 percent of that figure to be sold in Canada). Other models worth considering: the Mazda Miata ($27,695), or the BMW Z4 ($51,500).

The S2000 comes fully equipped with limited-slip differential; anti-lock brakes; AC; cruise control; leather seats; seat belt pretensioners; 16-inch wheels; high-intensity headlights; remote door locks with an engine immobilizer; a CD player with remote audio controls; power everything; and an air deflector.

Powered by a torquier 2.2L VTEC engine, the S2000 will reach 0–100 km/h in about the same time it takes to close the convertible roof—under 6 seconds. All this without straining the high-revving VTEC engine (the tachometer has an unbelievable 9000 redline). The upgraded 6-speed shifter has short throws helped by a direct link with the gearbox, rather than shift-by-wire units used in other cars. New standard features include larger wheels and wider tires.

Performance drivers will immediately discover that the S2000 excels at acceleration, braking, cornering, and shifting, due in large part to the car's powerful engine, anti-lock brakes, double wishbone body, rigid suspension, and electronically controlled rack-and-pinion steering, which enhances steering response without compromising stability.

Not everything is perfect, though. Rear visibility is cut a bit by the convertible top boot cover, the convertible top when it's up, and the rear plastic window *sans* defroster. The small trunk houses a temporary spare tire (shame!) and Honda doesn't include a passenger airbag shut-off switch (double shame!). The ride can be jolting, as with many convertibles: there are quite a few squeaks and rattles; tall drivers won't fit comfortably in the cockpit; and the sound system has Lilliputian controls and sounds cheap.

Service bulletins address erratic operation of the driver's window, a noisy convertible top, and the centre console lifting up in the back.

CIVIC ★★★★

RATING: Above Average. The Honda Civic is one of the best-performing, most dependable small cars money can buy. Nevertheless, its rating has suffered due to the serious safety-related complaints registered against the car over the past several years, and not corrected with its redesign of the 2001 model. **Strong points:** Good acceleration; smooth-shifting automatic transmission; great handling; comfortable ride; good front and rear visibility; lots of interior and trunk space; excellent resale value; and better than average reliability. **Weak points:** Suspension may be too firm for some; difficult rear access; rear seat room limited to two adults; and an unusually large number of performance- and safety-related complaints that include airbag and seat belt malfunctions, sudden acceleration, transmission breakdowns, chronic stalling, and complete brake failure. **New for 2004:** New front- and rear-end treatments; Si Sedan gets VTEC engine and larger tires.

OVERVIEW: The Civic is one of the most refined and competent subcompacts on the market today. Few larger and more expensive cars can match its quality, performance, and roominess.

Civic sedans can be found in three trim levels, much like the Accord's sedans: with a base DX, a mid-level LX, and a high-performance EX. In addition to the above trim levels, the HX coupe is offered to buyers who put fuel economy over highway performance.

DX and LX models use either a 5-speed manual or a 4-speed automatic transmission coupled to a 115-hp 4-cylinder engine, while the Si gets 12 more horses, with the emphasis on low-end torque, from its VTEC power plant. Carried-over HX versions come with a 117-hp VTEC engine mated to either a manual 5-speed or a CVT (continually variable transmission) gearbox.

Last year, Honda upgraded the suspension, steering, seats, and instrumentation. LX and HX models got a manual seat-height adjustment feature.

Cost analysis/best alternatives: New Civics in any form usually cost a few thousand dollars more than domestic offerings. However, they are easy to resell and always command premium prices that equal or surpass the extra money spent when first bought. Two other reasons to buy a 2004 Civic: prices are practically the same as last year, and dealers, feeling the competitive heat from companies like Hyundai, Mazda and Toyota, are willing to dicker. Naturally, the other cars worth considering are the Hyundai Accent, Mazda Protegé, and Toyota Echo. **Options:** Try to get a free extra set of ignition keys written into the contract; Honda's anti-start, theft-protection keys may cost as much as $150 a set. Steer clear of the standard-issue radio and Firestone or Bridgestone tires. **Rebates:** $500– $1,000 and low-rate financing early in 2004. **Delivery/PDI:** $1,010. **Depreciation:** Much slower than average. **Insurance cost:** High insurance costs, wrote one Elmvale, Ontario shopper: "I was choosing between a Saturn and a Civic and I simply couldn't afford the insurance costs for the Civic. They were almost three times the cost of the Saturn in our area." **Parts supply/cost:** Moderately priced and easily found at dealers and independent suppliers. **Annual maintenance cost:** Less than average. **Warranty:** Bumper-to-bumper 3 years/60,000 km; powertrain 5 years/100,000 km; rust perforation 6 years/unlimited km. **Supplementary warranty:** Not needed. **Highway/city fuel economy:** 5.7–7.4L/100 km with the base engine.

Quality/Reliability/Safety

Pro: Quality control: Better than average. **Reliability:** Also better than average, but worrisome in view of the problematic transmissions that can endanger your life or leave you stranded. **Warranty performance:** Much better than average.

Con: Service bulletin problems: Coolant in the engine oil pan; AC goes off in defrost mode; dashboard creaking noise; clicking noise while driving in Second gear, or when turning; noisy steering, particularly steering damper; airbag light stays on for no reason; instrument panel brightness doesn't stay set; erratic function of instrument gauges; rear suspension squeak; windshield cracks; suction cup marks on glass; faulty window run channel; warped or deformed windshield moulding; front air spoiler becomes detached from the bumper, or is loose, deformed, or shows gaps; and key beeper, light chime, and ceiling light don't work. **Owner-reported problems:** Oil leak from lower engine block crack; automatic transmission leaks, sluggish performance when it rains or when passing through a large puddle, and noisy engagement; transmission periodically wouldn't shift into Third or Fourth gear (torque converter replaced); and front strut leakage causes noise and difficult handling. Electrical system and fit-and-finish deficiencies include poorly mounted driver's seat; inoperable door locks and power windows; driver's side window won't roll back up; fan button has to be turned on before AC button will work; erratic fuel gauge; speedometer, and tachometer; interior light hums as it dims; lousy radio speakers; vehicle produces a high magnetic field (confirmed by a Gauss meter); water leaks through the door bottoms, from the tail light into the trunk and onto the driver-side footwell carpet; AC doesn't cool properly and AC condensate drips from under the glove compartment (heating core replaced); dashboard buzzing; loose, rattling door panels and door latches. Owners also report that fuel economy is only half what was advertised and Hybrid fuel savings are overrated. **NHTSA safety complaints/safety:** Hybrid unintended braking action when switching to electrical power can be dangerous; seat belts tighten up progressively when connected; child injured when he became entangled in unfastened rear centre shoulder belt, which retracted, cutting off his air; seat belts failed to lock up in a sudden panic stop; and many complaints that the airbags failed to deploy in a frontal collision:

> I hit the car in front of me going about 40 mph [64 km/h]. I do not think that my seat belt locked. My airbag did not deploy but my passenger-side airbag did. I hit the steering wheel very hard! I had to be transported to the hospital. I have severe chest contusions and neck and back sprain/strain. My insurance company believes my car may be a total loss but my airbag did not go off??? And because it did not go off, I was injured. Honda was contacted the next day. Nine days later, they sent an adjuster out to inspect my car (but not an engineer). I have been told there is no reason to send an engineer out. The adjuster seemed to be more concerned with investigating me than why the airbag did not deploy. I would like to know why it is that I do not have a front end to my car on the driver's side and my airbag did not go off, but there is no damage to the front passenger side and that airbag went off!!

Airbag warning light stays lit; front tie-rod broke, causing complete steering loss; sudden acceleration when AC or heater is engaged; sudden acceleration, particularly when braking; several incidents where cellular phone triggered sudden acceleration:

> My wife installed a Nokia 5110 to car kit, started engine, selected rear gear, and the engine started acceleration. My wife immediately shut the engine down. Now, she drives the car with the cellular phone closed. I think that a possible reason for this accident is the unintended triggering of the throttle to full open by the electromagnetic signal created by the car kit. So, this shows that the insulation of the throttle control system is weak in this model. It has to be much better insulated to screen the electromagnetic noise created by any other equipment.

Chronic stalling accompanied by steering wheel lock-up; steering lock-up while slowing for a traffic light, engine continued running; fuel leakage into the engine compartment while vehicle was underway; car has to warm up a few minutes before brake will work properly; vehicle rolls back when stopped on an incline with automatic transmission engaged; transmission surges forward when put into Reverse (blamed on transmission solenoid); automatic transmission will suddenly downshift in traffic; when accelerator pedal is tapped at less than 5 km/h, vehicle suddenly passes from Drive, to Neutral, to Reverse; transmission won't easily go into First gear; steering wheel wouldn't lock when parked; steering wheel shakes when turned sharply to the left or right; taller drivers' vision blocked by non-adjustable, windshield-mounted rear-view mirror; loose driver's seat; windshield cracked suddenly; difficult to see through bottom of windshield; with AC engaged at night, film covers rear windshield (engine head gasket failure?); exterior and interior lights dim to an unsafe level; and Firestone tire tread separation.

Road Performance

Pro: Acceleration/torque: The base engine is smooth and responsive. Transmission shifting is quiet and effortless with the manual and practically imperceptible with the automatic. The Si's 160-hp 1.6L 4-cylinder is a tire burner. **Routine handling:** Handling is excellent, and the ride, though a bit stiff, is among the best in the subcompact class. The Si's larger wheels add to its handling prowess. **Braking:** Disc/drum with base models and disc/disc with the SiR perform impeccably.

Con: The VTEC variant is noisy—get the VTEC option only if you enjoy constant gear shifting and intend to do a lot of highway driving, where it's most useful and less interactive. The HX gets its impressive fuel economy by sacrificing acceleration power through its tall gearing. **Transmission:** The CVT often feels like it's slipping in gear until you get used to its little quirks. **Emergency handling:** Some body roll when cornering at highway speeds. Skinny tires add to overall noise and harshness. **Steering:** Optional power steering could be more precise.

Comfort/Convenience

Pro: Driving position: Very good due to the height adjustable driver's seat; seat height is higher in the sedans than in previous years. A large cockpit area, along with interior refinements and the large window area, makes for excellent visibility and a feeling of spaciousness. **Controls and displays:** Excellent dashboard design, with easy-to-read gauges and accessible controls. **Climate control:** Heating and ventilation are first class. **Interior space/comfort F/R:** Spacious front seating, less roomy in the rear. **Cargo space:** Versatile and spacious cargo areas. **Trunk/liftover:** Trunk space is acceptable. **Quietness:** Engine/road noise has been cut considerably.

Con: Standard equipment: Though it's reasonably well appointed, the base Honda has few standard features in order to keep the base price down.

Interestingly, the front side airbag feature has a warning device to alert you to the fact that the passenger's head is resting on the airbag. I guess it's there so you can waken the passenger just prior to impact so that he or she is properly positioned. Go figure. Doors give a "tinny" sound when closed. **Entry/exit:** The small door openings and low seating make for difficult entry and exit, and restrict overall visibility. **Interior space/comfort F/R:** Both front and rear seats are uncomfortable on long trips. Trunk has a high sill, making for difficult loading.

Civic

List Price (firm)	**Residual Values** (months)			
	24	**36**	**48**	**60**
DX 2d: $16,100 (11%)	$12,000	$10,000	$8,000	$6,500
LX 2d: $18,500 (12%)	$13,000	$11,000	$9,000	$7,500
DX 4d: $16,100 (11%)	$12,000	$10,000	$8,000	$6,500
Si 2d: $20,800 (12%)	$14,000	$12,000	$10,000	$8,500
Si 4d: $21,500 (12%)	$16,000	$13,000	$11,000	$10,000
SiR 2d: $25,500 (14%)	$18,000	$15,000	$13,000	$11,000

Technical Data

Powertrain (front-drive)
Engines: 1.7L 4-cyl. (115 hp)
• 1.7L 4-cyl. (127 hp)
• 2.0L 4-cyl. (160 hp)
Transmissions: 5-speed man.
• 4-speed auto.

Dimension/Capacity
Passengers: 2/3
Height/length/width: 55.1/175/66.7 in.
Headroom F/R: 38.2/36.2 in.
Legroom F/R: 42.7/34.1 in.
Wheelbase: 103.2 in.
Turning circle: 37 ft.
Cargo volume: 12.9 cu. ft.
Tow limit: N/A
Fuel tank: 50L/reg.
Weight: 2,416 lb.

Safety Features/Crashworthiness

	Std.	**Opt.**
Anti-lock brakes	■	■
Seat belt pretensioners F/R	■	—
Side airbags	—	■
Traction control	—	—
Head restraints F/R	***	**
Visibility F/R	***	**
Crash protection (front) D/P		
2d	*****	*****
4d	*****	*****
Crash protection (side) F/R		
2d	***	****
Hatchback	****	****
4d	****	****
Crash protection (offset)	*****	—

ACCORD ★★★★

RATING: Above Average. This is a practically all-new car that has addressed the powertrain, performance, and comfort deficiencies of its predecessor in its redesigned 2003 model. Unfortunately, the reworked model has serious design and quality deficiencies that are unacceptable in any new vehicle, let alone a Honda Accord. **Strong points:** Excellent acceleration with both engines; new 5-speed automatic transmission works like a charm; well equipped with user-friendly instruments and controls, larger tires, and a sturdier chassis; easy handling; a comfortable ride; good craftsmanship; above-average reliability; and high resale value. **Weak points:** Factory-related defects make the interior stink, cause airbags to explode for no reason, and cause complete brake failure or sudden acceleration in Reverse! Coupe rear access and passenger room is still a bit tight; and windshield dash reflection may be distracting. **New for 2004:** Nothing significant following the 2003s' extensive improvements.

OVERVIEW: Having competed toe-to-toe for what seems like forever with the Toyota Camry and Ford Taurus/Sable, the 2003 Accord jumped leagues ahead of the competition last year with its additional power, larger interior, and impressive array of new safety, performance, and convenience features.

Although the sedan didn't get much longer 0.3 cm (0.1 in.), the car did get an extra 2.5 cm (1 in.) in width and wheelbase. Coupes were lengthened and widened by 2.5 cm (1 in.), and set on the same wheelbase as earlier models.

You want performance? Well, you can choose between two engines: a competent base 2.4L 160-hp 4-banger and a powerful 3.0L 240-hp V6 with 40 more horses than its predecessor. The sportier EX V6 arrived in mid-2003 with standard side curtain airbags, a 6-speed manual transmission, and 17-inch tires. You want ride comfort and responsive handling? The new Accord gives you that too, through a more refined suspension and steering set-up. What about space? Look out, Taurus/Sable—the Accord sedans are roomier than ever before, with interior dimension/capacity that provides more interior space than Ford's mid-sized duo.

Overall, the Honda Accord is smooth, quiet, mannerly, and predictable. Every time Honda has redesigned this line it not only caught up with the latest advances, but also went slightly ahead. Other strong points are ergonomics that prioritize comfort, easy driveability, and predictable handling.

Fast and nimble without a V6, this is the mid-sized sedan of choice for drivers who want maximum fuel economy and comfort along with lots of space for grocery hauling and occasional highway cruising. With the optional V6, the Accord is one of the most versatile mid-sized cars you can find. It offers something for everyone, and its top-drawer quality and high resale value mean there's no way you can lose money buying one.

Cost analysis/best alternatives: Get this year's identical version and try haggling for a better price; the redesigned 2003s are more glitch-prone than previously thought. If you go for a 2003, try to at least get a better-quality second-series version made in the spring of 2003. Some other vehicles worth considering: the BMW 3 Series, Mazda6, Nissan Altima, and Toyota Camry. **Options:** Choose the V6 for a smoother ride that gives you some grunt in reserve. The DVD navigation with voice control feature found on the EX and V6 coupe is a bit gimmicky, but easy to use and understand. Try to get a free extra set of ignition keys written into the contract; Honda's anti-start, theft-protection keys may cost as much as $150 a set. **Rebates:** Not likely. **Delivery/PDI:** $1,095. **Depreciation:** Slower than average. **Insurance cost:** Higher than average. **Parts supply/cost:** Good availability and moderately priced. **Annual maintenance cost:** Less than average. **Warranty:** Bumper-to-bumper 3 years/60,000 km; powertrain 5 years/100,000 km; rust perforation 5 years/unlimited km. **Supplementary warranty:** Not needed. Honda's base warranty and "goodwill" warranty extensions should sufficiently cover the Accord's first-year defects. **Highway/city fuel economy:** 6.9–9L/100 km with the base engine; 7.8–11.6L/100 km with the V6 automatic sedan.

Quality/Reliability/Safety

Pro: Curtain side airbags. **Reliability:** Better than average, yet an increase in transmission and brake failures is worrisome. Still, one would imagine that Honda has learned from its mistakes and isn't putting the same failure-prone components into its 2003s. **Warranty performance:** Above average. CAA surveys have shown that customer satisfaction is an impressive 88 percent, compared to 85 percent for both the Toyota Camry and Mazda 626. Keep in mind that Honda puts a "goodwill" clause in almost all of its service bulletins, allowing service managers to submit any claim to the company long after the original warranty has elapsed.

Con: Quality control: Bad news. The 2003 redesign has generated almost two hundred safety-related complaints, when a quarter of that number would be normal. Also, following the 2003's redesign, Honda promised us better performing, more durable transmissions. They lied. Fortunately, an extended warranty now covers tranny breakdowns up to eight years. **Owner-reported**

problems: "Rotten-egg" smell in the cabin; coolant in the engine oil pan; frequent hard starts; transmission shifts erratically; brake shuddering, grinding, and squealing; warped brake rotors; popping noise when accelerating; hole in AC condenser; defective CD changer; stereo speaker hums; windshield creaks in cold, dry weather; moon roof doesn't close all the way; pervasive rattling; windows rattle and tick; wrinkled, bubbling door window moulding; roof water leaks; and headliner sags in the rear. One owner found that falling acorns pock-marked the Accord's body, while his CRV escaped damage. **Service bulletin problems:** "Goodwill" campaigns to replace automatic transmissions and MICU units that regulate door locks, trunk alarm, etc; corrective action for coolant leakage into the engine oil pan; oil leaks at the cylinder head cover; a faulty air intake air breather pipe hose may cause the engine warning light to remain lit; engine clicking or ticking at idle; automatic transmission leaks on the cooler lines; hard starts; troubleshooting calipers, rotors, and pads, following complaints of excessive brake vibration (wait a minute; weren't these the same chronic problems present before Honda's much-vaunted 2003 brake upgrade?):

> My 2003 Accord has been in the Honda dealer with problems with the front brakes (warped rotors). After having the front rotors resurfaced and pads replaced twice it was found that the front calipers were hanging up. My car just turned 17,000 miles [27,200 km] and it has had problems with the front brakes since it had 8,000 miles [12,800 km]. At that time I was told by the dealer that I was "riding" my brakes and they would only take care of this problem for me at no cost just this "one time." I feel Honda is knows that 2003 Accords have brake problems and Honda puts the blame on everyone else except themselves. My vehicle has now been in the Honda dealer for two weeks because they are unable to locate new calipers. From talking with other 2003 Accord owners and reading information on other websites, [I know] I'm not alone with this problem....

Troubleshooting ABS brake light illumination; inaccurate gauges and odometer; excessive steering vibration; noisy power steering countermeasures; investigating owner reports of cracked windshields; doors don't unlock in cold weather; poor AC cooling; rear shelf rattling; dash or pillar creaking or clicking; roof water leak fix; exhaust pipe rust stains; and HomeLink remote system range is too short. **NHTSA safety complaints:** Sudden, unintended acceleration while car is in motion or when put into Reverse; axle suddenly snapped; front and side curtain airbags deployed for no reason; airbag failed to deploy during collision; severe pulling to the right; power-steering groan believed to be caused by the steering pump; console overheats and smells burnt; keys overheat in the ignition; complete brake failure; ABS and traction control lights stay lit and corrective parts aren't easily found; frequent Michelin tire blowouts; and rear vision is obstructed by head restraints, high rear deck, and roof pillar.

Road Performance

Pro: Emergency handling: Impressive. The ride is firm and well controlled, except for a rear suspension that may be too soft for some. **Steering:** Sporty and very responsive, with plenty of road feel. **Acceleration/torque:** Excellent with both the 4- and 6-cylinder engine. Both engines give sparkling performance with plenty of low-end torque and minimal noise. **Transmission:** The 5- and 6-speed manual and automatic transmissions work very well, with smooth and light clutch action. **Routine handling:** Excellent handling in town and on the highway. It's also amazingly smooth and quiet. There isn't a trace of vibration throughout the operating range, including at idle. Excels in smoothness when passing over freeway expansion joints and potholes, where it damps out the jolts better than most cars in its class. The emphasis on comfort also dominates the Accord's ride and handling. Although its chassis is as good as any, Honda's new suspension settings aren't as firm as those of the other cars in this class. **Braking:** Disc/drum and disc/disc brakes perform quite well with little fading.

Con: Hard cornering compromised a bit by softer rear suspension. Standard rear drum brakes are out of character for a car with upscale pretensions.

Comfort/Convenience

Pro: Standard equipment: A nice array of standard features on the base model, like a tilt/telescope steering column. Everything is right where it should be, easily seen and easily reached. **Driving position:** Excellent driving position and forward visibility. Very comfortable, sporty front seats are well padded and supportive. **Controls and displays:** Clear gauges and instruments. User-friendly controls and instruments. **Climate control:** Very easy to access and comprehend; quiet, efficient, and easily calibrated. **Entry/exit:** Easy entry and exit, though the doors could be a bit wider. **Interior space/comfort F/R:** Ample interior room for average-sized adults. **Cargo space:** Better than average. **Trunk/liftover:** Large trunk on sedans and versatile hatchback design; and a low liftover.

Con: Rear room for three is a bit tight. Sedans lose a little rear-seat legroom and coupes have a bit less rear headroom and legroom. Rear seat cushions lack thigh support. Locating the power sunroof controls left of the steering column on the dash is counterintuitive. **Quietness:** Worse than average, highlighted by constant interior rattling and creaking.

Accord

List Price (firm)	**Residual Values** (months)			
	24	**36**	**48**	**60**
DX Sedan: $23,900 (13%)	$19,500	$16,500	$13,500	$10,500
Accord LX Coupe: $25,300 (13%)	$20,500	$17,500	$14,500	$11,500
Accord EX V6: $32,900 (15%)	$25,000	$22,000	$19,000	$16,000

Technical Data

Powertrain (front-drive)
Engines: 2.4L 4-cyl. (160 hp)
• 3.0L V6 (240 hp)
Transmissions: 5-speed man.
• 5-speed auto.
• 6-speed auto.
Dimension/Capacity (coupe)
Passengers: 2/3
Height/length/width:
55.7/187.6/71.3 in.
Headroom F/R: 37.5/36.1 in.
Legroom F/R: 43.1/31.9 in.
Wheelbase: 105.1 in.
Turning circle: 36.1 ft.
Cargo volume: 13.2 cu. ft.
Tow limit: N/A
Fuel tank: 65L/reg.
Weight: 3,265 lb.

Safety Features/Crashworthiness

	Std.	Opt.
Anti-lock brakes	■	—
Seat belt pretensioners F/R	■	—
Side airbags	■	—
Traction control	■	—
Head restraints F/R	****	***
Visibility F/R	*****	***
Crash protection (front) D/P	*****	*****
Crash protection (side) F/R	*****	*****
Crash protection (offset)	*****	—

ODYSSEY ★★★★

With Toyota's new Sienna, the Odyssey begins playing catch-up.

RATING: Above Average, but not as well built as the redesigned 2004 Toyota Sienna. On both minivans, quality has slipped over the years, as evidenced by the disturbingly frequent reports of safety- and performance-related failures. Of particular concern are airbag malfunctions; automatic sliding door failures; engine failures (Toyota: oil sludge; Honda: V6 engine oil leaks); transmission breakdowns and erratic shifting (now covered by Honda's latest goodwill warranty); and sudden brake loss. **Strong points:** Strong engine performance;

carlike ride and handling; easy entry/exit; second driver-side door; quiet interior; most controls and displays are easy to reach and read; lots of passenger and cargo room; low step-up facilitates entry/exit; an extensive list of standard equipment, and a willingness to compensate owners for production snafus. **Weak points:** A high base price; front-seat passenger legroom is marginal due to the restricted seat travel; you can't slide your legs comfortably under the dash; some passengers bump their shins on the glove box; third-row seat is only suitable for children; the narrow back bench seat provides little legroom, unless the middle seats are pushed far forward, inconveniencing others; radio control access is blocked by the shift lever, and it's difficult to calibrate the radio without taking your eyes off the road; power-sliding doors are slow to retract; some tire rumble, rattles, and body drumming at highway speeds; premium fuel is required for optimum performance; a decline in quality; poor head restraint crashworthiness rating; and rear-seat head restraints impede side and rear visibility. The storage well won't take any tire larger than a "space saver"—meaning you'll carry your flat in the back. **New for 2004:** Nothing important.

OVERVIEW: No longer simply an Accord masquerading as a minivan, the Ontario-built Odyssey is longer, wider, taller, and more powerful than ever before. It has a lean look, yet the interior is wide enough and long enough to accommodate a 4' x 8' sheet of plywood laid flat. Sliding doors are offered as standard equipment and, if you buy the EX version, they will both be power assisted.

You can only get one engine—a powerful 240-hp V6 that easily handles most driving chores. Odyssey also offers an upgraded, fully independent suspension, front and rear AC, second-row captain's chairs that can be shoved side by side to create a bench seat, and a convenient third-row bench seat, which folds into the floor when not in use.

Cost analysis/best alternatives: If you can find one, a discounted 2003 model would give you all the performance and convenience features you'll need. But, if you can only find a more expensive 2004, don't dismay—you'll get the difference back come trade-in time. Although a fully loaded EX costs $3,000 more than the base model, you *do* get traction control, power rear doors, remote entry, a power driver's seat, fore and aft adjustment for the middle seats, a more refined climate control system, and an upgraded sound system. I suggest you bite the bullet and buy your Odyssey before the second round of price increases are announced early next year. There is room for some price negotiation, with savings in the $1,500–$2,000 range, but make sure you have a specific delivery date spelled out in the contract, along with a protected price, in case there's a price increase while you're waiting for delivery. The reworked 2004 Nissan Quest is also worth considering if you plan mostly light-duty urban commuting. However, if you want better handling and reliability, the closest competitor to the Odyssey is Toyota's Sienna minivan. The Pontiac Montana and Chevrolet Venture from GM are good front-drive choices—they're only a bit less reliable, but are still acceptable. If you're looking for lots of towing "grunt," then the rear-drive GM full-sized vans would be good buys. **Options:** Traction control is a good idea. Remember, if you want the gimmicky video entertainment and DVD navigation system, you also have to spring for the expensive (and not-for-everyone) leather seats. **Rebates:** Honda has pulled back on its incentives due to higher exchange rates and the unabated popularity of its Odyssey. Alberta is the most difficult region in which to find a bargain; don't expect any substantial incentives in that province before the fall of 2004. **Delivery/PDI:** $1,095. **Depreciation:** Slower than average. **Insurance cost:** Higher than average. **Annual maintenance cost:** Average. **Parts supply/cost:** Supply is better than average because the Odyssey uses many generic Accord parts. **Warranty:** Bumper-to-bumper 3 years/ 60,000 km; powertrain 5 years/100,000 km; rust perforation 5 years/ unlimited km. **Supplementary warranty:** Not needed. **Highway/city fuel economy:** 8.6–12.9L/100 km. Premium fuel is recommended, but the Odyssey will run on regular, with a slight loss of power (about five horses).

Quality/Reliability/Safety

Pro: Reliability: Reliability is better than average. **Warranty performance:** Comprehensive base warranty that's usually applied fairly, with lots of "wiggle room" that the service manager can use to apply "goodwill" adjustments for post-warranty problems. **Safety:** Honda claims it's the only automaker to offer head restraints and three-point seat belts in all seven seating positions. Plus, the body structure has been extensively reinforced to protect occupants from all kinds of collisions, including rollovers.

Con: Owners complain that "goodwill" refunds aren't extended to all model years with the same defect; recall repairs take an eternity to perform; and dealers exhibit an arrogant, uncaring "take it, or leave it" attitude. **Quality control:** Honda still hasn't tackled its engine and automatic transmission failures as effectively as has Toyota. Another notable and hazardous shortcoming: failure-prone sliding doors. Imagine, they open when they shouldn't, won't close when they should, catch fingers and arms, get stuck open or closed, are noisy, and frequently require expensive servicing. The Check Engine light may stay lit due to a defective fuel-filler neck. There's a fuel sloshing noise when accelerating or coming to a stop, and the transmission clunks or bangs when backing uphill or when shifted into Reverse. There are also reports of rattling and chattering when put into forward gear. Owners note a loud wind noise and vibration from the left side of the front windshield, along with a constant vibration felt through the steering assembly and front wheels. Passenger doors may also require excessive force to open. And owners have complained of severe static electricity shocks when exiting. Other potential problem areas are frequent and high-cost front brake maintenance and trim and accessory items that come loose, break away, or malfunction. Automatic transmission failures continue to be a major shortcoming. **Owner-reported problems:** Engine oil leaks and timing belt breakage. Transmission breakdowns; when shifting into

Fourth gear, transmission suddenly downshifts to First gear, or engine almost stalls out and produces a noise like valve clattering. There's a transmission gear whine at 90 km/h or when in Fourth gear (transmission replaced); front-end clunking caused by welding breaks in the front subframe; loud fuel splashing sound in the fuel tank when coming to a stop; vehicle pulls to the right when underway; premature front brake wear; excessive front brake noise; sliding side door frequently malfunctions; electrical glitches; and accessory items that come loose, break away, or won't work. Plastic interior panels have rough edges and are often misaligned. **Service bulletin problems:** Fuel may leak from gas tank; engine oil leaks; engine runs roughly; hard starts; power-steering pump noise; front steering damper noise; exhaust rattles or buzzes upon acceleration; squealing from rear quarter windows; front door howls in strong crosswind; dash or front strut creaking or ticking; broken or shattered tailgate glass; deformed windshield moulding; inoperative centre-mount brake light; warning lights may blink on and off for no reason; and troubleshooting an off-centre steering wheel. **NHTSA safety complaints/safety:** Fire ignited in the CD player; passenger seatbacks collapsed when vehicle was rear-ended; child injured from a side collision because second-row seat belt failed to hold her in due to gap caused by door attachment; rear seat belts lock for no reason; defective speed sensor caused vehicle to suddenly lose power when merging into traffic; vehicle suddenly shut down in traffic; hard starts and engine misfiring; many automatic transmission malfunctions; faulty steering causes wander; sudden brake failure; driver often shocked when touching door handle; fuel spits out when refuelling; inaccurate fuel gauge readings; and sliding door closed on driver's hand.

Road Performance

Pro: Emergency handling: Predictable and well controlled. **Acceleration/torque:** Quiet yet powerful V6 engine performance with plenty of torque throughout the power band. In fact, only the Toyota Sienna exceeds the Odyssey in powertrain refinement and mid-range grunt. **Transmission:** Usually smooth and quiet shifting (note exception below), though a bit slow to downshift at full throttle. **Routine handling:** Better handling than most of the competition, with a short turning circle, making it quite nimble; akin to driving a large, quiet, sporty family sedan. **Braking:** Excellent braking now that rear discs have been added. Even with a full load and after repeated stops, they perform quite well—most of the time.

Con: Steering: A bit heavy at low speeds; sometimes feels over-boosted at highway speeds. Column-mounted shift lever tends to slide past Drive into Third gear when shifted out of Park. Automatic transmission clunks and jolts the occupants when shifted into Reverse. The taut suspension is also a bit jolting over bumps, but the ride smoothes out as the load is increased.

Comfort/Convenience

Pro: Standard equipment: Well appointed. **Driving position:** Comfortable, commanding driving position. Additional padding to the front seat cushions adds to driving comfort. **Controls and displays:** Easy to reach and read. **Climate control:** Quiet and efficient. Each seat has its own AC and heating vent. **Entry/exit:** Low step-up. The Odyssey provides carlike room in the front two rows of seats; getting in is no more difficult than climbing into an Accord. **Interior space/comfort F/R:** It's easy to strap a child seat into the Odyssey because there's far less lifting and reaching. The bench seat in the centre row is more practical than the captain's chairs, which can't be removed. It's more like a seven-seater station wagon than a minivan. All passengers have their own reading light, air vent, and power rear vent window controls. **Trunk/liftover:** There's space enough for groceries, and the edging around the storage area will help keep your apples from rolling around. **Cargo space:** Lots of little storage areas. A handy storage tray between the front seats folds down when not in use. If you need to expand the cargo area, the third seat folds flat into the floor. **Quietness:** The interior is quiet in most situations.

Con: Driving position: The driver's seat has backrest bolstering that is too padded for some, not enough for others (try before you buy). Styling isn't to everyone's taste: colour and model selection are limited, carpeting looks cheap, and some find the Odyssey looks too buslike. **Interior space/comfort F/R:** Front-seat passenger legroom is marginal due to the restricted seat travel; you can't slide your legs comfortably under the dash. Some passengers bump their shins on the glove box. Third row seat is only suitable for children; the narrow back bench seat provides little legroom, unless the middle seats are pushed far forward, inconveniencing others. **Controls and displays:** Access to radio controls is blocked by the shift lever and it's difficult to calibrate the radio without taking your eyes off the road. Power outlet, located at the base of the centre console, is awkward to access. Sealed centre windows. Power-sliding doors are slow to retract. Huge interior is slow to warm up and rear side windows fog up with a full passenger load. **Climate control:** AC controls and pictograms on the EX aren't very user-friendly and are confusing to adjust. **Cargo space:** Stowing the "hideaway" third seat exposes metal sidewall anchors that can easily damage cargo. The storage well won't take any tire larger than a "space saver"—meaning you'll carry your flat in the back. **Quietness:** Some tire rumble and body drumming at highway speeds.

Odyssey

List Price (negotiable)	**Residual Values** (months)			
	24	**36**	**48**	**60**
Odyssey LX: $32,200 (18%)	$25,500	$21,500	$17,500	$14,500
Odyssey EX: $35,500 (18%)	$26,500	$22,500	$18,500	$15,500

Technical Data

Powertrain (front-drive)
Engine: 3.5L V6 (240 hp)
Transmission: 5-speed auto.
Dimension/Capacity
Height/length/width:
68.5/201.2/75.6 in.
Headroom F/R1/R2: 41.2/40/38.9 in.
Legroom F/R1/R2: 41/40/38.1 in.
Wheelbase: 118.1 in.
Turning circle: 37.7 ft.
Passengers: 2/2/3
GVWR: N/A
Cargo volume: 146.1 cu. ft.
Tow limit: 3,500 lb.
Ground clear.: 6.4 in.
Fuel tank: 76L/prem.
Weight: 4,310 lb.

Safety Features/Crashworthiness

	Std.	Opt.
Anti-lock brakes (4W)	■	—
Seat belt pretensioners F/R	■	—
Side airbags	■	—
Traction control	■	—
Head restraints F/R	**	**
Visibility F/R	*****	*
Crash protection (front) D/P	*****	*****
Crash protection (side) F/R	*****	*****
Crash protection (offset)	*****	—

Hyundai

Hyundai quality is incredibly good and decidedly better than what Detroit and recently designed Honda and Toyota models offer. If you still have your doubts, check out the most recent J.D. Power survey reports or click onto NHTSA's Internet consumer complaints database (*www-odi.nhtsa.dot.gov/cars/problems/complain*). For my part, I'll add that I bought my wife a 2001 Elantra three years ago and have only had to replace the CD changer—for free. Hyundai doesn't know from whom or where I bought the car.

Hyundais are now backed by a fairly comprehensive warranty, and they're reasonably priced—more reasons why Canadians are giving Hyundai products more consideration and the company's North American sales are an unprecedented success.

Unfortunately, Hyundai's executives aren't as reliable as their cars; they've had a 10-year problem with telling the truth. During that period, the company misled over 1.3 million owners as to their vehicles' true horsepower—in some cases boosted by almost 10 percent. Hyundai admitted its deception in September 2002, over six months after a Vancouver Competition Bureau investigator received an anonymous tip that Hyundai was breaking the law with impunity.

Hyundai has beat being charged with misleading advertising by offering buyers various forms of compensation that run the gamut of a $500 refund to an extended warranty.

Who says crime doesn't pay? Ask the boys in Ottawa.

On a more positive note, Hyundai's complaint ratio has dropped considerably over the past few years, concomitant with the company's ditching its failure-prone Excel and upgrading the Sonata. Government-collected factory- and safety-related complaints have fallen to a level where they're less frequent than what you'll find recorded by Honda and Toyota owners.

ACCENT ★★★★★

RATING: Recommended, if you're using your Accent primarily as a fuel-sipping urban econocar. Think of it as a more refined Metro/Sprint from South Korea, built for light commuting chores. **Strong points:** Reasonably priced and well appointed; adequate engine and automatic transmission performance in most situations; comfortable driving position with good visibility and height-adjustable driver's seat; cheap on gas; average PDI and freight fee; and a low base price. **Weak points:** Primitive manual transmission; passing power comes up short, especially when equipped with the automatic transmission; mediocre braking; excessive engine, road, and wind noise; and uncertain long-term reliability, particularly with the automatic transmission. **New for 2004:** Nothing important.

OVERVIEW: This front-drive 4-cylinder entry-level sedan is one of the cheapest small cars sold in North America. Carrying a homegrown 1.6L 4-cylinder engine coupled to a 5-speed manual or 4-speed automatic transmission, the Accent offers bare-bones motoring without sacrificing basic amenities, including AC, a height-adjustable driver's seat with lumbar support, and split-fold rear seats.

Cost analysis/best alternatives: It makes little difference whether you buy a 2003 or 2004 model, since they are identical; shop in order to take advantage of the new year rebates, instead. Other vehicles worth considering: the Honda Civic, Nissan Sentra, and Toyota Echo and Corolla. **Options:** An automatic transmission and power steering are essential. **Rebates:** Expect $1,500 rebates and zero percent financing on unsold 2003 models. **Delivery/PDI:** $795. **Depreciation:** Slower than average, now that fuel prices have soared. **Insurance cost:** Average. **Parts supply/cost:** Parts aren't hard to find and they're reasonably priced. **Annual maintenance cost:** Average. **Warranty:** Bumper-to-bumper 3 years/60,000 km; powertrain 5 years/100,000 km; rust perforation 5 years/unlimited km. **Supplementary warranty:** A good idea, since Hyundai powertrains can be problematic. **Highway/city fuel economy:** 6.5–8.1L/100 km.

Quality/Reliability/Safety

Pro: Quality control: Average. Repairs are straightforward because of a fairly simple design, and parts are less expensive than average. **Reliability:** Above average. **Warranty performance:** Above average.

Con: Service bulletin problems: Clicking noise from speakers. **Owner-reported problems:** Engine may burn oil; engine cooling system and cylinder head gasket failures; automatic transmission gears grind when shifting, slip, and feel like they're in no gear at all; front end shakes when passing over a hill; premature ignition coil failure; Check Engine light stays lit; premature front brake wear and excessive noise when braking; wheel bearings, fuel system, and electrical components may also fail prematurely. **NHTSA safety complaints/ safety:** Airbags failed to deploy and excessive shaking when brakes are applied.

Road Performance

Pro: Although the 1.6L engine is no pocket rocket, it performs adequately in city traffic. **Emergency handling:** Slow, but predictable. **Steering:** The non-power steering is precise and transmits plenty of road feedback at higher speeds. **Transmission:** Surprisingly, this is one car where the automatic transmission performs better than the manual gearbox. **Routine handling:** Above average. The Accent rides comfortably and handles responsively.

Con: Acceleration/torque: The manual transmission is hard to shift correctly due to its balky linkage and long lever movements. Uneven pavement makes for a busy, jittery ride that's accentuated as passengers are added. Steering without the power-assist option takes a lot of effort around town and when parking. **Braking:** Unacceptably long stopping distances (100–0 km/h: 42 m).

Comfort/Convenience

Pro: Driving position: Very good. Multiple front seat adjustments and a well-appointed interior make for a pleasant driving environment with a good view of the road. **Controls and displays:** Well laid-out dashboard and controls. **Climate control:** Efficient, easy-to-understand climate control system. **Entry/exit:** Front and rear access is impressively easy with the sedan; more problematic with the hatchback. **Interior space/comfort F/R:** Room for two adults in the rear. Plenty of front headroom. Fairly comfortable but very firm and narrow front seats. **Cargo space:** Reasonable amount of luggage space, but the hatchback is more versatile with its folding rear seatbacks. **Trunk/liftover:** Average-sized trunk with a low liftover.

Con: Standard equipment: Few standard features, and there aren't a lot of options to choose from. Limited front legroom for tall passengers; rear headroom is inadequate for tall occupants. Sedans have a small trunk opening. **Quietness:** Excessive engine, road, and wind noise.

Accent

List Price (negotiable)	**Residual Values** (months)			
	24	**36**	**48**	**60**
Accent GS: $12,895 (11%)	$8,000	$6,500	$5,500	$4,500
Accent GL: $14,195 (12%)	$9,000	$7,500	$6,500	$5,500
Accent GSi: $14,995 (12%)	$10,000	$8,500	$7,500	$6,500

Technical Data

Powertrain (front-drive)
Engine: 1.6L 4-cyl. (104 hp)
Transmissions: 5-speed man.
• 4-speed auto.
Dimension/Capacity
Passengers: 2/3
Height/length/width:
54.9/166.7/65.7 in.
Headroom F/R: 38.9/38 in.
Legroom F/R: 42.6/32.7 in.
Wheelbase: 96.1 in.
Turning circle: 32.5 ft.
Cargo volume: 11.8 cu. ft.
Tow limit: N/A
Fuel tank: 45L/reg.
Weight: 2,350 lb.

Safety Features/Crashworthiness

	Std.	**Opt.**
Anti-lock brakes	—	—
Seat belt pretensioners F/R	■	—
Side airbags	—	■
Traction control	—	—
Head restraints F/R	***	***
Visibility F/R	*****	*****
Crash protection (front) D/P	****	****
Crash protection (side) F/R	***	**

ELANTRA ★★★★

RATING: Above Average. **Strong points:** Peppy engine, with plenty of passing power; smooth-shifting transmission; good ride quality; plenty of interior room; reasonably priced; low PDI and freight fee; well appointed; and nimble handling. **Weak points:** Lots of wind noise; difficult rear entry/exit; below-average crashworthiness scores. **New for 2004:** Minor styling changes and the engine gains CVVT (continuously variable valve timing) for smoother performance and less power (five horses).

OVERVIEW: This Italian-designed, front-drive, conservatively styled subcompact is only marginally larger than the discontinued Excel. Nevertheless, its 2.0L 138-hp 4-cylinder power plant supplies the much-needed power that you won't find with many other cars in this price range. The ride and handling are also quite good, due mainly to the Elantra's longer wheelbase and more sophisticated suspension.

Available as a four-door sedan or a GT hatchback and sedan, the Elantra is loaded with standard equipment, including standard AC and intermittent wipers, full centre console, dual remote control outside mirrors, remote fuel-filler door and trunk release, reclining bucket seats, and tinted glass. Upgrading to the GT will get you leather seating, four-wheel disc brakes, a high-performance suspension, larger alloy wheels, and remote keyless entry.

Cost analysis/best alternatives: I like the styling changes, but I'm worried about the new CVVT engine cutting power and having first-year reliability problems. For these reasons, I'd choose the 2003 version, particularly, if it's sufficiently discounted. Other cars worth considering are the Honda Civic, Nissan Sentra, and Toyota Echo, Corolla, or Matrix. **Options:** The GT version is a bargain, when one totes up the cost of its standard features if purchased separately. **Rebates:** $2,000 rebates and zero percent financing. **Delivery/PDI:** $795. **Depreciation:** Slower than average. **Insurance cost:** A bit higher than average. **Parts supply/cost:** Easy-to-find and reasonably priced parts, with heavy discounting by dealers. **Annual maintenance cost:** Average. **Warranty:** Bumper-to-bumper 3 years/60,000 km; powertrain 5 years/

100,000 km; rust perforation 5 years/unlimited km. **Supplementary warranty:** A wise investment. **Highway/city fuel economy:** 7.1–10.8L/100 km.

Quality/Reliability/Safety

Pro: Quality control: Above average. **Reliability:** Overall reliability has been quite good. **Warranty performance:** Above average. Hyundai staffers usually deal with customer claims in a fair and efficient manner.

Con: Owner-reported problems: Chronic stalling; excessive brake noise; passenger-side scraping noise when underway; wind howls in the interior when encountering a crosswind; while driving, a humming noise emanates from the corners of the windshield; tire thumping; window defrosting takes too long; rainwater seeps in under the door. **Service bulletin problems:** Automatic transmission sticks in Second gear; harsh shifts into Drive or Reverse; rough-running engine may require an upgraded fuel pump; corrosion in the front door wiring connector; and repair tips for an electro-chromatic mirror wire short. **NHTSA safety complaints/safety:** Engine surging while on the highway; vehicle suddenly lost all power; complete loss of brakes; headlights will read "dim" but will actually be on high; distracting windshield glare; and seat belts fail to lock.

Road Performance

Pro: Acceleration/torque: Good acceleration with plenty of low-end torque. **Transmission:** Both transmissions perform well, although the automatic shifts frequently when the engine is pushed. **Emergency handling:** Better than average. **Steering:** Accurate and responsive. **Routine handling:** Independent suspension helps make for a pleasant ride, with exceptional handling and control. **Braking:** Brakes are usually quite efficient (see below), with little fading after successive stops.

Con: Braking: Brakes are difficult to modulate.

Comfort/Convenience

Pro: Standard equipment: Reasonably well equipped. **Driving position:** Good driving environment with excellent all-around visibility and comfortable front seats. **Controls and displays:** The attractive interior includes an easy-to-read dashboard with large analogue gauges that are complete and clear, and climate and radio controls that are effective and easy to operate. **Climate control:** The climate control system works without a hitch. **Interior space/comfort F/R:** Plenty of interior room, fore and aft. **Entry/exit:** Good up front; difficult access to the rear. **Cargo space:** Average. **Trunk/liftover:** Average, with a low liftover.

Con: Firm rear seats are uncomfortable for long trips. **Quietness:** Excessive high-speed tire, engine, and road noise.

Elantra

List Price (negotiable)	**Residual Values** (months)			
	24	**36**	**48**	**60**
Elantra GL: $15,625 (13%)	$9,000	$7,000	$5,500	$4,500
Elentra GT: $19,025 (14%)	$12,000	$10,000	$7,500	$6,000

Technical Data

Powertrain (front-drive)
Engine: 2.0L 4-cyl. (138 hp)
Transmissions: 5-speed man.
• 4-speed auto.
Dimension/Capacity
Passengers: 2/3
Height/length/width:
56.1/177.1/67.7 in.
Headroom F/R: 39.6/38 in.
Legroom F/R: 43.2/35 in.
Wheelbase: 102.8 in.
Turning circle: 36 ft.
Cargo volume: 12.9 cu. ft.
Tow limit: N/A
Fuel tank: 55L/reg.
Weight: 2,698 lb.

Safety Features/Crashworthiness

	Std.	**Opt.**
Anti-lock brakes	—	■
Seat belt pretensioners F/R	■	—
Side airbags	—	■
Traction control	—	■
Head restraints F/R	****	***
Visibility F/R	*****	*****
Crash protection (front) D/P	****	*****
Crash protection (side) F/R	*****	****
Crash protection (offset)	*	—

SONATA ★★★★

RATING: Above Average. **Strong points:** Well equipped, stylish, and reasonably priced; plenty of V6 power; good handling; comfortable ride; and much-improved quality control. **Weak points:** Some powertrain glitches; 4-cylinder engine overwhelmed by a full load and hilly terrain; uncomfortable rear seating; excessive cabin noise. **New for 2004:** No important changes.

OVERVIEW: Give Hyundai credit: it's building more refined cars, and last year's engine and suspension improvements make the Sonata one of the best mid-sized sedans to come out of South Korea to date. It returns this year with a rigid structure, based on the Kia Optima/Magentis platform, a double-wishbone front suspension (like the Accord) and independent multi-link rear suspension, and more attractive styling. This mid-sized sedan still lacks the handling and ride refinements to put it in the same league as the redesigned Honda Accord and Nissan Altima, but it has become a credible alternative to most other cars in that niche—and it's priced thousands of dollars less. Sonata's recent upgrades make it quite attractive when compared to the second-tier mid-sized sedans represented by the aging and less reliable Ford Taurus and Sable and Chrysler Cirrus and Stratus.

Like Hyundai's other models, the Sonata is loaded with standard features that are optional or not found on competing makes. Some of those standard features include traction control, an automatic transmission with a semi-manual mode, ABS, AC, power steering, and a 2.4L 147-hp 4-cylinder engine. It also features an optional 2.7L 170-hp V6.

Hyundai has again put the accent on safety, giving buyers a lot more safety features for their money than competing models. Standard features include: side airbags, Seat belt pretensioners F/R, an integrated rear child safety seat, and a "smart" passenger-side airbag that won't deploy if the passenger weighs less than 30 kg (66 lb.).

Cost analysis/best alternatives: Buy a discounted 2003 model, since it offers practically the same equipment and performance as the 2004. Other cars worth considering are the Honda Accord, Mazda6, Nissan Altima, and Toyota Camry. **Options:** Choose the V6 engine and 16-inch wheels for better performance and handling. If you get the 4-banger engine, keep in mind that good fuel economy means putting up with an engine that's rougher and noisier than the V6. Be wary of the sunroof; it eats up a lot of headroom and has a history of leaking. **Rebates:** Expect $3,000 rebates and zero percent financing on all leftover models. **Delivery/PDI:** $795. **Depreciation:** Average. **Insurance cost:** Average. **Parts supply/cost:** Easy to find and relatively inexpensive. Large engine compartment for easy servicing. **Annual maintenance cost:** Higher than average. **Warranty:** Bumper-to-bumper 3 years/60,000 km; powertrain 5 years/100,000 km; rust perforation 5 years/unlimited km. **Supplementary warranty:** Not needed. **Highway/city fuel economy:** 7.7–11.4L/100 km with the 2.4L and 7.9–12.3L/100 km with the V6.

Quality/Reliability/Safety

Pro: Quality control: Above average quality control. Repairs are straightforward because of the Sonata's simple design. **Reliability:** Despite some powertrain deficiencies, no serious reliability problems have taken these cars off the road. **Warranty performance:** Above average.

Con: Owner-reported problems: Erratic engine performance (poor idling and stalling); vehicle loses power when going uphill; transmission leaks, clunks, and hangs up in gear; and excessive vibration when the brakes are applied. **Service bulletin problems:** Harsh shifts into Drive or Reverse; sticks in Second gear. **NHTSA safety complaints/safety:** Driver-side seat belt buckle released and rear-view mirror is too big and obstructs the view.

Road Performance

Pro: Routine handling: Above average. Hyundai has changed the suspension setting to provide a more comfortable ride and better handling. **Emergency handling:** Also above average, particularly with 16-inch wheels. **Steering:** Very good. Predictable, with lots of road feedback. **Acceleration/torque:** Reasonably good acceleration, but the base engine's insufficient torque makes the optional V6 a prerequisite for highway cruising, especially over hilly terrain. **Transmission:** The 4-speed automatic/manual transmission works smoothly and is well adapted to the engine's power range. **Braking:** Disc/disc; average, with some brake fade.

Con: Ride quality deteriorates as load increases; excessive body lean in turns; Transmission sometimes shifts harshly and malfunctions (see "Quality/ Reliability/Safety," above).

Comfort/Convenience

Pro: Standard equipment: Well equipped for the price. **Driving position:** Very good. Great visibility and front seats are fairly supportive and are set high off the floor. **Controls and displays:** Well laid-out dashboard and controls. Easy-to-read gauges. **Climate control:** Good heating-defrosting-ventilation system. **Entry/exit:** Easy front and rear access. **Interior space/comfort F/R:** Up front, this is one of the roomiest mid-sized vehicles around. **Cargo space:** Many storage bins and a good-sized trunk with a low liftover.

Con: Trunk/liftover: Rear headroom is tight for tall passengers and the low, short rear seat cushions force occupants into a painful knees-to-chin position. Trunk has a small opening. **Quietness:** Excessive wind, road, and suspension noise.

Sonata

List Price (very negotiable)	**Residual Values** (months)			
	24	**36**	**48**	**60**
GL: $22,395 (17%)	$16,500	$14,000	$11,000	$8,000
GL V6: $23,795 (16%)	$18,000	$15,000	$12,000	$9,000
GLX: $27,395 (18%)	$21,500	$17,000	$14,000	$11,000

Technical Data

Powertrain (front-drive)
Engines: 2.4L 4-cyl. (138 hp)
• 2.7L V6 (172 hp)
Transmissions: 4-speed auto./man.
Dimension/Capacity
Passengers: 2/3
Height/length/width:
56/186.9/71.7 in.

Headroom F/R: 39.3/37.6 in.
Legroom F/R: 43.3/36.2 in.
Wheelbase: 106.3 in.
Turning circle: 34.4 ft.
Cargo volume: 14 cu. ft.
Tow limit: N/A
Fuel tank: 65L/reg.
Weight: 3,217 lb.

Safety Features/Crashworthiness

	Std.	Opt.
Anti-lock brakes	—	■
Seat belt pretensioners F/R	■	—
Side airbags	■	—
Traction control	—	■
Head restraints F/R	***	***
Visibility F/R	*****	*****
Crash protection (front) D/P	****	****
Crash protection (side) F/R	****	****
Crash protection (offset)	***	—

TIBURON ★★★

RATING: Average. **Strong points:** Well equipped; impressive 2.0L performance with a manual gearbox; and an exceptional ride and handling. **Weak points:** Base 2.0L engine's passing power is seriously handicapped by the automatic transmission. Restricted rear visibility; uncomfortable seating; not for six-footers or portly occupants. No crashworthiness data. **New for 2004:** Nothing significant.

OVERVIEW: The Tiburon (it means shark in Spanish) is a stylish, two-door spin-off of the Elantra sedan that delivers sports car thrills in a compact coupe. It's powered by a base 136-hp 2.0L 4-banger, an optional 172-hp 2.7L V6 mated to a 4-speed electronically controlled automatic, or a 5- or 6-speed manual transmission. The Tiburon handles well, with gas-charged shocks inside coil springs at all four corners, and front MacPherson struts. Steering is light and responsive with a minimum of body flex. Standard brakes consist of discs up front and drums in the rear; the FX package includes four-wheel discs and ABS.

Cost analysis/best alternatives: Get an identical, discounted 2003 model. Other vehicles worth considering: the Acura RSX, Ford Mustang, Honda Civic Si, Mitsubishi Eclipse, and Toyota Corolla or Matrix. **Options:** Get the V6 engine and better-gripping tires. **Rebates:** $2,000 rebates and low percent financing on all models. **Delivery/PDI:** $800. **Depreciation:** Average. **Insurance cost:** Higher than average. **Parts supply/cost:** Parts aren't hard to find and they're less expensive than average. Repairs are straightforward because of a fairly simple design. **Annual maintenance cost:** Average. **Warranty:** Bumper-to-bumper 3 years/60,000 km; powertrain 5 years/100,000 km; rust perforation 5 years/unlimited km. **Supplementary warranty:** Not needed. **Highway/city fuel economy:** 7.4–10.5L/100 km with the base engine.

Quality/Reliability/Safety

Pro: Quality control: Above average. **Reliability:** Better than average. **Warranty performance:** Above average. **Safety:** Impressive headlight illumination and four-wheel disc brakes.

Con: Owner-reported problems: Power shutdown caused by a faulty fuel regulator; slipping clutch; faulty sunroof won't open or close; inoperative power windows and misaligned doors; premature front brake wear and excessive brake noise; hard starting; some electrical glitches; gas cap pops off when fuel tank falls below a quarter full; and paint flaking. **Service bulletin problems:** Engine RPM fluctuation; harsh shifts into Drive or Reverse; sticks in Second gear; headlight dimming; and troubleshooting sunroof malfunctions. **NHTSA safety complaints:** Hood popped up while car was underway; under-hood fire; airbag deployed when it shouldn't have, didn't deploy when it should have; and the rear spoiler is distracting and cuts rearward vision.

Road Performance

Pro: The 2.0L gives respectable power (0–100 km/h: 9.1 seconds), but the V6 is the preferred power plant for its reserve power. **Emergency handling:** Very good. Well controlled, with minimal body roll. **Steering:** The light, responsive steering is precise and predictable. **Transmission:** The 5-speed manual transmission shifts smoothly and is well adapted to the engine's power range. The ride is reasonably soft on good roads. **Braking:** Disc/disc; acceptable, though not impressive.

Con: Acceleration/torque: The 4-speed automatic transmission shifts roughly under full throttle and is noisy at high revs. **Routine handling:** Uneven pavement makes for a busy, jittery ride that's accentuated as passengers are added. Poor braking on wet pavement.

Comfort/Convenience

Pro: Attractive styling and a well-appointed interior. **Controls and displays:** Well laid-out dashboard and controls. **Climate control:** Efficient, easy-to-understand climate control system. **Interior space/comfort F/R:** Comfortable seatbacks. Multiple front-seat adjustments. Plenty of front headroom. **Cargo space:** Average. **Trunk/liftover:** Reasonable amount of luggage space, but the hatchback is more versatile with its folding rear seatbacks. **Standard equipment:** Fairly well equipped for a vehicle in this price range.

Con: Rear view is obstructed by the Tiburon's high tail and wide pillars. Tiny radio controls and temperature controls aren't within easy reach. **Entry/exit:** Front and rear access isn't easy. **Interior space/comfort F/R:** Low, firm, narrow, and short seat cushions give little thigh support. Limited front legroom for tall passengers, and rear headroom is tight. Barely enough back-seat room for two adults. **Trunk/liftover:** High liftover to the cargo area. **Quietness:** Plenty of engine, road, and wind noise.

Tiburon

List Price (very negotiable)	**Residual Values** (months)			
	24	**36**	**48**	**60**
Base: $20,495 (15%)	$13,000	$11,000	$9,000	$8,000
SE: $22,895 (17%)	$15,000	$13,000	$11,500	$10,500

Technical Data

Powertrain (front-drive)
Engines: 2.0L 4-cyl. (138 hp)
• 2.7L V6 (172 hp)
Transmissions: 5-speed man.
• 6-speed man.
• 4-speed auto.
Dimension/Capacity
Passengers: 2/2
Height/length/width:
52.4/173/69.3 in.
Headroom F/R: 38/34.4 in.
Legroom F/R: 43.1/29.9 in.
Wheelbase: 99.6 in.
Turning circle: 37 ft.
Cargo volume: 14.8 cu. ft.
Tow limit: N/A
Fuel tank: 55L/reg.
Weight: 3,025 lb.

Safety Features/Crashworthiness

	Std.	**Opt.**
Anti-lock brakes	—	■
Seat belt pretensioners F/R	■	—
Side airbags	■	—
Traction control	■	—
Head restraints F/R	*****	****
Visibility F/R	*****	**

XG350 ★★★★

RATING: Above Average. **Strong points:** Adequate engine power, automatic transmission shifts smoothly, lots of standard features at a reasonable retail price, comfortable ride; good steering, handling, and braking; plenty of front-seat room and comfort; fit and finish improved this year. Three-year reliability scores have been remarkably good. **Weak points:** Sometimes jerky acceleration due to the car's heavy weight and power-sapping automatic transmission; body lean and front-end plowing when pushed hard; limited rear seat headroom and footroom for tall occupants; obstructed rear corner visibility; some wind noise and tire roar intrudes into the interior; split rear seatbacks don't fold flat. **New for 2004:** Re-styled front and rear ends, deck lid uses hydraulic pistons; larger front disc brakes; and a full-sized spare tire on an alloy wheel.

OVERVIEW: This $32,995, front-drive, four-door, five-passenger sedan has been on the market for three years, and is the closest Hyundai comes to putting out a luxury car. XG350 designates the car's engine configuration, a 194-hp 3.5L V6, and two trim levels are offered: the base XG350 and the XG350L.

Well appointed and offering plenty of passenger space up front, the GX350's V6 engine is teamed with a 5-speed automatic transmission, with manual override that offers fairly good acceleration.

Cost analysis/best alternatives: Get the 2004 model; its brake upgrade alone is worth the extra money. Other vehicles worth considering: the Honda Accord, Nissan Altima, and Toyota Avalon and Camry. **Options:** The L trim level isn't worth the extra money. **Rebates:** Look for $3,000 rebates, deep discounting, and low percentage financing. **Delivery/PDI:** $1,065. **Depreciation:** Average. **Insurance cost:** Higher than average. **Parts supply/cost:** Mechanical parts aren't hard to find and they're less expensive than average. Body parts are sometimes expensive and in short supply. Repair is straightforward because of a fairly simple design. **Annual maintenance cost:** Average. **Warranty:** Bumper-to-bumper 3 years/60,000 km; powertrain 5 years/100,000 km; rust perforation 5 years/unlimited km. **Supplementary warranty:** Not necessary. **Highway/city fuel economy:** 8.4–13.9L/100 km.

Quality/Reliability/Safety

Pro: Quality control: Impressive. **Reliability:** Better than average. **Warranty performance:** Above average.

Con: Owner-reported problems: Automatic transmission sometimes shifts erratically and radio and AC/heater display is hard to see in daylight. **Service bulletin problems:** Harsh shifts into Drive or Reverse and sticks in Second gear. **NHTSA safety complaints:** Stalling and engine surging.

Road Performance

Pro: The 194-hp 3.5L V6 is no tire burner, but it is competent and performs most driving tasks smoothly. **Emergency handling:** Very good. Well controlled, with minimal body roll. **Steering:** The light, responsive steering is precise and predictable. **Transmission:** The 5-speed automatic transmission with a manual shiftgate usually shifts smoothly and is well adapted to the engine's power range (see below). The ride is reasonably soft on good roads. **Braking:** Disc/disc; much better stopping ability with less fading after repeated application.

Con: Acceleration/torque: The automatic transmission sometimes shifts roughly under full throttle. This problem has been around since the car's debut, although to a lesser extent following each model-year upgrade. **Routine handling:** Uneven pavement makes for a busy, jittery ride that's accentuated as passengers are added. Poor braking on wet pavement.

Comfort/Convenience

Pro: An impressive array of standard features and the interior is nicely garnished in upscale-feeling materials. Seats are reasonably supported. **Controls and displays:** Dashboard design and instrument cluster are easy to comprehend and controls are sensibly sized and easily accessed. **Climate control:** Hassle-free, simple settings are very user-friendly. **Interior space/comfort F/R:** Comfortable seating for tall occupants up front, with multiple front seat adjustments. **Cargo space:** Average. **Trunk/liftover:** Good amount of luggage space and lots of little storage areas. **Standard equipment:** Well equipped.

Con: Rear corners are hard to see. **Entry/exit:** Problematic rear access. **Interior space/comfort F/R:** Limited rear headroom and legroom for tall passengers. Barely enough back seat room for two adults and a small child. **Trunk/liftover:** Trunk lid hinges can damage luggage. Split rear seatback doesn't fold flat, limiting pass-through width. **Quietness:** V6 whines a bit under full-throttle acceleration, frameless door glass increases wind whistling, especially in a strong highway crosswind; and there's some tire noise.

XG350

List Price (very negotiable)	**Residual Values** (months)			
	24	**36**	**48**	**60**
Base: $32,995 (17%)	$22,000	$19,000	$16,000	$13,000

Technical Data

Powertrain (front-drive)
Engine: 3.5L V6. (194 hp)
Transmission: 5-speed auto.

Dimension/Capacity
Passengers: 2/3
Height/length/width: 55.9/192/72 in.
Headroom F/R: 39.7/38 in.
Legroom F/R: 43.3/37.2 in.
Wheelbase: 108.3 in.
Turning circle: 36.1 ft.
Cargo volume: 14.5 cu. ft.
Tow limit: N/A
Fuel tank: 70L/reg.
Weight: 3,651 lb.

Safety Features/Crashworthiness

	Std.	Opt.
Anti-lock brakes	■	—
Seat belt pretensioners F/R	■	—
Side airbags	■	—
Traction control	■	—
Head restraints F/R	***	***
Visibility F/R	*****	***
Crash protection (offset)	****	—

Infiniti

Unlike Toyota's Lexus division, which started out with vehicles akin to your dad's fully loaded Oldsmobile, Nissan's Infiniti alter ego has historically stressed performance over comfort and opulence and offered buyers lots of high-performance features at very reasonable prices. But, the company got greedy during the mid-90s: its Infiniti lineup became more mainstream and lost its price and performance edge, particularly after de-contenting the Q45, resurrecting its embarrassingly incompetent G20, and dropping the J30 and J30t.

Now the company is returning to its roots and is building a more balanced roster of luxury and high-performance luxury vehicles. For example, this year Infiniti has dropped its QX4 sport-utility, put a merciful end to its gutless, entry-level G20, introduced a rear-drive FX35 SUV, and will add all-wheel-drive to its G35.

All Infinitis come fully equipped and offer owners the prestige of driving a reliable and nicely styled luxury car with lots of standard performance and safety features. Head restraints across the entire lineup are rated "Good" by IIHS and NHTSA-rated crashworthiness is fairly good.

Although Infinitis are sold and serviced by a small dealer network across Canada, this doesn't compromise either the availability or quality of servicing, since any Nissan dealer can carry out most non-warranty maintenance work.

I35 ★★★★

RATING: Above Average. But do you want to spend this much money on a fully loaded year-old Maxima? **Strong points:** Fully equipped with lots of innovative safety features; powerful engine and transmission matchup; a comfortable ride, impressive handling with the Sports Package, head restraints are rated "Good" by IIHS; front- and side-crashworthiness rated four stars by NHTSA; and first-class quality control. **Weak points:** This upscale mid-sized car doesn't do anything wrong, it just needs some of the 2004 Maxima's performance features to enhance its overall performance and handling. **New for 2004:** Returns unchanged, despite the fact that its Maxima kissing cousin returns this year substantially improved.

OVERVIEW: The I35's $39,700 base price makes it the world's most expensive Maxima, incorporating many of that car's chassis and drivetrain components, including a 255-hp 3.5L V6 engine, an enhanced suspension system for better handling and increased ride comfort, standard 17-inch wheels, an upgraded interior and braking system, front seat belt pretensioners, and xenon headlights.

There's only one model offered. A Sports Package is available for an extra $3,000. It comes with a Vehicle Dynamic Control stability system and other performance goodies to provide sportier performance, but don't look for a manual transmission; it was dropped several years ago.

Cost analysis/best alternatives: Get the cheaper, practically identical 2003, unless you spy a better equipped, cheaper 2004 Maxima. Other vehicles worth considering: the Acura TL, BMW 3 Series, and Lexus ES 300. **Options:** The Sports Package isn't worth an extra $3,000. A rear spoiler is a useless option. **Rebates:** $3,000+ rebates, and low-interest financing; discounts will become even more generous in mid-2004 as the competition heats up. **Delivery/PDI:** $1,150. **Depreciation:** Slower than average. **Insurance cost:** Higher than average. **Parts supply/cost:** Moderately priced parts are easily found in the Maxima parts bin. **Annual maintenance cost:** Should be much lower than average. **Warranty:** Bumper-to-bumper 4 years/100,000 km; powertrain 6 years/ 100,000 km; rust perforation 7 years/unlimited km. **Supplementary warranty:** Not needed. **Highway/city fuel economy:** 8.3–12.1L/100 km.

Q45 ★★★★

RATING: Above Average. A nice compromise between luxury and performance, but no match for the better-performing European competition. **Strong points:** Well appointed and stylish; impressive acceleration for a car this size; smooth, quiet engine performance; pleasant riding; provides plenty of head and legroom; and brakes well. There's also a plethora of high-tech gadgets that will both challenge and frustrate you, while providing dealership service personnel with a secure future. Head restraints and offset crash ratings given a "Good" score by IIHS. **Weak points:** Outrageously overpriced and over-engineered; slow transmission kick-down; average cornering ability; road groove wandering; some road and wind noise; trunk space is compromised by trunk lid hinges that can crush luggage; cheap-looking flip-up door handles; no driver's left footrest; undersized rear-view mirrors; get used to oncoming drivers flashing their lights in reaction to your powerful xenon headlights; no NHTSA front-or side-crashworthiness data; and requires premium fuel. **New for 2004:** Relatively unchanged.

OVERVIEW: Over the years, Infiniti has shifted the Q45's emphasis from performance to luxury and comfort. In fact, when the engine was downsized a few years ago from a 4.5L to a 4.1L, the Q45 designation no longer made sense. Now, equipped with a 340-hp, twin-cam 4.5L V8, Infiniti returns to its performance roots—at a price.

This $74,900 luxury sedan comes fully loaded with lots of convenience, performance, and safety features that add to the car's complexity, drive up the cost of servicing, can be difficult to use, and may have as many drawbacks as advantages. These features include: 5-speed automatic transmission with a manual shift feature; four-wheel disc ABS with brake assist, traction control, and anti-skid capability; front side and curtain side airbags; front seat belt pretensioners; an in-dash information screen; and voice-activated audio, climate, and navigation systems, and large, multi-lens projector high-intensity discharge headlights, guaranteed to tick off other drivers.

Standard safety features that are a good idea include side airbags for both the driver and right front passenger, dual-locking shoulder belts, and front Seat belt pretensioners F/R that activate in a crash to reduce belt slack.

Cost analysis/best alternatives: A heavily discounted 2003 version is the better buy, particularly in view of the fact that these identical cars are selling poorly and unsold leftovers are piling up. If you really want a 2004, wait for more generous rebates and additional discounting in the new year. Or better yet, consider buying a BMW 5401, Lexus GS 430, or Mercedes-Benz E430. **Options:** Auto-adjusting shock absorbers don't have much effect. Forget about the laser-guided "intelligent" cruise control option due out later this year. It's unproven and its failure can have disastrous consequences. Premium package's rear-view TV system is just a video gimmick—get used to turning your head. The Sports Package only adds better tires and more wood trim. **Rebates:** $3,000 rebates and low-interest financing. **Delivery/PDI:** $1,250. **Depreciation:** Much slower than average. **Insurance cost:** Much higher than average. **Parts supply/cost:** Parts may be more expensive than other cars in this class, and they're available only from Nissan or Infiniti dealers. **Annual maintenance cost:** Much less than average. **Warranty:** Bumper-to-bumper 4 years/ 100,000 km; powertrain 6 years/100,000 km; rust perforation 7 years/ unlimited km. **Supplementary warranty:** Not necessary. **Highway/city fuel economy:** 8.8–13.6L/100 km.

M45 ★★★

RATING: Average. This is the Q45's $13,000-cheaper, high-performance alter ego. Other cars you may wish to consider: Acura RL V6, BMW 540i, Mercedes-Benz E430, and the Audi A6 4.2. **Strong points:** Well appointed; superior acceleration and handling; easy entry/exit; head restraints lower into seatback; predicted low depreciation; and head restraints rated "Good" by IIHS. **Weak points:** Slow downshifts when passing or merging; cruise control of limited usefulness; complicated, counterintuitive controls; confuses "sportiness" with a jolting ride; rear pillars hinder rear visibility; not much storage space; excessive tire and road noise; and requires premium fuel and delivers disappointing fuel economy. **New for 2004:** Returns relatively unchanged.

OVERVIEW: The M45 is a rear-drive luxury sedan that rides on a four-wheel independent suspension and carries the Q45's powerful and smooth 340-hp V8. It's essentially a small Q45 that bridges the gap between the Q and the G35. The M45 shares the Q's drivetrain, is longer than the G35, and is set on a shorter wheelbase. Due to its lighter curb weight and potent engine, it can do 0–100 km/h in under six seconds, or a half-second faster than the Q45.

The M45 gives shoppers the thrills of a Q45 for about $62,000. It comes packed with all the techno-goodies one could want. Of particular note is its 5-speed automatic transmission, incorporating a convenient, though not very impressive, manual shift mode, a sport suspension, Vehicle Dynamics Control stability system, active head restraints, and four-wheel disc brakes supplemented by brake assist and electronic brakeforce distribution.

This isn't a pretty car. In fact, it's rather bland-looking, has an excessive rear overhang, and sports a front grille that says Ford Crown Victoria.

G35 ★★★★

RATING: Above Average. This is the one Infiniti with the most to offer from a price/quality perspective and gives hope that Infiniti has learned its lesson that the public won't stand for price-gouging, no matter how pretty the package. **Strong points:** Smooth, responsive automatic/manual transmission works imperceptibly with the sedan's powerful 260-hp V6; handles well with good ride quality; carries a fairly large trunk; and head restraints are rated "Good" by IIHS. **Weak points:** Harsh ride with the Sport Suspension; poorly located power seat controls aren't easily accessible and are vulnerable to cupholder spillovers; difficult rear-seat access; seat cushions may be too firm for long trips; limited rear visibility (coupes); and the trunk is hindered by a small opening. **New for 2004:** Returns relatively unchanged.

OVERVIEW: The $39,600 ($1,250 PDI and freight) G35 is Infiniti's latest rear-drive luxury sports sedan. It arrived in mid-2002 and was joined in 2003 by a bit wider and shorter coupe sporting a slightly more powerful engine (20 extra horses), a 6-speed manual transmission, larger tires, and smooth aerodynamic styling. All for a $45,200 base price.

The sedan doesn't come with a manual transmission and acceleration is good, though not overwhelming. The coupe is the speed leader in this lineup. Nevertheless, the G35 has more passenger room than a BMW 540i, Lexus 300, or Cadillac CTS, beats the Lexus IS 300 equipped with a manual transmission in a 0–100 km/h run, and costs less than most of the competition.

Kia

Do or die time

After going bankrupt in 1998, Kia was bought by Hyundai (it's also partly owned by Ford through its Mazda affiliation) and now sells two small cars, a mid-sized sedan, a minivan, and a sport-utility. Unlike GM's unstable Daewoo, Kia has the money and backing to build a solid, stable dealer organization in Canada, but it'll take a great deal of Hyundai's time and money.

Kia has posted impressive sales over the past few years, thanks to low prices and a comprehensive base warranty. It sold 1,417 cars in 1999, followed by 13,343 in 2000, 26,013 in 2001, and 20,014 in 2002. It is expected Kia will sell more than 30,000 units by year-end.

Despite these respectable sales, all of the consumer and government feedback I've seen paints a very poor picture of Kia's quality control, particularly when it comes to automatic transmission reliability (a Daewoo bugaboo, too). Apparently, the company uses a longer base warranty as a substitute for quality control and simply pays consumer claims, instead of improving the product.

But this is do or die time.

Hyundai says it will no longer treat the company with benign neglect and will, henceforth, turn its attention to strengthening Kia's dealer network and improving quality and servicing. Undoubtedly, it has the money to invest in Kia and can continue to share platforms and key components to put out cheap and fuel-efficient small cars. Still, there's that nagging suspicion that if warranty costs for Kia defects or labour unrest threaten Hyundai's overall profits, Kia will be dumped before you can say Pony, Stellar, or Excel.

Kia's Amanti luxury sedan will appear early in 2004.

In the meantime, let's be positive and look at what Kia's doing right this year. Firstly, it's introducing the Amanti, the company's first luxury sedan. Additionally, the Rio and Megantis get a slight freshening and extra options and the company is expanding its dealer network throughout Canada.

RIO ★★★

RATING: Average. If you do buy a Rio, get an identical, discounted 2003 version. Keep in mind, though, that the Hyundai Accent, Honda Civic, or Toyota Echo are a better buy. **Strong points:** Cheap, cheap, and cheap. Highly manoeuvrable in city traffic. Given an "Acceptable" rating for head restraint protection, and a four-star rating for frontal crash protection. **Weak points:** Weak and noisy engine performance; busy, harsh ride; highway handling is slow and imprecise; limited passenger room and problematic entry/exit; tire thumping; small audio controls and missing remote trunk release; low-budget interior materials; small door openings and limited rear headroom and legroom; trunk's small opening doesn't take bulky items and doesn't offer a pass-through for large objects; optional tilt-steering wheel doesn't tilt much; poor body construction; chronic stalling caused by a defective engine control module; earned a two-star rating for side-impact protection; and depends upon a small dealer network that may complicate servicing and warranty performance. **New for 2004:** Driver's seat lumbar support.

OVERVIEW: A bit smaller than the Sephia and Spectra, the base Rio S sells for $12,650 (plus $930 freight and PDI) and is essentially a spin-off of the Ford Aspire, sold from 1995–97. Base Rios are equipped with a 104-hp 1.6L 4-cylinder engine teamed with a 5-speed manual transmission. A 4-speed automatic transaxle is one of the few options available. **Highway/city fuel economy:** 6.8–8.6L/100 km.

Owners report problems with the automatic transmission, electrical and fuel systems; chronic stalling; power loss due, in part, to faulty engine control modules; brakes, and seat belts. **Service bulletin problems:** N/A. **NHTSA safety complaints/safety:** Rio safety defects mirror those reported by Sephia owners. Defective fuel line ignited an under-hood fire, hood flew up and broke front windshield; side airbag failed to deploy; vehicle disengages from Overdrive due to a missing transmission control modulator; transmission jumps out of gear when brakes are applied or is slow shifting; sudden loss of steering; brakes stick and pedal goes to the floor without stopping vehicle; brakes are noisy; excessive shaking and vibration; vehicle swerves all over the road; no start caused by faulty computer; chronic stalling; front seat belts ride high up on the neck and rear seat belt shreds or jams; rear seat belts don't lock when a child safety seat is installed; steering binds and grinds when turned; premature tire wear; vehicle assembled without a horn; various electrical problems, including clock spring failure, cause the Check Engine light and airbag warning lamp to remain lit; fuel light constantly stays lit.

SPECTRA ★

RATING: Not Recommended. The front-drive, five-passenger Spectra RS is about the size of a Toyota Corolla and sells for $14,995 plus $930 transport/ PDI. Consider getting a Honda Civic, Hyundai Elantra, or Toyota Echo. **Strong points:** Inexpensive; hatchback head restraints rated "Good," frontal crashworthiness earned four stars, while side crash protection received three stars. **Weak points:** Crude and weak engine performance; harsh ride; poor handling; noisy engine; lots of road and wind noise; subpar fit and finish; offset crash-protection rated "Poor;" and a small dealer network that may complicate servicing and warranty performance. **New for 2004:** A tachometer.

OVERVIEW: Launched several years ago, the Spectra is really a sportier, more upscale version of the Sephia, with two fewer doors and different front-end styling. It comes with a twin-cam 124-hp 1.8L 4-cylinder engine hooked to a standard 5-speed manual or an optional 5-speed automatic transmission, set on a four-wheel independent suspension. Other standard features with the base models: front Seat belt pretensioners F/R; power steering; an AM/FM/CD sound system; and split-fold rear seatbacks.

On the road, the Spectra is a modest performer, accelerating from 0–100km/h in about 9 seconds with a manual and two seconds more with an automatic transmission. There's little power left for passing or merging with traffic. **Highway/city fuel economy:** Unimpressive: 7.1–10.3L/100 km. Kia's products have a very poor safety and quality track record and Hyundai's purchase of the company hasn't improved quality much, if one uses NHTSA's complaint records as a guideline. **Service bulletin problems:** Cold engine hard starts and steering wheel noise. **NHTSA safety complaints/safety:** Airbags failed to deploy; sudden automatic transmission failure; chronic stalling; vehicle loses power when radio and AC are turned on; faulty heater; snow enters and freezes the doors shut; seat belts aren't long enough; defective passenger door latch; many complaints relating to sudden windshield cracking, window air leaks, wipers failing intermittently, and a defective trunk release.

MAGENTIS ★★★

RATING: Average. Consider buying a competitor with more refinement and a proven history for reliability and good quality control. Three good choices: a Honda Accord, Nissan Maxima, or Toyota Camry. **Strong points:** Nicely appointed; good V6 performance; good overall visibility; plenty of front headroom; better-than-average fuel economy with regular fuel; nice ride quality and handling; spacious trunk; and a bit better built that Kia's other models. **Weak points:** Weak 4-cylinder engine; poorly performing 4-speed automatic transmission; mediocre braking; average rear headroom; considerable body lean when turning; excessive wind and tire noise; trunk has small opening; and the small dealer network may complicate servicing and warranty performance.

OVERVIEW: Sold in the States as the Optima, this front-drive, five-passenger sedan was launched several years ago in Canada as the Magentis. The entry-level model sells for $22,250 (LX), $25,750 gets you an LX V6, and $28,750 is the asking price for a high-end EX V6, plus a $930 transport/PDI fee for all models.

Essentially a Hyundai Sonata without traction control, the Magentis comes with a twin-cam 138-hp 2.4L 4-cylinder or optional 170-hp 2.7L V6 hooked to a 4-speed automatic transmission (V6s have the Sonata's separate gate for manual shifting and four-wheel ABS). Other standard features: front Seat belt pretensioners F/R; a tilt steering column; independent double wishbone front suspension and independent multi-link rear suspension; 60/40 split-fold rear seat; and tinted glass.

Performance and reliability are expected to be similar to the Sonata. Prudent shoppers should stay away from this car until it has a confidence-inspiring track record. **Service bulletin problems:** Nothing reported, yet. **NHTSA safety complaints/safety:** N/A.

AMANTI ★

RATING: Not Recommended during its first year on the market. Try a Toyota Avalon instead. **Strong points:** Fully loaded with safety, performance and convenience features; competent V6 performance; good overall visibility; headroom and legroom beat out such key competitors as the Toyota Avalon, Chrysler Concorde and Buick LeSabre; comfortable ride; predictable handling; and a spacious trunk. **Weak points:** The 6-cylinder engine, on par with the Toyota Camry's V6, may be overwhelmed by the car's heft and not have a great deal of reserve power for passing or merging. Expect numerous first-year production glitches, exacerbated by Kia's weak dealer network.

OVERVIEW: This large front-drive sedan is a $22,500 (U.S.) Hyundai XG350 clone, sold elsewhere as the Kia Opirus. It carries a standard 195-hp 3.5L V6 coupled to a 5-speed automatic transmission with a sequential manual mode. Ride comfort is enhanced by an independent double wishbone front suspension with coil springs and a stabilizer bar. At the rear, there's an independent multi-link suspension with coil springs and a stabilizer bar. Overall styling is reminiscent of a Mercedes-Benz E-Class.

Amanti is no slouch in offering standard active and passive safety features, including anti-lock brakes; electronic braking distribution; brake assist; traction control; electronic stability control; and eight "smart" airbags.

Lexus

Ho-hum, another year of record sales for Toyota and its Lexus spin-off. Lexus has now become a luxury automaker on its own merits, even though it started out selling dressed-up Camrys as upscale models (only the ES 300 fits that description now). But so did Acura and Infiniti, right? Unlike Infiniti, though, Lexus has generally become the epitome of luxury and comfort, with a small dab of performance thrown in. No matter how often car enthusiast magazines say that drivers want "road feel," "responsive handling," and "high-performance" thrills, the truth of the matter is that most drivers simply want to travel from point A to point B in safety and comfort, without interruption, in cars that are more than fully equipped Camrys or warmed-over, gadget-laden Maximas.

Although these luxury imports do set advanced benchmarks for quality control, they're not engineering perfection, as a recent spate of engine failures from sludge buildup will attest. And, yes, cheaper luxury cars from Acura, Nissan, and Toyota give you almost as much comfort and reliability, but without as much cachet and resale value.

Technical service bulletins show that these cars are affected by some minor body fit and trim glitches. Most owners haven't heard of these problems, because Lexus dealers have been particularly adept at fixing many of them before they become chronic.

In summary, the Lexus lineup is geared more to the luxury cruiser crowd, which prefers comfort over performance thrills. All models come with a full array of standard safety features.

Lexus holds a winning hand, and won't make any major changes to its existing line except for a slight re-styling and engine upgrade for the ES 300 and a redesigned GS 300/GS 430 for the 2005 model year.

ES 330 ★★★★

RATING: Above Average. A near-luxury sedan that's slowly morphing into a baby LS 430. **Strong points:** Good acceleration; pleasantly quiet ride; quiet running; top-quality components; impressive crashworthiness ratings; favourable accident injury claim data; and excellent quality control and warranty performance. **Weak points:** Primarily a four-seater; three adults won't sit comfortably in the rear. Headroom is just adequate for tall occupants. Muted steering feel; automatic transmission is reluctant to kick down for passing, hesitates, shudders, and constantly feels like it's slipping; manual shifting system isn't very user-friendly; and overall handling isn't as nimble as its BMW or Mercedes rivals. Traction control is no longer a standard feature; requires the anti-skid option. Trunk space is limited (low liftover, though) and rear corner visibility is hampered by the high rear end; requires premium fuel. **New for 2004:** A 3.3L V6 engine upgrade accompanied by a name change to the ES 330, and refreshed styling. Other new features include: optional Adaptive Variable Suspension, power adjustable pedals with memory setting, a DVD navigation system, a 10-way power adjustable driver's seat with memory, and rain-sensing wipers.

OVERVIEW: A Camry clone with more standard luxury features, this $43,800 entry-level Lexus front-drive carries a 225-hp V6 mated to an electronically controlled 5-speed automatic transmission that handles the 3.3L engine's horses effortlessly—without sacrificing fuel economy. All ES 330s feature dual front and side airbags, anti-lock brakes, double-piston front brake calipers, 60/40 split-fold rear seats, and one of the rarest features of all: a conventional spare tire.

Cost analysis/best alternatives: Here's the deal: the upgraded 2004s are selling for about what the 2003s cost, but the freight/PDI fee is still usurious and should at least be halved. Other choices: an all-dressed Camry, the Acura TL, Audi A6, BMW 3 Series, Toyota Avalon, and Volvo S40 or S60. **Options:** Anti-skid Variable Stability Control is worthwhile. I don't recommend the Adaptive Variable Suspension option; it's mostly a gimmick with little functional improvement. **Rebates:** Look for discounting, $3,000 rebates, and zero percent financing. Toyota has cut back on what were once generous leasing programs. **Delivery/PDI:** $1,185. **Depreciation:** Depreciation is much lower than average. **Insurance cost:** Much higher than average. **Parts supply/cost:** Average availability, and parts are moderately priced. **Annual maintenance**

cost: Below average. **Warranty:** Bumper-to-bumper 4 years/80,000 km; powertrain 6 years/110,000 km; rust perforation 6 years/unlimited km. **Supplementary warranty:** May be needed to cover automatic transmission malfunctions. **Highway/city fuel economy:** 7.5–11.6L/100 km.

Quality/Reliability/Safety

Pro: Quality control: Except for the powertrain, most mechanical and body components are top-quality. **Reliability:** Usually very reliable (see below). **Warranty performance:** Excellent; the Lexus division has actually been more generous than Toyota in interpreting its warranty and post-warranty responsibilities.

Con: Owner-reported problems: Computer glitch caused powertrain to jerk and drag; when slowing to a stop or coming out of a turn, transmission hesitates or feels like it's slipping (a similar problem affected the 2002s, Lexus has said for months that a fix is in the works); some trim items and interior fit and finish deficiencies. **Service bulletin problems:** Windshield or window creaking; rear interior door trim panel rattle; gauges wash out in direct sunlight; gas cap may stick; and free seat belt extender offer. **NHTSA safety complaints/safety:** Chronic hesitation when accelerating (see below):

> Transmission is unpredictable. Sometimes it hesitates, sometimes it lurches. It shudders at times when shifting. I've had the car in to the dealer for two computer-download fixes. The first lasted a day making the car seem like the demo I test drove. The car was returned and another fix by computer was done with no change to the quirky, unsafe condition of this transmission. My wife was nearly broadsided in cross traffic pattern as the car hesitated as she attempted to turn onto a side street. I have had four similar situations. These problems are at speeds up to 40 mph [64 km/h]. These problems do not seem to occur if you stomp on the accelerator from a stopped position. The general manager at the agency has driven the car "around the block" and experienced none of these problems. Yet the service department admits it's not the way this car should operate and no further manufacturer update/correction is available. They say they are working on it. The manager says per Toyota this is the normal way this car drives and they won't do anything to satisfy my complaint. These two statements are a contradiction. I've been advised by one of the techs to use premium fuel because it would help the engine/ transmission work better. I've done this since that conversation and the performance is still substandard and in stop-and-go traffic still unpredictable and dangerous.

Airbag light comes on for no reason; and turn signals aren't loud enough to signal they are activated.

Road Performance

Pro: Emergency handling: Very good, predictable, and easy to master. **Acceleration/torque:** Fast acceleration and lots of torque, especially with this year's power upgrade. **Transmission:** Shifts smoothly in all gear ranges. **Routine handling:** Capable handling, with little body roll. **Braking:** Disc/disc; better-than-average braking.

Con: Considerable body lean when cornering under power. Transmission kick-down when passing is slower than what one would expect. Steering is still a bit numb and over-assisted.

Comfort/Convenience

Pro: Standard equipment: Although the ES 330 is a Camry V6 twin, it does have its own styling and more standard features. **Driving position:** Better than average. **Controls and displays:** User-friendly and fairly complete instrumentation. **Climate control:** Efficient and easy to adjust. **Entry/exit:** Easy access. **Interior space/comfort F/R:** Comfortable seating with plenty of headroom and legroom. **Trunk/liftover:** Modest-sized trunk has a low liftover. **Quietness:** Quiet, with little road or engine noise.

ES 330

List Price (firm)	**Residual Values** (months)			
	24	**36**	**48**	**60**
ES 330: $43,800 (20%)	$36,500	$32,500	$27,500	$22,500

Technical Data

Powertrain (front-drive)
Engine: 3.3L V6 (225 hp)
Transmission: 5-speed auto.
Dimension/Capacity
Passengers: 2/3
Height/length/width:
57.3/191.1.2/71.3 in.
Headroom F/R: 38.5/37.4 in.
Legroom F/R: 42.2/35.6 in.
Wheelbase: 107.1 in.
Turning circle: 36.8 ft.
Cargo volume: 14.5 cu. ft.
Tow limit: N/A
Fuel tank: 70L/prem.
Weight: 3,460 lb.

Safety Features/Crashworthiness

	Std.	Opt.
Anti-lock brakes	■	—
Seat belt pretensioners F/R	■	—
Side airbags	■	—
Traction control	■	—
Head restraints F/R	***	***
Visibility F/R	*****	**
Crash protection (front) D/P	*****	*****
Crash protection (side) F/R	*****	****
Crash protection (offset)	*****	—

IS 300 ★★★★

RATING: Above Average. **Strong points:** Nice array of standard features; good acceleration, handling, and braking; pleasant riding; and first-class workmanship. **Weak points:** Automatic transmission manual E-shifter is awkward to use; excessive road noise and tire thump; seat cushions don't provide adequate thigh support; cramped rear seating; trunk hinges eat up trunk space and may damage luggage; and car requires premium fuel. **New for 2004:** Automatic locking doors.

OVERVIEW: Targeting BMW's 3 Series, Lexus's latest rear-drive sports-compact sedan comes with similar power and features to the BMW 328I—for a few thousand dollars less ($37,775; $39,450 with an automatic transmission). The only engine is a 215-hp 3.0L inline-six (borrowed from the GS 300), mated to a 5-speed automatic transmission, incorporating Lexus's E-shift feature for manual shifting. Other important features: four-wheel, ABS-equipped disc brakes, 17-inch wheels, performance tires, front seat belt pretensioners, and traction control.

Just a bit narrower than the BMW 3 Series, the IS 300 is a competent performer both for routine tasks and in emergency situations. Fortunately, it hasn't been afflicted by the ES 300's chronic powertrain defect. **Service bulletin problems:** CD changer improvements and free seat belt extender offer. **NHTSA safety complaints/safety:** Airbags failed to deploy.

Cost analysis/best alternatives: Forego the automatic locking doors and get last year's model for a cheaper price. Other cars worth considering are the Acura TL, Audi A4, BMW 3 Series, and Infiniti I35. **Options:** Limited-slip differential is a good idea. Don't waste money on the sunroof, heated seats, or leather upholstery. **Rebates:** None. **Delivery/PDI:** $1,500. **Depreciation:** Depreciation predicted to be much lower than average. **Insurance cost:** Higher than average. **Parts supply/cost:** Average availability, though parts may be expensive until independent resellers come on stream. **Annual maintenance cost:** Below average. **Warranty:** Bumper-to-bumper 4 years/80,000 km; powertrain 6 years/110,000 km; rust perforation 6 years/unlimited km. **Supplementary warranty:** Not necessary. **Highway/city fuel economy:** 8.9–13.1L/100 km.

GS 300, GS 430 ★★★★★

RATING: Recommended. **Strong points:** Excellent high-performance powertrain set-up; pleasantly quiet ride; superb handling and braking; and exceptional quality control and warranty performance. Standard side airbags and Vehicle Stability Control (VSC) anti-skid system. **Weak points:** Primarily a four-seater with limited headroom for six-footers; high window line impedes rear visibility; poor fuel economy (premium fuel); some instruments hidden by the steering wheel; trunk and fuel-door releases hidden at the base of the dash;

and dash-mounted GPS navigator; climate control and entertainment system tries to do too much. No crashworthiness data available. **New for 2004:** Nothing significant until the redesigned 2005 model debuts.

OVERVIEW: Redesigned in 1999 and revamped in 2002, these rear-drives represent the middle range of the Lexus product lineup. Starting at $61,700, the 300 carries an inline, 220-hp 3.0L 6-cylinder engine and the $69,500 430 uses a 300-hp 4.3L V8. Both engines have VVT-i (Variable Valve Timing with intelligence), a feature that continually changes the engine timing to achieve peak horsepower with low emissions and high fuel economy. Other innovative features include VSC, dual side airbags, and dual-zone climate controls, plus oodles of standard safety features. **Service bulletin problems:** Free seat belt extender. **NHTSA safety complaints/safety:** N/A.

Cost analysis/best alternatives: Inasmuch as the 2004s return without any significant changes, buy a discounted 2003, if you can find one. Other cars worth considering are the Acura RL, BMW 5 Series, and Mercedes-Benz E-Class. **Options:** 17-inch tires. Stay away from the in-dash navigator; it complicates the calibration of the sound system and climate controls. **Rebates:** $3,500 rebates, zero percent financing, generous leasing terms, and discounting on the MSRP by about 5 percent. Discounting and rebates will become more attractive as the redesigned models' launching date approaches. **Delivery/PDI:** $1,520. **Depreciation:** Much slower than average. **Insurance cost:** Much higher than average. **Parts supply/cost:** Parts aren't easily found outside of the dealer network; prices tend to be on the high side. **Annual maintenance cost:** Less than average. **Warranty:** Bumper-to-bumper 4 years/ 80,000 km; powertrain 6 years/110,000 km; rust perforation 6 years/ unlimited km. **Supplementary warranty:** Not necessary. **Highway/city fuel economy:** *GS 300:* 9.0–13.3L/100; *GS 430:* 9.4–12.9L/100 km.

Road Performance

Pro: Acceleration/torque: Hot rod acceleration with lots of torque (0–100 km/h: 6.1 seconds with the V8). **Braking:** Disc/disc; exceptional braking (100–0 km/h: 37 m).

GS 300, GS 430

List Price (firm)	**Residual Values** (months)			
	24	**36**	**48**	**60**
GS 300: $61,700 (25%)	$46,500	$40,500	$35,500	$30,500
GS 430: $69,500 (25%)	$49,000	$44,000	$39,000	$35,000

Technical Data

Powertrain (rear-drive)
Engines: 3.0L 6-cyl. (220 hp)
• 4.3L V8 (300 hp)

Headroom F/R: 39/37.4 in.
Legroom F/R: 44.5/34.3 in.
Wheelbase: 110.2 in.

Transmissions: 5-speed auto.
• 5-speed auto.

Dimension/Capacity

Passengers: 2/3
Height/length/width: 56.7/189.2/70.9 in.
Turning circle: 37.1 ft.
Cargo volume: 14.8 cu. ft.
Tow limit: N/A
Fuel tank: 70L/prem.
Weight: 3,685 lb.

Safety Features/Crashworthiness

	Std.	Opt.
Anti-lock brakes	■	—
Seat belt pretensioners F/R	■	—
Side airbags	■	—
Traction control	■	—
Head restraints F/R	*****	*****
Visibility F/R	*****	**
Crash protection (offset)	*****	—

LS 430, SC 430

RATING: Recommended. **Strong points:** Impressive V8 acceleration; a quiet, comfortable ride; convertible is agile with no body flexing, well appointed with curtain side airbags, high-intensity headlights, and Vehicle Stability Control anti-skid system; first-class construction and high-quality mechanical components; and better-than-average warranty performance. **Weak points:** 5-speed automatic transmission on the 2003; automatic transmission is slow to downshift when passing (a common Lexus trait); mind-numbing, complex navigation system controls and voice-activation system; sedan handling is no match for the European competition; convertible's 18-inch tires make for a jittery, wandering ride over rough pavement; and premium fuel required. No crashworthiness data available. **New for 2004:** *LS 430:* A new 6-speed automatic transmission (with a manual-shift gate), adaptive headlights that swivel during turning manoeuvres, front knee airbags, a tire pressure monitor and upgraded shocks, steering, and brakes. The re-styled front end adds 2.5 cm to overall length. *SC 430:* Nothing new before the 2006 model's redesign.

OVERVIEW: The $83,200 LS 430 rear-drive luxury compact sedan is Lexus' response to the Audi A8 and BMW 7 Series. It carries a 290-hp, twin-cam V8 engine hooked to a 6-speed automatic transmission. It's chock-full of the latest techno-gadgets to assure your comfort and safety. The $86,800 SC 430 is Lexus' first convertible, filling the void left by the discontinued SC 400. It comes equipped with the LS 430's V8 engine (with 10 additional horses) and 5-speed automatic transmission. A power-retractable metal roof is housed in the trunk (think of the old Ford Skyliner), forcing you to use the small rear seat for cargo. Capable of reaching 100 km/h in less than six seconds, the SC 430's double-wishbone suspension provides both a comfortable ride and precise, responsive handling. **Service bulletin problems:** Free seat belt extender offer; ticking or rattling from the SC's convertible top; and procedures to determine if run-flat tires can be repaired. **NHTSA safety complaints/safety:** Premature failure of Dunlop tires; seat belt tightened around child's neck and had to be cut away; and the fuel tank can be easily damaged by the placement of the tire jack.

Cost analysis/best alternatives: The upgraded 2004 LS 430 is worth this year's increase. Since the 2004 SC 430 returns practically unchanged, look for a cheaper 2003. Other cars worth considering are the Audi A8, BMW 7 Series, and Mercedes-Benz S-Class. Nissan's Q45 doesn't have the build quality or performance to match the LS 430. Other vehicles you may wish to consider: the BMW Z8 and Mercedes-Benz CLK. **Options:** Euro-tuned suspension is a worthwhile, no-charge option that enhances handling and gives a firmer ride. Ultra Luxury package contains an expensive array of unnecessary features. Convertible rear spoiler not needed but run-flat tires are a good idea, considering the SC 430's limited storage space. **Rebates:** $4,000, plus zero percent financing and favourable lease deals. **Delivery/PDI:** $1,520. **Depreciation:** Much lower than average. **Insurance cost:** Higher than average. **Parts supply/cost:** Predicted average availability and moderately expensive parts. **Annual maintenance cost:** Average. **Warranty:** Bumper-to-bumper 4 years/ 80,000 km; powertrain 6 years/110,000 km; rust perforation 6 years/ unlimited km. **Supplementary warranty:** Not necessary. **Highway/city fuel economy:** *LS 430:* 8.8–12.8L/100 km; *SC 430:* 9.3–13.1L/100 km.

Mazda

Back to the future

Mazda has come full circle.

Following its debut almost four decades ago as a small automaker selling inexpensive econocars and a rotary-engine-powered 1968 R100 coupe, followed by the RX-7 roadster (1978–2002), Mazda went upmarket and almost went bankrupt in 1994. Then Ford stepped in to turn the company around (Mazda was a major Ford supplier), and two years later bought a controlling interest of the company's shares. This takeover raised expectations that Mazda would soon turn profitable through better management and an exciting new array of cars, trucks, and minivans.

That didn't happen. But it's about to.

First, Mazda returns this year with its Wankel rotary engine powering a brand new RX-8 sports car. The Wankel is a truly innovative engine, first bought by General Motors from bankrupt NSU, and originally destined to power the Chevrolet Vega, until the General got cold feet and sold the technology to Mazda.

Stay away from this year's RX-8. Allow at least a year for the correction of design and production defects, wider distribution of parts, better servicing, and more reasonable prices.

Additionally, a new small car, the Mazda3, will replace the long-running Protegé, bringing the company back to its compact car roots.

Finally, the MPV minivan will get a slight re-styling and be more competitively priced to gain market share in the profitable minivan niche. The Miata sportster along with the Mazda6, and Tribute and B-series trucks return unchanged.

Everyone agrees that Mazda makes high-quality, competent vehicles. Unfortunately they have always been eclipsed by Honda and Toyota because there was never enough of a performance or price difference to attract buyers to Mazda. Now, with new products anchoring both ends of its 2004 lineup, the automaker has a unique opportunity to lure back customers shopping for vehicles that are reliable, fuel efficient, affordable, and on the cutting-edge of technology.

PROTEGÉ (2003) 

RATING: Recommended. New wagon entries look like winners. MP3 is basically a "boom box" on wheels, geared to young drivers. **Strong points:** Good engine performance (manual transmission); comfortable ride; plenty of interior room; good fuel economy; few safety-related complaints; and quality workmanship. **Weak points:** Limited highway passing power is further compromised by the automatic transmission; there is excessive engine, road, and wind noise; ride is a bit harsh in the SE version, and fold-down seatbacks can't be locked. **New for 2004:** Replaced by the all-new Mazda3.

OVERVIEW: The Protegé is Mazda's least costly model. It shares platforms with the discontinued Escort and Tracer, but keeps its own sheet metal, engine, and interior styling. Powered by a standard, fuel-efficient 105-hp 1.6L engine mated to a manual 5-speed transmission, and offering an optional Miata-based 130-hp 2.0L power plant, the Protegé is one of the roomiest and most responsive small cars around.

The Protegé5 wagon carries the same 2.0L engine boosted to 140 horses and some extra performance features, like a better-tuned suspension and high-performance tires to enhance handling.

The MP3's main claim to fame is its powerful and complicated trunk-housed sound system that incorporates a huge woofer for the omnipresent stereo system. This dramatically reduces the trunk's usefulness and, unfortunately, draws attention away from the MP3's other useful features, like a suspension system with sportier tuning and peppy 4-cylinder engine.

The $27,000 MazdaSpeed Protegé is basically a high-performance sedan customized by Mazda Japan and equipped with a turbocharged 2.0L 4-cylinder engine and manual transmission. Other features include a limited-slip differential, four-wheel disc brakes, a heavy-duty clutch, low-profile tires, and a more responsive suspension.

Cost analysis/best alternatives: Why recommend a car that's being replaced this year? Because the Protegé is a fine little car and there are plenty of reasonably priced 2003 leftovers available that will get cheaper as dealers make room for the new Mazda3. Other cars worth considering are the Honda Civic, Hyundai Accent, Toyota Corolla and Matrix, and VW Golf and Jetta. Steer clear of the Mazda3 during its first year on the market. **Options:** None needed. **Rebates:** $2,500–$3,000 rebates. **Delivery/PDI:** $895. **Depreciation:** Average. **Insurance cost:** Average. **Parts supply/cost:** Mazda parts costs are a bit more reasonable since the company cut prices several years ago, however, company insiders tell me they're frustrated by dealers not passing on the price reductions to their customers. Fortunately, parts aren't difficult to find from independent suppliers. Plus, Ford dealers or independents can easily service these cars if you're not satisfied with your Mazda dealer's servicing. In fact, most Ford dealers have parts that can be used on Mazda products because the discontinued, but very much alive, Ford Escort uses lots of Mazda mechanical components. This knowledge may be useful if you find your Mazda dealer jacking up parts prices or providing lousy service (same thing applies to Ford dealers who give poor servicing to Escorts and Tracers). **Annual maintenance cost:** Relatively inexpensive, but be careful. Some scummy dealers have been boosting scheduled maintenance fees to almost double what they should be. If this occurs, contact Mazda for the name of another dealer and let CBC TV's *Marketplace* program know of your experience (they ran a recent exposé of Mazda fee-gouging). **Warranty:** Bumper-to-bumper 3 years/80,000 km; powertrain 5 years/100,000 km; rust perforation 5 years/unlimited km. **Supplementary warranty:** Not necessary. **Highway/city fuel economy:** *2.0L and auto.:* 7.3–9.6L/100 km.

Quality/Reliability/Safety

Pro: Quality control: Very good. Few quality control complaints. **Reliability:** Better than average. **Warranty performance:** Better than average, though a lot depends on the dealer going to bat for you. ABS is standard on the ES and optional on the LX.

Con: Owner-reported problems: Some engine rattling, transmission lock-up when shifting into Fifth gear; premature wearout of the front brakes, MacPherson struts, and rear shock absorbers. A few complaints of poor fit and paint defects, seat belt malfunctions, and electrical system glitches. **Service bulletin problems:** Cold engine rattling; wind noise around doors; a "rotten-egg" exhaust smell; and musty, mildew-type odours emanating from the AC.

AC - Musty/Mildew Type Odours

Bulletin No.: 07-001/03

1998–2002 626; 1998–2003 Protegé; 2002–03 Protegé5; 2000–03 MPV; 1998 MPV; 1998–2002 Millenia; and 1999–2003 Miata.

A musty/mildew type odour may come from the vents when the air conditioning (AC) system is operating. It is most noticeable when the AC is first turned ON. This odour is the result of mould growth in the AC evaporator/cooling unit, which is caused by condensation, dust, and pollen within the cooling unit. This condition is usually worse during high humidity conditions. "Mazda Air Cooling Coil Coating" is available to encapsulate the mould to reduce odours. If the product is properly applied, it can effectively reduce the musty/mildew odour for up to three years.

NHTSA safety complaints/safety: Airbags failed to deploy; transmission jumped from Drive to Neutral while vehicle was underway; random hard shifting at moderate speed; heavy, nauseating fumes entered the vehicle; brake pedal pushed almost to the floor before brakes work, and they produce excessive noise; inadequate high beam illumination; erratic speedometer movement; driver sprained knee when got caught by the handle underneath seat; passenger seat belt won't disengage, forcing occupant to "slither" out of the seat; passenger seat belt broke; seat belt case and release button broke while trying to release belt.

Road Performance

Pro: Emergency handling: Comfortable, firm, no-surprise ride doesn't deteriorate as passenger load increases or when cornering at high speed. **Steering:** Predictable, well controlled, and transmits good road feedback. **Acceleration/torque:** Excellent overall performance with either engine. **Transmission:** Both transmissions shift smoothly. The 5-speed manual has a handy "Hold" feature that allows the driver to select and hold any of the three lower gears through the shift lever. Two other advantages: the automatic transaxle can be driven away in Second gear, an asset on snow and ice; and you can lock out Overdrive with the push of a button, allowing you to traverse hilly terrain or urban traffic without constant shifting. **Routine handling:** Handling is quite nimble.

Con: The automatic transmission robs the base engine of much-needed power. Some body roll in turns. SE suspension is a bit too firm for some. MazdaSpeed transmission lever set a bit too low for some and has a slippery grip. **Braking:** Average braking without ABS.

Comfort/Convenience

Pro: Driving position: Better than average. The driver and front passenger are pampered with firm, comfortable front bucket seats and multiple-seat steering wheel adjustments. Great all-around visibility. **Controls and displays:** Clear and well laid-out controls. **Climate control:** Efficient, quiet heating and defrosting, although controls are not as user-friendly as they could be. **Entry/exit:** Large doors make for easy access to the front and rear. **Interior space/comfort F/R:** One of the roomiest small cars on the market, although the rear seating is better suited to two average-sized adults. **Cargo space:** Lots of interior space to store large and small items, particularly with the wagon. **Trunk/liftover:** Trunk space is expanded with locking, folding rear seatbacks. A low liftover facilitates loading.

Con: Standard equipment: Cheap-looking, sober interior. The sunroof drastically reduces headroom. Even with a tilt steering column, the steering wheel may be too low for some tall drivers. Others may find the gearshift lever a bit too far away when shifting into Fifth gear. **Quietness:** Excessive engine, road, and wind noise intrusion into the interior.

Protegé

List Price (very negotiable)	**Residual Values** (months)			
	24	**36**	**48**	**60**
SE: $15,795 (13%)	$11,000	$9,000	$7,500	$6,000
LX: $16,750 (13%)	$12,000	$9,500	$8,000	$6,500
ES: $18,995 (14%)	$13,000	$10,500	$9,000	$7,500

Technical Data

Powertrain (front-drive)
Engines: 1.6L 4-cyl. (103 hp)
- 2.0L 4-cyl. (130 hp)
- 2.0L 4-cyl. (140 hp)
- 2.0L 4-cyl. (170 hp)

Transmissions: 5-speed man.
- 4-speed auto.

Dimension/Capacity
Passengers: 2/3
Height/length/width: 55.5/174/67.1 in.
Headroom F/R: 37.9/36.7 in.
Legroom F/R: 42.2/35.6 in.
Wheelbase: 102.8 in.
Turning circle: 38 ft.
Cargo volume: 13.1 cu. ft.
Tow limit: N/A
Fuel tank: 50L/reg.
Weight: 2,350 lb.

Safety Features/Crashworthiness

	Std.	Opt.
Anti-lock brakes	—	■
Seat belt pretensioners F/R	■	—
Side airbags	—	■
Traction control	—	—
Head restraints F/R	***	**
Visibility F/R	*****	*****
Crash protection (front) D/P	*****	****
Crash protection (side) F/R	***	****
Crash protection (offset)	***	—

MAZDA3 ★

RATING: Not Recommended during its first year on the market. **Strong points:** Good powertrain set-up; easy, predictable handling; good steering feedback; spacious, easy-to-load trunk; and better-than-average workmanship. **Weak points:** Mazda has a history of automatic transmission and fit and finish deficiencies; high deck cuts rear visibility; Mazda dealers may overcharge for scheduled maintenance. **New for 2004:** Everything.

OVERVIEW: Scheduled to arrive in dealer showrooms in early January, the front-drive Mazda3 is a totally new entry-level small car that replaces the Protegé. With considerable engineering help from Ford and Volvo, these cars use a platform that will also serve future iterations of the Ford Focus and Volvo S40.

Powered by 148-hp 2.0L and 160-hp 2.3L 4-cylinder engines coupled to either a 5-speed manual or a 4-speed Sport mode automatic transmission, the car offers spirited acceleration and smooth, sporty shifting. Handling is enhanced with a highly rigid body structure, front and rear stabilizer bars, multi-link rear suspension, and four-wheel disc brakes. Interior room is also quite ample with the car's relatively long 263.9-cm (103.9-in.) wheelbase (5 cm ger than the present Protegé), extra width, and straight sides, which maxi-e headroom, legroom, and shoulder room.

Priced at $16,195 for the GX and $17,695 for a GS, the Mazda3 is a competent competitor to the Honda Civic, Nissan Sentra, and Toyota Corolla. At $21,345, the GT appears to be outgunned by more proven competitors from Honda and Toyota.

MAZDA6 ★★★★

RATING: Above Average. **Strong points:** Adequate powertrain set-up; good overall handling with responsive steering; impressive braking; better-than-average workmanship. **Weak points:** Manual transmission shifting isn't all that smooth; hard riding over uneven pavement; some body lean in turns; torque steer (pulling) to the right when accelerating; excessive road noise intrudes into the cabin; rear spoiler blocks rear visibility; trunk opening isn't conducive to loading large objects; Mazda has a history of automatic transmission and fit and finish deficiencies; and watch out for scheduled maintenance overcharges. **New for 2004:** Next spring, a wagon and hatchback will join the lineup.

OVERVIEW: Blame it all on Nissan. They kicked off the mid-sized family sedan war by revamping their Altima a few years ago to offer a sweet combination of high performance, a capacious interior, and clean, aerodynamic styling. Then Toyota reworked its Camry, and last year, Honda brought out their seventh-generation Accord. Now it's Mazda's turn, so, *voila!* We have the Mazda6.

The Mazda6 is offered with two power plants: an impressive 160-hp inline 4-cylinder that equals the Accord's entry-level engine and a 220-hp V6 that trumps the Camry's 192 horses, but comes up a bit short when compared with the 240 horses unleashed by the Accord and Altima. Either engine can be hooked to a 5-speed manual or automatic transmission that also offers a semi-manual "Sport Shift" feature. Early testers of the manual tranny say the MazdaSpeed manual is far more precise and doesn't get hung up when shifting into Fifth gear.

Don't get the idea that this is a warmed-over 626. It's set on an entirely new platform and carries safety and convenience features never seen by its predecessor, like two-stage airbags and a chassis engineered to deflect crash forces away from occupants. Wider than the Accord, the Mazda6's interior allows for a comfortable ride, carries an unusually large 17.6 cubic foot trunk, and will offer hatchback and wagon versions in late 2003. Highway/city fuel economy for the base engine and a manual transmission is 6.7–9.6L/100km; with an automatic: 7.5–10.5L/100km; the V6 and manual: 8.1–12.1L/100km; and 8.1–12.3L/100km with the GT and an automatic transmission.

During its first year on the market, there have been few owner complaints. However, some owners mention that interior clunks are omnipresent and a number of driveability concerns that include poor engine and transmission performance, and a "rotten-egg" smell that pervades the interior.

One major Mazda6 drawback may be price—not enough difference to lure Honda, Nissan, and Toyota buyers away.

Mazda6

List Price (negotiable)	**Residual Values** (months) 24	36	48	60
GS: $24,295 (15%)	$18,000	$15,000	$12,000	$9,000
GS V6: $28,195 (16%)	$20,000	$17,000	$14,000	$10,500
GT: $28,195 (16%)	$21,000	$18,000	$15,000	$11,000
GT V6: $31,995 (17%)	$23,000	$19,500	$16,000	$12,500

Technical Data

Powertrain (front-drive)
Engines: 2.3L 4-cyl. (160 hp)
• 3.0 V6 (220 hp)
Transmissions: 5-speed man.
• 4-speed auto.
Dimension/Capacity
Passengers: 2/3
Height/length/width:
56.7/186.8/70.1 in.
Headroom F/R: 38.7/37.1 in.
Legroom F/R: 42.3/36.5 in.
Wheelbase: 105.3 in.
Turning circle: 38.7 ft.
Cargo volume: 17.6 cu. ft.
Tow limit: N/A
Fuel tank: 68L/reg.
Weight: 3,045 lb.

Safety Features/Crashworthiness

	Std.	Opt.
Anti-lock brakes	■	—
Seat belt pretensioners F/R	■	—
Side airbags	—	■
Traction control	■	■
Head restraints F/R	***	**
Visibility F/R	*****	*****
Crash protection (front) D/P	*****	*****
Crash protection (side) F/R	***	****
Crash protection (offset)	*****	—

MX-5 (MIATA) ★★★★★

RATING: Recommended. An exceptional, reasonably priced roadster. **Strong points:** Well-matched powertrain provides better-than-expected acceleration with both the manual and automatic transmission; nice handling; impressive braking; good fuel economy; no safety-related and few performance-related complaints; and a high resale value. **Weak points:** All the things that make roadsters so much "fun," like limited passenger and cargo room (not much room for six-footers); a firm ride; excessive road, wind, and engine noise; difficult entry/exit; and restricted rear visibility with the top up. **New for 2004:** A limited number of turbocharged MazdaSpeed versions will be introduced early in 2004.

OVERVIEW: Now in its fourteenth model year, the Miata is a stubby, lightweight, rear-drive, two-seater convertible sports car that combines new technology with old British roadster styling reminiscent of the Triumph, Austin-Healy, and Lotus Elan. It comes in a variety of packages: the base model; the racy Club Sport, which comes *sans* AC and power steering; the LS; and the top-of-the line SE. All cars come with a folding softtop and heated glass rear window. A removable hardtop is available as an option.

It's amazing how well the Miata is put together, considering that it isn't particularly innovative and most parts are borrowed from Mazda's other models. For example, the 142-hp 1.8L twin-cam 4-cylinder engine is borrowed from the Protegé, and the rear suspension is a copy of the discontinued RX-7s. A 5-speed manual gearbox is standard fare and a 4-speed automatic is optional, as is the 6-speed on the LS. ABS is an option only with the LS and SE versions. Built on a shortened 323 platform, the Miata is shorter than most other sports cars, nevertheless, this is a fun car to drive, costing much less than other vehicles in its class.

Cost analysis/best alternatives: Get a 2004, since prices have changed little (about $200) and 2003 leftovers are quite rare. Other cars worth considering are the BMW Z4 and Toyota MR2 Spyder. No, the Honda S2000 isn't a credible alternative; check out its rating. **Options:** Power steering, and a limited-slip differential are only useful to the lazy and cautious; the 6-speed manual transmission doesn't offer much more than the 5-speed; and anti-lock brakes haven't proven to be that effective. **Rebates:** Some discounting and zero percent financing. **Delivery/PDI:** $895. **Depreciation:** Much slower than average. **Insurance cost:** Higher than average. **Parts supply/cost:** Parts are easy to find, but often cost more than average. **Annual maintenance cost:** Less than average. **Warranty:** Bumper-to-bumper 3 years/80,000 km; powertrain 5 years/100,000 km; rust perforation 5 years/unlimited km. **Supplementary warranty:** Not needed. **Highway/city fuel economy:** 7.5–10.5L/100 km.

Quality/Reliability/Safety

Pro: Quality control: Excellent workmanship and exceptional quality. **Reliability:** Nothing reported that would take these cars out of service. **Warranty performance:** Average, although a lot depends on dealer servicing. If you must carry an infant, the Miata has a factory-installed airbag cut-off switch.

Con: Owner-reported problems: Some transmission clutch shudder and front-end vibration at cruising speeds; interior tends to overheat due to the close proximity of the catalytic converter to the centre console. **Service bulletin problems:** Tips on eliminating clutch chatter from manual transmissions; and preventing musty, mildew-type odours emanating from the AC. **NHTSA safety complaints:** Headlights give poor, unsettling illumination when driven on hilly roads. **Safety:** Shoulder belts may chafe your neck; seat belt's low anchor causes the belt to pull down against the shoulder.

Road Performance

Pro: Emergency handling: No surprises. Performs emergency manoeuvres predictably and almost as quickly as the Porsche Boxster. Exceptionally responsive, with minimal body roll; the rigid chassis gives the car a solid feeling. **Steering:** Steering is crisp and predictable. **Acceleration/torque:** Brisk acceleration with a good amount of low-end torque (0–100 km/h: 8.5 seconds). **Transmission:** Easy, precise throws with the manual and smooth, quiet shifting with the automatic. The 6-speed gearbox is helpful in keeping the noise level down, but there's little performance difference between it and the 5-speed. **Routine handling:** Lightness and 50/50-weight distribution make it an easy car to toss around corners without tossing your cookies. **Braking:** Disc/disc; impressive braking performance on dry pavement with little brake fading after successive stops (100–0 km/h: 31 m).

Con: Although the ride is a bit choppy, it's not as harsh as previous versions. The rear end tends to swing out a bit when cornering under speed.

Comfort/Convenience

Pro: Standard equipment: The base model is well equipped. The round dash air vents heighten the sports car image, giving the cockpit a 1960s British roadster allure. The convertible top is easily lowered from inside or outside of the car, and the optional hardtop is quite practical and easy to install. An innovative wind block flips up behind the rear seats. **Controls and displays:** Gauges are simple to comprehend and well positioned. **Climate control:** Efficient and quiet heating and defrosting system is easy to adjust. **Entry/exit:** Not difficult once you get used to stepping down into your Miata. **Cargo space:** Storage areas include a relatively large locking glove compartment and centre console bin (with cupholder), net pouches behind the seats, and small map pockets in each door. However, storage space for large objects is practically nonexistent, and there's almost no room behind the rear seats. **Trunk/liftover:** A low liftover makes for easy loading, although the trunk itself is quite small and shallow.

Con: Driving position: Not for six-footers. The steering wheel is set too close to the driver, isn't height-adjustable, and blocks your view of the ignition switch and power mirror control. Seats are set too low for short drivers. Unusually small inside door releases are mounted too far back to be accessed comfortably. With the top in place, rear visibility is obstructed by the wide rear panels, the small rear window, and small side-view mirrors. Small bucket seats give marginal lateral support and could use a bit more padding. **Interior**

space/comfort F/R: Interior is a bit small for tall passengers, who must sit bolt upright when pushing their seat all the way back. Insufficient lower back and thigh support. Inadequate headroom and legroom for tall adults. If you have large thighs, you may need a month at Weight Watchers to fit them between the steering wheel and seat cushion. **Quietness:** Better get used to road, tire, and engine noise that increases as the Miata picks up speed.

MX-5 (Miata)

List Price (firm)	**Residual Values** (months)			
	24	**36**	**48**	**60**
GX: $27,895 (17%)	$21,000	$17,000	$14,000	$12,500
GS: $30,100 (17%)	$23,000	$18,500	$15,500	$13,500
GT: $33,165 (18%)	$24,500	$19,500	$16,500	$14,500

Technical Data

Powertrain (rear-drive)
Engine: 1.8L 4-cyl. (142 hp)
Transmissions: 5-speed man.
• 6-speed man.
• 4-speed auto.

Dimension/Capacity
Passengers: 2/0
Height/length/width: 48.3/155.3/66.1 in.
Headroom: 37.1 in.
Legroom: 42.8 in.
Wheelbase: 89.2 in.
Turning circle: 30.2 ft.
Cargo volume: 5.1 cu. ft.
Tow limit: N/A
Fuel tank: 48L/reg.
Weight: 2,443 lb.

Safety Features/Crashworthiness

	Std.	Opt.
Anti-lock brakes	—	■
Seat belt pretensioners F/R	■	—
Side airbags	—	—
Traction control	—	■
Head restraints F/R	***	**
Visibility F/R	*****	*
Crash protection (front) D/P	—	—
Convertible	****	*****
Crash protection (side) F/R	—	—
Convertible	***	—

RX-8 ★

RATING: Not Recommended during its first year on the market. **Strong points:** An incredibly quiet, smooth engine that loves to chase the redline; engine's front midship layout assures excellent handling; reinforced chassis reduces body-flexing squeaks, rattles, and leaks; well-appointed interior; and good rear-seat access. **Weak points:** No-haggle pricing; inevitable first-year production glitches, mechanics are inexperienced; the 4-speed automatic dramatically reduces horsepower and doesn't mesh well with the engine;

multi-link suspension and 18-inch tires make for a slightly harsh ride. **New for 2004:** Everything.

OVERVIEW: The rotary is back! After an absence of almost nine years Mazda has brought back its latest iteration of the Wankel-inspired rotary engine in its RX-8 rear-drive, four-door sports coupe. And, this time, the price is right—$12,000 less than what the $48,795 RX-7 sold for in 1995.

The RX-8 automatic is the less powerful of the two models offered, but it still has 197 hp pushing just a little over 1,360 kg (3,000 lb.) Managing that horsepower is a 4-speed Sport AT automatic transmission, which also gives you the option of shifting through the gears yourself with special "upshift" and "downshift" paddles on the steering wheel. The Sport Package upgrades the automatic's handling ability to the level of the 6-speed model.

The more expensive 6-speed has all the performance and handling features available. This includes the 238-hp RENESIS engine, a limited-slip differential, sport-tuning for the suspension and 18-inch wheels housing larger front brakes.

Safety features for both cars include 4-wheel ventilated disc brakes with ABS, side air bags and side air curtains. Highway/city fuel economy for the 4-speed automatic transmission is 8.8–12.9L/100 km; the 6-speed manual transmission is rated at 9.2–12.8L/100 km.

RX-8

List Price (firm)	**Residual Values** (months)			
	24	**36**	**48**	**60**
GS: $36,795 (17%)	$28,000	$24,000	$19,000	$14,500
GT: $39, 595 (18%)	$30,000	$26,000	$21,000	$16,500

Technical Data

Powertrain (rear-drive)
Engine: 1.3L rotary. (197-238 hp)
Transmissions: 4-speed auto.
• 6-speed man.

Dimension/Capacity
Passengers: 2/2
Height/length/width:
52.8/174.3/66.1 in.
Headroom F/R: 38.2/36.8 in.
Legroom F/R: 42.7/32.2 in.
Wheelbase: 106.4 in.
Turning circle: 34.8 ft.
Cargo volume: 5.1 cu. ft.
Tow limit: N/A
Fuel tank: 60L/prem.
Weight: 3,053 lb.

Safety Features/Crashworthiness

	Std.	**Opt.**
Anti-lock brakes	■	—
Seat belt pretensioners F/R	■	—
Side airbags	■	—
Traction control	■	—
Head restraints F/R	***	**
Visibility F/R	*****	****

MPV ★★★★

RATING: Above Average. Mazda has accomplished an amazing turnaround in the past two years; its MPV is now smaller, sportier, and more nimble than its more space/comfort-oriented counterparts. **Strong points:** Well appointed (lots of gadgets); an adequate engine; smooth-shifting 5-speed automatic transmission; comfortable ride; easy handling; good driver's position; responsive steering; and innovative storage spots. Without a doubt, the MPV manages its limited interior space far better than the competition. Remarkably few factory-related defects reported at NHTSA or in the service bulletin database. **Weak points:** Smaller than most of the competition; don't believe for a minute that the MPV will hold seven passengers in comfort—six is more like it. Elbow room is at a premium, and it takes a lithe figure to move down the front- and middle-seat aisle. Excessive engine and road noise. Transport and preparation fee is excessive. Dealer servicing and head office support have been problematic in the past. **New for 2004:** Refreshed interior and exterior styling of no real consequence, except for the driver's seat additional lumbar support. Engine won't be re-powered until 2006.

OVERVIEW: Manufactured in Hiroshima, Japan, this small minivan offers a number of innovative features, like "theatre" seating (rear passenger seat is slightly higher) and a third seat that pivots rearward to become a rear-facing bench seat—or folds into the floor for picnics or tailgate parties. Another feature unique among minivans is Mazda's Side-by-Slide removable second-row seats, which move fore and aft as well as side-to-side while a passenger is seated. Rarely seen sliding door crank windows are standard on the entry model and power assisted on the LS and ES versions.

The MPV uses a new 3.0L Duratec V6 found in the Taurus. That means horsepower is up 35 to 200 and a similar number of lb.-ft. are also available, though some refinements produce a lower torque peak—3000 rpm versus 4400 rpm—giving the Mazda engine better pulling power at lower speed. Making better use of that power is a new 5-speed automatic transmission, which should cut fuel consumption a bit.

One immediate benefit: the 3.0L is able to climb hills without continuously downshifting and Mazda's "slope control" system automatically shifts to a lower gear when the hills get very steep.

Torque is still less than the Odyssey's 3.5L or the 3.8L in top-of-the-line Chryslers, though comparable with lesser Chrysler products and GM's Venture and Montana. The engine operates more smoothly than its predecessor, though it's not as quiet as Honda's 3.5L V6. The suspension has been firmed up to decrease body roll, enhancing cornering ability, which produces a sportier ride than other minivans. This firmness may be too much for some.

I've been tougher on the MPV than has *Consumer Reports* and others, due to the dealer price-gouging, poor servicing, and the minivan's small size, underpowered drivetrain, and reliability problems (all too common on pre-2000 models). Presently, it looks like most of these concerns have been dealt with, although the Ford-built 3.0L Duratec may still be problematic with high mileage.

Cost analysis/best alternatives: Go for a discounted 2003 model, since it's practically identical to the 2004 version. Other minivans worth considering are the Honda Odyssey, GM front-drive minivans, Nissan Quest, and the Toyota Sienna. **Options:** Beware: dealers pretend Mazda accessories like a roof rack don't include the labour charge. The upgraded stereo that comes with the Sports Package may not include a cassette player. The higher trim levels don't offer much of value, except for power windows and door locks and an ignition immobilizer/alarm/keyless entry system. An integrated child safety seat isn't available. Rear AC and seat height-adjustment mechanisms are recommended convenience features. **Rebates:** Expect $3,000 rebates and zero percent financing. **Delivery/PDI:** $1,095. **Depreciation:** Faster than average. **Insurance cost:** Higher than average. **Parts supply/cost:** Likely to be back ordered and cost more than average, despite Mazda's warning to dealers to keep prices in check. **Annual maintenance cost:** Average; any garage can repair these minivans. **Warranty:** Bumper-to-bumper 3 years/80,000 km; powertrain 5 years/100,000 km; rust perforation 5 years/unlimited mileage. **Supplementary warranty:** An extended warranty is worth having, particularly in view of the fact that powertrain problems have plagued these vehicles in the past. **Highway/city fuel economy:** 9.6–13.6L/100 km.

Quality/Reliability/Safety

Pro: Quality control: Recent models feature improved workmanship and more rugged construction.

Con: Reliability: Expect some transmission failures, ABS malfunctions, and premature wearout of the front and rear brakes. **Warranty performance:** Inadequate base warranty. Mediocre past servicing. **Owner-reported problems:** There have been an extraordinary low number of engine complaints on 2001–03 models; except for a half-dozen complaints of "rotten-egg" exhaust, stalling and surging, and some oil leakage. Owners report that the electronic computer module (ECU), automatic transmission drive shaft, upper shock mounts, front 4X4 drive axles and lash adjusters, AC core, and radiator fail within the first three years. Cold temperatures tend to "fry" the automatic window motor, and the paint is easily chipped and flakes off early, especially around the hood, tailgate, and front fenders. Premature brake caliper and rotor wear and excessive vibration/pulsation are chronic problem areas (repairs are needed about every 12,000 km). Premature paint peeling afflicting white-coloured MPVs is quite common. **Service bulletin problems:** "Rotten-egg" exhaust smell; AC musty, mildew odours; wind noise around doors; and cargo net hooks detach. **NHTSA safety complaints/safety:** Airbags failed to deploy; fuel smell emanating from the AC; sudden loss of power at highway speeds; stalling upon deceleration; engine oil leakage; sudden stalling; sliding doors don't lock in place; brake caliper bolt fell off, causing vehicle to skid; shifter obscures dash and is easily knocked about; excessive drivetrain vibration at 90 km/h; child's neck tangled in seat belt; seat belts lock up when vehicle is moving; seat belts are too short; difficult keeping child safety seat secured properly due to the anchor placement; rear visibility obstructed by high seatbacks.

Road Performance

Pro: Acceleration/torque: Ford's Duratec V6 (essentially the same engine that powers the Mercury Cougar) has plenty of power for most driving situations. **Transmission:** Smooth-shifting 5-speed automatic is much improved over earlier versions. **Routine handling:** Very good, particularly in close quarters, due to the minivan's small size. Better-than-average ride quality. **Emergency handling:** Much better than average; the ES corners especially well. **Steering:** Predictable and responsive. **Braking:** Disc/drum; much better than average.

Con: Little low-end torque, insufficient power with a full load. Some brake fade after successive stops.

Comfort/Convenience

Pro: Standard equipment: Lots of standard features, but the 180-watt, nine-speaker audio system that stores up to six CDs is particularly impressive. **Driving position:** Driver sits comfortably high with excellent forward visibility and lots of headroom. **Controls and displays:** Standard tilt steering wheel houses auxiliary audio controls. Instrumentation is easy to read, with large gauges, buttons, and knobs. Controls are within easy reach. **Climate control:** Simple and easily accessed radio and heating/defrosting/ventilation system. **Entry/exit:** Easy entry/exit up front. **Interior space/comfort F/R:** Without a doubt, the MPV manages its limited interior space far better than the competition. All seats are well bolstered with plenty of headroom in all seating positions. Middle-row seating converts easily from bench, with an aisle on each side, to buckets, with an aisle down the middle. Third-row seat flips and folds into a floor well. Side doors have windows that drop down, rather than swing out, which tend to be rattle-prone. Walk-through passage. **Cargo space:** Adequate; large door bins. Seats can be easily removed for additional cargo room and they weigh 6 kg (13 lb.) less than the Odyssey's seats. **Trunk/liftover:** Easy to load and unload.

Con: The sliding side door openings are narrow and doors only operate manually. **Interior space/comfort F/R:** Limited elbow room; it takes a lithe figure to move down the front- and middle-seat aisle. Won't hold seven passengers in comfort. Entry/exit compromised by second-row seats that don't flip forward and side doors that don't open as widely as longer minivans. The flip and fold third-row seat, when flipped 90 degrees backward for tailgate parties, puts the passenger-side occupant dangerously close to the hot exhaust pipe and leaves adult legs dangling uncomfortably. **Quietness:** Excessive engine growl when accelerating.

MPV

List Price (negotiable)	**Residual Values** (months) 24	36	48	60
GX: $26,595 (18%)	$17,000	$14,000	$12,000	$9,500
GS: $29,995 (18%)	$18,500	$15,500	$13,500	$11,000
GT: $35,995 (20%)	$21,500	$18,500	$15,000	$13,000

Technical Data

Powertrain (front-drive)
Engine: 3.0L V6 (200 hp)
Transmissions: 5-speed auto.
Dimension/Capacity
Passengers: 2/2/3
Height/length/width: 68.7/187.8/72.1 in.
Headroom F/R1/R2: 41/39.3/38 in.
Legroom F/R1/R2: 40.8/37/35.6 in.
Wheelbase: 111.8 in.
Turning circle: 33 ft.
Cargo volume: 127 cu. ft.
GVWR: 5,229 lb.
Tow limit: 3,000 lb.
Fuel tank: 75L/reg.
Weight: 3,796 lb.

Safety Features/Crashworthiness

	Std.	Opt.
Anti-lock brakes (2W; 4W)	■	■
Seat belt pretensioners F/R	■	—
Side airbags	■	—
Traction control	—	■
Head restraints F/R	**	*
Visibility F/R	*****	***
Crash protection (front) D/P	*****	*****
Crash protection (side) F/R	*****	*****
Crash protection (offset)	***	—

Mitsubishi

Adrift in red ink

Mitsubishi is in serious financial difficulty and has been shaken by large-scale defections of its top managers. The company got into this mess because it sold so many sub-prime loans in the States that were never paid; its product mix was lacklustre, and management lacked direction. But the worst may be over, judging by Mitsubishi's innovative new models, reduced prices, and expansion into Canada. With DaimlerChrysler owning 37.3 percent, the automaker launched its Canadian dealer network in 2002 by piggybacking its dealers onto Chrysler's existing dealer body. This is the company's second attempt to gain a foothold in Canada since 1997, when it put expansion on hold after diverting its energies to the American market. Canada's 65 Mitsubishi dealers will be located mostly in Ontario and Quebec.

Mitsubishi will sell the following eight models in Canada: the compact Lancer ES ($15,998); the mid-sized Galant DE ($23,498), Eclipse RS ($23,998), and Eclipse Spyder GS ($34,998); and the Montero XLS ($41,987), Montero Sport ES AWD ($32,497), Endeavour ($33,998), and Outlander LS ($24,197) sport-utilities. Buyers will be charged $895 for PDI and freight.

All things considered, Mitsubishis are fairly good buys—in the States. In Canada we are at the mercy of dual dealerships that won't have adequately trained mechanics or large parts inventories, and the cars will face a more challenging environment.

LANCER ★★★☆

RATING: Average; the high-performance O•Z Rally is Above Average. Unfortunately, Chrysler sales and service may not be as good as the car. **Strong points:** Awesome acceleration with the O•Z Rally's turbocharged engine; top-rated frontal and offset crashworthiness; reasonably good workmanship. **Weak points:** Base engine is a bit horsepower-challenged, especially with the automatic transmission; expect some parts delays; and side crashworthiness is below average for the driver. **New for 2004:** The arrival of a Sportback wagon and the Evolution AWD.

OVERVIEW: These entry-level front-drive econocars offer an incredible choice of vehicles that run the gamut from the cheap and mundane ES, to the high-performance Evolution AWD, to a sizzling street racer called the O•Z Rally. The Evolution has over double the horsepower (271) of the base ES (120). Evolution prices haven't been released yet in Canada.

Owner complaints on the 2003 models have focused on automatic transmissions that lurch, jerk, and slip; clutch failure; broken axles; sudden power loss, particularly with Evolutions (see *forums.evolutionm.net/showthread.php?s=&threadid=34867*); frequent paint defects that include swirls, scratches, and peeling; and rear-bumper support rusting.

Lancer

List Price (negotiable)	**Residual Values** (months) 24	36	48	60
Base: $15,998 (12%)	$11,000	$9,000	$7,500	$6,000
O-Z Rally: $20,098 (12%)	$13,000	$11,000	$9,500	$8,000

Technical Data

Powertrain (front-drive/AWD)
Engines: 2.0L 4-cyl. (120 hp)
• 2.4L 4-cyl. (160 hp)
• 2.0L 4-cyl. turbo (271 hp)
Transmissions: 5-speed man.
• 4-speed auto.

Headroom F/R: 38.8/36.7 in.
Legroom F/R: 43.2/36.6 in.
Wheelbase: 102.4 in.
Turning circle: 32.8 ft.
Cargo volume: 11.3 cu. ft.
Tow limit: N/A

• AWD

Dimension/Capacity

Passengers: 2/3

Height/length/width: 54.1/177.6/66.7 in.

Fuel tank: 50L/reg.

Weight: 2,646 lb.

Safety Features/Crashworthiness

	Std.	Opt.
Anti-lock brakes (4W)	—	■
Seat belt pretensioners F/R	■	—
Side airbags	—	■
Traction control	—	—
Head restraints F/R	*****	*****
Visibility F/R	*****	*****
Crash protection (front) D/P	****	****
Crash protection (side) F/R	**	****
Crash protection (offset)	*****	—

GALANT

RATING: Above Average, thanks to this year's improvements. **Strong points:** Competent 4- and 6-cylinder power plants; excellent handling and steering feedback; generally good crashworthiness scores; fairly good workmanship. **Weak points:** No manual transmission to make full use of the car's spirited engines; problematic rear seat entry exit; first-year glitches likely to plague this redesigned model. Expect slow parts delivery. **New for 2004:** More powerful engines, handling enhancements, a larger interior, and sleeker styling characterize this year's Galant. The manual transmission has been dropped.

OVERVIEW: The Galant, Mitsubishi's most popular car, is a mid-sized sedan that has been considerably improved this year with the addition of a better-performing engine that is smoother and has about 15 extra horses. Handling is better, too, thanks to the revised suspension, front and rear stabilizer bars, upgraded power steering, four-wheel disc brakes, and larger, 16-inch wheels. Higher trim levels get standard ABS brakes.

The 2.4L 4-banger is a standard feature with the DS and LS models, while the LS-V6 and GTS use a smooth-running V6 engine that's torquier than the Altima's and is mated to a semi-manual gearbox. The GTS comes with performance tires and larger, 17-inch wheels. This year's GTZ comes with standard V6 power and firmer shock absorber valving and higher spring rates.

Overall quality control is impressive, generating few owner complaints. Some of the problems that have been mentioned are: "rotten-egg" exhaust smell; engine hesitation; transmission and brake malfunctions; noisy, squeaking front brakes; the AC may not provide an even flow of cool air through the lower vents; and a loose driver's seat.

Technical service bulletins cover rear suspension rattling, seat adjustments, window glass freezing to the moulding, and tips on reducing brake noise.

Galant

List Price (negotiable)	**Residual Values** (months)			
	24	**36**	**48**	**60**
DE: $17,857 (13%)	$12,000	$9,000	$7,500	$6,000
ES: $18,717 (13%)	$13,000	$10,000	$8,500	$7,000
LS: $21,317 (14%)	$15,000	$12,000	$10,000	$8,000
ES V6: $20,137 (14%)	$14,500	$11,500	$9,500	$8,500
LS V6: $23,347 (15%)	$16,000	$13,000	$11,000	$9,000
GTZ: $24,637 (15%)	$17,000	$14,000	$12,000	$10,000

Technical Data

Powertrain (front-drive)
Engines: 2.4L 4-cyl. (160 hp)
• 3.8L V6 (230 hp)
Transmission: 4-speed man./auto.

Dimension/Capacity
Passengers: 2/3
Height/length/width:
54.1/177.6/66.7 in.
Headroom F/R: 38/37.7 in.
Legroom F/R: 43.5/36.3 in.
Wheelbase: 103.7 in.
Turning circle: 38.7 ft.
Cargo volume: 14.6 cu. ft.
Tow limit: N/A
Fuel tank: 62L/prem.
Weight: 3,031 lb.

Safety Features/Crashworthiness (2003)

	Std.	**Opt.**
Anti-lock brakes (4W)	■	—
Seat belt pretensioners F/R	■	—
Side airbags	■	—
Traction control	—	—
Head restraints F/R	**	**
Visibility F/R	*****	*****
Crash protection (front) D/P	****	****
Crash protection (side) F/R	***	****
Crash protection (offset)	***	—

ECLIPSE, SPYDER ★★★★

RATING: Above Average; the biggest negative is that you have to go mainly through Chrysler dealers to buy and service your Eclipse. **Strong points:** V6 engines have plenty of grunt throughout their power range and the base 4-cylinder handles most driving chores quite well. Handling is exceptionally good on all models, but the sportier GT and Spyder are better than the rest. **Weak points:** Base 15-inch tires are disappointing; so-so steering with the base sedan; excessive engine and road noise; limited rearward visibility (Spyder); difficult rear entry/exit; rear seating adequate only for children or small adults; sun washes out dash-mounted readouts; and a weak dealer network. **New for 2004:** The addition of a six-speaker audio system; the deletion of traction control.

OVERVIEW: These are beautifully styled, reasonably priced sporty coupes that perform as well as they look. Whether you choose the hardtop or the Spyder convertible, there are three engine choices available: a 147-hp 2.4L 4-banger (RS and GS), a 200-hp 3.0L V6 (GT), and a 210-hp variant of the same power plant (GTS). A manual 4-speed is standard, however, an automatic 4-speed is available with the RS; a semi-manual 4-speed automatic may be purchased with the GS, GT, and GTS models.

Although these cars are loaded with performance and convenience features, only the GTS gets standard side airbags, anti-lock brakes and better-performing 17-inch wheels.

Highway/city fuel economy varies between 7.6–10.1L/100 km with the base 2.4L and 8.3–11.3L with the 3.0L V6.

Eclipse, Spyder

List Price (negotiable)	**Residual Values** (months)			
	24	**36**	**48**	**60**
RS: $23,998 (14%)	$18,000	$14,000	$11,000	$8,000
Spyder GS: $34,998 (18%)	$24,000	$20,000	$16,500	$13,500

Technical Data

Powertrain (front-drive)
Engines: 2.4L 4-cyl. (147 hp)
• 3.0L V6 (200 hp)
Transmissions: 5-speed man.
• 4-speed auto.

Dimension/Capacity
Passengers: 2/3
Height/length/width: 51.6/176.8/68.9 in.
Headroom F/R: 37.9/34.9 in.
Legroom F/R: 42.3/30.2 in.
Wheelbase: 100.8 in.
Turning circle: 35.4/40 ft.
Cargo volume: 16.9/7.2 cu. ft.
Tow limit: N/A
Fuel tank: 62L/reg.
Weight: 2,910 lb.

Safety Features/Crashworthiness

	Std.	Opt.
Anti-lock brakes (4W)	—	■
Seat belt pretensioners F/R	■	—
Side airbags	—	■
Traction control	—	—

Head restraints F/R	**	*
Visibility F/R	****	*
Crash protection (front) D/P	****	****
Crash protection (side) F/R	***	—

Nissan

Nissan, or should I say, Renault's Nissan division, has risen from the dead after having been "written off" by most industry pundits a scant six years ago. Fortunately, Renault had faith in Nissan's future and promptly bought a controlling interest in the company. This action allowed the French automaker to send in its own management teams, and turn Nissan's fortunes around dramatically.

Today, amid record 2003 sales and profits, Nissan is showcasing both performance and styling with its sporty 350Z, redesigned Quest minivan, Murano cross-platform SUV, and Titan full-sized pickup

Nissan, like Mazda, makes dependable cars, trucks, and minivans that, until recently, flew under the public's radar screen due to their small volume and bland styling. But, the secret has gotten out, and shoppers are now flocking to Nissan showrooms to buy highly styled and reasonably priced models that combine good fuel economy with cutting-edge technology.

350Z ★★★★★

RATING: Recommended. These sporty coupes are so popular, they're often sold at more than list price, so don't expect much dealer haggling. If you want to get a better price, delay your purchase until mid-2004 when prices are expected to stabilize and dealer stocks are replenished. Plus, by then early production glitches will likely have been corrected. **New for 2004:** A new convertible and upgraded interior trim.

OVERVIEW: I remember driving the 240Z around Montreal in the early 1970s and loving the car, despite its iffy brakes ("*Non, monsieur, the brakes are fine, it's our customers who are at fault.*") and biodegradable body. Then the car put on weight over the years and became less and less reliable and sporty before it was axed in 1996.

Now, we return to the Z car's roots with last year's launching of the 350Z. A sporty, rear-drive, 287-hp V6-equipped, two-passenger hatchback and convertible, the Z is priced at $45,400 (the 1996 300ZX Turbo sold for $60,698!)

The new Z car is loaded with innovative features never seen on the original 240Z. These include a Continuously Variable Valve Timing Control System, front and rear ABS-equipped disc brakes, electronic brake force distribution, a drive-by-wire-throttle, a 6-speed manual transmission, and a unique carbon-fibre driveshaft that's said to give better acceleration and provide added crash safety. There are also many optional performance features, including a 5-speed automatic transmission (more of a non-performance option), traction control, a limited-slip differential, larger wheels and grippier tires, and a vehicle stability system.

Crashworthiness as tested by NHTSA is much better than average: five-star frontal crash protection for both the driver and passenger and four-star side-impact protection for the driver.

Oh, and don't forget the lockable luggage floor box—big enough to carry one *small* briefcase.

SENTRA ★★★★★

RATING: Recommended. Unlike many bare-bones economy cars, entry-level Sentras offer dependable motoring with lots of comfort features. **Strong points:** All three engines provide plenty of power, a comfortable ride, and easy handling. Good quality control, few safety-related or performance-related defects reported by owners. On top of that, the cars are reasonably priced. The 2.5L-equipped SE-R versions provide much better acceleration, handling, and ride. **Weak points:** Difficult rear access; poor highway handling in wet weather; and average crashworthiness. **New for 2004:** The SE-R gets new front and rear styling and a standard trip computer; V versions add Brembo brakes.

OVERVIEW: Nissan's entry-level small sedans come in three trim levels: the XE and GXE, housing a perky 126-hp 1.8L 4-cylinder engine, and the sporty SE-R, carrying a 165-hp 2.5L 4-banger with 10 additional horses in the SE-R Spec V version. A manual 5-speed transmission is standard; a 6-speed manual or a 4-speed automatic is optional. The exterior design is part Altima and part Maxima.

The 175-hp SE-R V comes with a limited-slip differential, a sport-tuned suspension, and 17-inch wheels to compete against the Honda Civic SiR and Mazda's high-performance spin-offs. Four-wheel disc brakes are a standard feature.

Cost analysis/best alternatives: Get the 2004 XE or GXE models; although they're not much different from the 2003s, they aren't much more expensive, either. Sentra's engine and body dimension improvements have made it a good competitor to the Honda Civic and Toyota Corolla. Other worthwhile cars to consider are the Hyundai Elantra and the Mazda compacts. **Options:** If you do a lot of highway driving, spend the extra money for the SE-R's stronger engine, enhanced suspension, and better-grade, larger tires. **Rebates:** $2,000 to $3,000 rebates and low financing programs. Additional discounting very likely by early summer. **Delivery/PDI:** $868. **Depreciation:** Average. **Insurance cost:** Average for the base models, much higher than average for the SE-R. **Parts supply/cost:** Inexpensive parts can be found practically anywhere. SE-R parts will likely be more difficult to find and relatively expensive. **Annual maintenance cost:** Less than average. **Warranty:** Bumper-to-bumper 3 years/ 80,000 km; powertrain 5 years/100,000 km; rust perforation 5 years/ unlimited km. **Supplementary warranty:** Not needed. **Highway/city fuel economy:** *1.8L and auto.:* 6.6–8.9L/100 km; *2.5L:* not yet tested.

Quality/Reliability/Safety

Pro: Quality control: Sentras are almost trouble-free. First-class body assembly. **Reliability:** Overall reliability is better than average. **Warranty performance:** Above average.

Con: Owner-reported problems: Clutch, exhaust, fuel system, and electrical problems are fairly common after the first three years. Front brake pads and rotors wear out quickly; excessive bouncing and vibration caused by prematurely worn struts; doors vibrate noisily; passenger-side window leaks; rear bumper may fall off; excessive wind noise around the windshield moulding. **Service bulletin problems:** Anti-theft system may make for hard starts or no-starts; troubleshooting tips for a lit MIL light; turn signals may be too fast; AC may operate erratically and have a sticking case door. **NHTSA safety complaints/safety:** Airbag failed to deploy; ABS brake failure; brakes lock up at low speed, particularly on wet roads; sudden, unintended acceleration; stalling; while driving, windshield wipers, turn signals, headlights, horn, and hazard lights failed; Continental tire tread separation; one complaint that the gas and brake pedals are set too far apart.

Road Performance

Pro: Acceleration/torque: Both the 1.8L and 2.5L 4-cylinder engines provide plenty of power for all driving situations. **Transmission:** Both manual and automatic transmissions shift very smoothly. **Routine handling:** Good manoeuvrability around town. Firm but well-mannered ride on most roads. **Steering:** Predictable, with little under-steer. Much better with the SE-R. **Braking:** Above average; minimal fading after successive stops.

Con: Slow steering response with the base models. SE-R version has sportier handling, but ride comfort is compromised. **Emergency handling:** XE and GXE high-speed handling is a bit sloppy, and emergency handling is sluggish. Brakes tend to lock up on wet roads.

Comfort/Convenience

Pro: Standard equipment: Average for an entry-level small car. Better than average for the SE-R. **Driving position:** Spartan but acceptable instruments and controls. Standard tilt wheel and height-adjustable seat creates a comfortable driving position. **Controls and displays:** Easy-to-read dash gauges and convenient controls. Neatly finished interior. **Climate control:** Efficient and uncomplicated heating, defrosting, and ventilation. **Interior space/comfort F/R:** Comfortable, supportive seats. Up front, there's good all-around headroom and legroom. In the rear, you have to move the front seats all the way forward to be comfortable. **Cargo space:** Average for this size of car. **Trunk/liftover:** Of average size, but a bit smaller than the Honda Civic or Ford Focus trunk; has a low liftover.

Con: Entry/exit: Small rear doors make access a bit difficult. **Interior space/comfort F/R:** Don't believe Nissan's claim of five-passenger seating: the four-door sedan only accommodates four people comfortably. All seats could use extra padding. **Quietness:** Lots of wind and tire noise.

Sentra

List Price (very negotiable)	**Residual Values** (months)			
	24	**36**	**48**	**60**
Sedan 1.8: $15,598 (12%)	$11,000	$9,000	$7,000	$5,500
SE-R: $21,498 (13%)	$14,000	$12,000	$10,000	$8,000
Spec V: $21,998 (13%)	$15,000	$13,000	$11,000	$9,000

Technical Data

Powertrain (front-drive)
Engines: 1.8L 4-cyl. (126 hp)
• 2.5L 4-cyl. (165 hp)
• 2.5L 4-cyl. (175 hp)
Transmissions: 5-speed man.
• 6-speed man.
• 4-speed auto.
Dimension/Capacity
Passengers: 2/3
Height/length/width:
55.5/177.5/67.3 in.
Headroom F/R: 39.9/37 in.
Legroom F/R: 41.6/33.7 in.
Wheelbase: 99.8 in.
Turning circle: 38 ft.
Cargo volume: 11.6 cu. ft.
Tow limit: 1,000 lb.
Fuel tank: 50L/reg./prem.
Weight: 2,548 lb.

Safety Features/Crashworthiness

	Std.	Opt.
Anti-lock brakes (4W)	—	■
Seat belt pretensioners F/R	■	—
Side airbags	■	—
Traction control	—	■
Head restraints F/R	***	**
Visibility F/R	*****	***
Crash protection (front) D/P	****	****
Crash protection (side) F/R	**	—
Crash protection (offset)	***	—

ALTIMA ★★★

RATING: Average. Downgraded for poor quality control and problematic handling. **Strong points:** Scintillating V6 acceleration; flawless automatic transmission operation; good braking; well laid-out instruments and controls; and better-than-average interior room and craftsmanship. **Weak points:** V6 acceleration overpowers this car and causes excessive rear-end instability and steering pull to one side. The 4-cylinder engine isn't as refined as the competition and is noisy when pushed. Brakes tend to lock up on wet roads. Interior appointments lack panache; limited rear headroom; snug rear seating for three adults; obstructed rear visibility; dashboard reflects into windshield and dash gauges wash out in sunlight; and parts are often back ordered. Quality problems appear as these cars age. **New for 2004:** Nothing much.

OVERVIEW: Shaped like a Passat, with the heart of a Maxima, Nissan's front-drive, mid-sized sedan stakes out the territory occupied by the Honda Accord, Mazda 626, and Toyota Camry. The car's powerful 175-hp 2.5L 4-cylinder engine is almost as powerful as the competition's V6 power plants, and the optional 245-hp 3.5L V6 has few equals among cars in this price and size class. And, when you consider that the Altima is much lighter than most of its competitors, it's obvious why this car produces sizzling (and sometimes uncontrollable) acceleration with little fuel penalty. Four-wheel independent suspension strikes the right balance between a comfortable ride and sporty handling. With the V6 option, drivers get 17-inch tires for more grip.

Cost analysis/best alternatives: Although the 2003 models are practically identical to the 2004s, there aren't many around, and those that are for sale aren't discounted by much. So, stick with a 2004, preferably one equipped with a V6, and console yourself with the fact that you'll get most of your money back come trade-in time. Although 2004 prices haven't been boosted much, dealers are sneaking in extra profit from inflated PDI/freight charges. Other cars worth considering are the redesigned Mazda6, Honda Accord, and Toyota Camry. **Options:** Watch out for option-loading after you agree to a reasonable base price. Canny Nissan sales agents pretend that some options must be purchased, or that some can't be bought without others included. **Rebates:** $3,000 rebates and zero percent financing on fully loaded models. **Delivery/PDI:** $1,035. **Depreciation:** Much slower than average, especially the much-coveted V6-equipped model. **Insurance cost:** Higher than average. **Parts supply/cost:** Parts may be hard to find and dealer prices will be steep until independent suppliers enter the market. **Annual maintenance cost:** Average. **Warranty:** Bumper-to-bumper 3 years/80,000 km; powertrain 5 years/ 100,000 km; rust perforation 5 years/unlimited km. **Supplementary warranty:** Not needed. **Highway/city fuel economy:** *4-cyl. and auto.:* 7.3–10.1L/ 100km; *V6 and auto.:* 8.2–11.2L/100 km.

Quality/Reliability/Safety

Pro: Quality control: Good body assembly and powertrain components. **Warranty performance:** Average. Nissan staffers aren't particularly generous in giving out post-warranty "goodwill."

Con: Reliability: Lots of powertrain and body deficiencies as these vehicles age. **Owner-reported problems:** Many reports of engine surging, stalling, and hard starting, possibly due to a defective engine crank sensor or throttle switch (on national back order); annoying and hazardous dash reflection onto the windshield; electrical glitches; and excessive brake wear, noise, and pulsations. Snow builds up in the small wheel wells, making steering difficult and causing excessive shimmy; clutch failure when downshifting into Fourth gear; noisy, failure-prone rear shocks; ABS warning light stays lit; and ignition noise in radio speakers. **Service bulletin problems:** Engine won't crank; low power, poor running, and MIL light stays lit; inoperative AC/warm air flows from vents; poor heater performance; AC temperature isn't adjustable (see below):

> AC compressor does not disengage when the manual selector switch is moved from the defrost position. This is covered by a service bulletin NTB03-048 dated 5-6-03 from Nissan concerning this problem. This is noted as a change in operation from earlier models and I feel it is dangerous. The only way to disengage the compressor after using the defroster is to stop the car, move the selector switch to any function that is not defroster related, turn off the ignition and restart the car. If you don't there is at least a 10 percent loss in mileage due to the fact that the AC compressor is operating and the temp selector must be adjusted up or the occupants of the vehicle will freeze.

AC drain hose may leak into interior; howling noise when clutch is released; rear suspension and radio speaker noise; sunroof wind noise and water leaks; water leakage on front floor area; wind noise from doors; sunroof won't close at highway speeds; faulty airbag warning light; wheel cover squeak, click; buzzing noise when engine reaches 2500 rpm; automatic transmission slips in Reverse and won't brake when in D1 range, also makes a clicking noise; and headliner rattle. **NHTSA safety complaints/safety:** Safety defects become more common as these cars age. Over 120 complaints already recorded on the 2002s. Dealerships said to be aware of Altima's tendency to catch fire: fire erupted after collision; fire ignited due to a faulty fuel injection system on another occasion, and also ignited while cruising on the highway; airbags failed to deploy; vehicle was idling and then suddenly went into Reverse and accelerated as groceries were unloaded from the trunk (dash indicator showed car in "P"); driver run over by his own car when it slipped into Reverse; sudden acceleration when brakes are applied; transmission slips and engine hesitates when accelerating; chronic stalling; windshield distortion; exhaust pipe hanger pin catches debris that may ignite; crankshaft position sensor failure; tail lights constantly fail; battery suddenly blew up; seat belt fails to retract; and instrument panel gauges wash out in sunlight.

Road Performance

Pro: Acceleration/torque: Impressive acceleration with powerful 4- and 6-cylinder engines. **Transmission:** Both transmissions usually perform flawlessly (see above). **Steering:** Precise and responsive with lots of feedback. **Emergency handling:** Handles sudden corrections quite well. **Routine handling:** Good manoeuvrability around town, and highway cruising is quite comfortable. Suspension handles rough pavement without jostling passengers. **Braking:** Above average.

Con: Acceleration is hard to modulate; often accompanied by tire-spinning and excessive torque steer (twisting, or pulling to the right). The 4-cylinder engine tends to be buzzy and less refined than the competition. Be wary of the automatic transmission slipping into Reverse while you fool around in the trunk and of brakes locking up on wet highways.

Comfort/Convenience

Pro: Standard equipment: Fairly well appointed. **Controls and displays:** All controls are easy to understand and access, and instruments are easy to read and understand. **Climate control:** Good heating, defrosting, and ventilation. **Interior space/comfort F/R:** Up front, you'll find comfortable, supportive seats and good all-around headroom and legroom. **Cargo space:** Acceptable. Trunk pass-through allows for storage of long items. **Trunk/liftover:** Good-sized trunk with a low sill to facilitate loading. **Quietness:** Little road and tire noise gets into the passenger compartment.

Con: Driving position: Rear roof pillars and high-tail styling obstruct rear/side visibility. **Entry/exit:** Rear seat access is a bit difficult to master. **Interior space/comfort F/R:** Despite this year's larger interior, the small cabin provides limited rear headroom.

Altima

List Price (firm)	Residual Values (months)			
	24	36	48	60
2.5 S: $23,798 (17%)	$17,000	$14,000	$12,000	$10,000
2.5 SL: $29,498 (18%)	$21,000	$18,000	$14,000	$12,000
3.5 SE: $29,098 (18%)	$22,000	$19,000	$15,500	$12,500

Technical Data

Powertrain (front-drive)
Engines: 2.4L 4-cyl. (175 hp)
• 3.5L V6 (245 hp)
Transmissions: 5-speed man.
• 4-speed auto.
Dimension/Capacity
Passengers: 2/3
Height/length/width:
57.9/191.5/70.4 in.
Headroom F/R: 40.8/37.6 in.
Legroom F/R: 43.9/36.4 in.
Wheelbase: 110.2 in.
Turning circle: 41 ft.
Cargo volume: 15.6 cu. ft.
Tow limit: N/A
Fuel tank: 60L/reg.
Weight: 3,050 lb.

Safety Features/Crashworthiness

	Std.	Opt.
Anti-lock brakes	—	■
Seat belt pretensioners F/R	■	—
Side airbags	—	—
Traction control	—	—
Head restraints F/R	**	**
Visibility F/R	****	**
Crash protection (front) D/P	****	****
Crash protection (side) F/R	***	****
Crash protection (offset)	**	—

MAXIMA ★★★

RATING: Average; more horsepower than its competitors, but not backed up with the technical refinements needed for safe and pleasurable driving. The Maxima has been downgraded due to this year's redesign and the sharing of the Altima's defective DNA. Wait at least six months while Nissan corrects the inevitable factory-induced defects sure to afflict what has become a luxurious Altima. **Strong points:** Impressive powertrain performance; comfortable ride; and above-average reliability. **Weak points:** Tires are easily spun and there's excessive pulling to the right when accelerating; the shiftgate indents aren't user-friendly; 18-inch tires produce high-speed tire whine; dashboard reflection on windshield; increased chassis dimensions don't translate into a roomier interior, tall occupants may find rear seating a bit cramped; small trunk opening limits what luggage you can carry; premium fuel required; redesign will likely hurt quality, and lead to an unusually high number of safety-related failures. **New for 2004:** Redesigned and based on the Altima platform, which increases interior room and improves handling.

OVERVIEW: The front-drive mid-sized Maxima soldiers on as Nissan's luxury flagship, one step removed from the best-selling Altima. The 265-hp V6 coupled to a 6-speed manual or 4- or 5-speed automatic transmission is a real dazzler. The Maxima comes with an impressive array of standard features, and a host of performance and safety features, like large front brakes with full brake assist, a power driver's seat, xenon headlights, and 18-inch wheels.

Cost analysis/best alternatives: Since its price hasn't increased this year, the 2004 luxo-Altima is a tempting buy. But are you ready for its handling problems and likely first-year production glitches? It's a sure bet the 2004 Maxima won't be made as well as its predecessor, but this drawback can be attenuated by getting a second-series model. Other cars worthy of consideration are the 200-hp Acura TSX and 260-hp 3.2TL Type S, BMW 3 Series, automatic-equipped versions of the 240-hp Honda Accord V6, 215-hp Lexus IS300, and 210-hp Lexus ES300, Mazda6 GT V6 5-speed, and the 225-hp Toyota Camry SE-V6. **Options:** Traction control wouldn't be a bad idea if you tend to spin your tires when accelerating. On the other hand, consider swapping the 6-speed manual transmission for a 4-speed Jatco automatic with a manual shift mode at the same trim level, or even for a 5-speed Aisin automatic found in the 3.5SL model. **Rebates:** $2,500 and low-interest financing by mid-2004. **Delivery/PDI:** $972. **Depreciation:** Average. **Insurance cost:** Higher than average. **Parts supply/cost:** Parts cost and availability are likely to be a problem until independent suppliers can fill the pipeline. **Annual maintenance cost:** Predicted to be higher than average. **Warranty:** Bumper-to-bumper 3 years/ 80,000 km; powertrain 5 years/100,000 km; rust perforation 5 years/ unlimited km. **Supplementary warranty:** Not needed. **Highway/city fuel economy:** (2003) *V6 and auto.:* 8.1–12.1L/100 km.

Quality/Reliability/Safety

Pro: Quality control: Better than average. IIHS hasn't crash-tested a new Maxima, but the Altima upon which it is based merits a "Good," its top rating. NHTSA gives the Altima four out of five stars in frontal crashes, but only three for driver-side impacts.

Con: Reliability: Maximas have had a better reliability history than Altimas; keep your fingers crossed. **Safety**: This is the scary part (see Altima rating). **Warranty performance:** Average. **Owner-reported problems:** Scattered reports of automatic transmission, brake, and electrical system malfunctions (MIL lamp stays lit, for example), as well as paint defects and assorted body squeaks, clunks, and rattles. Too soon to predict how the Altima platform will hold up. **Service bulletin problems:** Hesitation upon acceleration with vehicles equipped with a manual transmission; lack of power; no-starts; driver's seat won't move forward or backward, inoperative left side speakers; and ABS light stays lit. **NHTSA safety complaints:** Vehicle underway at 100 km/h when suddenly the steering wheel locked up and the brakes failed; while vehicle was in motion, the front wheels locked up, causing extensive undercarriage damage; vehicle suddenly swerved out of control; vehicle suddenly accelerated in Reverse when put into Drive; sudden acceleration upon start-up (a faulty air control valve is suspected); airbags fail to deploy in a collision; unable to control engine speed with the accelerator pedal; vehicle stalls without warning in cold weather (suspect the computer module); fuel leaks from seal when over-filled; excessive front-end vibrations; strong "bleach-type" odour permeates the interior; many complaints that the headlights are poorly designed, placing the high beams too high for adequate visibility; drivers complain they can't see between the high and low beams; trunk lid and latch are hazardous when raised; sunroof opens and closes on its own; steering wheel overheats in direct sunlight; and the driver-side windshield washer may not work in cold weather.

Road Performance

Pro: Acceleration/torque: The powerful and smooth standard V6 engine provides substantial power without excessive noise. **Transmission:** The smooth-shifting 4-speed automatic transmission delivers power smoothly and efficiently. **Routine handling:** Very good, with lots of control. The ride is fairly firm but comfortable, due to the refined suspension that handles rough roads well if the car is lightly loaded. Very little brake fade after successive stops. **Emergency handling:** Takes practice (see below). **Steering:** Fairly good when not accelerating too quickly or coming out of a turn.

Con: Acceleration is hard to modulate. It's often accompanied by tire spinning and excessive torque steer (twisting, or pulling to the right). The power steering has a strong return-to-centre motion which means you have to fight the steering wheel to stay on track when coming out of a turn. The 12-m turning circle is very tight. The 6-speed manual transmission is a bit hard-shifting and noisy. Steering feels increasingly over-assisted as the Maxima's speed picks up. **Braking:** Extended stopping distance.

Comfort/Convenience

Pro: Controls and displays: High-tech luxury interior with emphasis on electronic gadgets; power seat controls are user friendly. Impressive theft deterrent system sounds an alarm, flashes the lights, and disables the car if the vehicle is disturbed. **Driving position:** Very good. Firm, supportive, and comfortable front seating; good forward visibility. **Climate control:** Efficient and quiet. **Entry/exit:** Easy front access. **Interior space/comfort F/R:** Adequate front head and legroom. **Quietness:** Much improved this year, with little wind and road noise invading the passenger compartment.

Con: The longitudinal Skyview glass roof running the length of the car's top is ugly (a conventional sunroof is optional). It doesn't open, and uses sliding panels to shutter the glass panes. Rear visibility is blocked by the new arched roofline and rear seat entry/exit can be difficult. Front and rear headroom has been reduced by 2 cm and 0.8 cm, respectively. Front seats could use additional rearward travel for tall occupants. **Interior space/comfort F/R:** Only two adults will fit comfortably in the rear, particularly in the Elite version, which provides two rear bucket seats in place of the usual bench. **Trunk/liftover:** Small trunk opening keeps out large objects.

Maxima

List Price (negotiable)	**Residual Values** (months)			
	24	**36**	**48**	**60**
SE: $34,500 (20%)	$27,000	$24,000	$21,000	$17,000

Technical Data

Powertrain (front-drive)
Engine: 3.5L V6 (265 hp)
Transmissions: 6-speed man.
- 5-speed auto.
- 4-speed auto.

Dimension/Capacity
Passengers: 2/2
Height/length/width: 58.3/193.5/71.7 in.
Headroom F/R: 40/37.3 in.
Legroom F/R: 43.9/36.5 in.
Wheelbase: 111.2 in.
Turning circle: 40 ft.
Cargo volume: 15.5 cu. ft.
Tow limit: N/A
Fuel tank: 70L/prem.
Weight: 3,463 lb.

Safety Features

	Std.	Opt.
Anti-lock brakes	■	—
Seat belt pretensioners F/R	■	—
Side airbags	—	■
Traction control	■	■
Head restraints F/R	***	***
Visibility F/R	*****	**

Note: The new Maxima hasn't been crash-tested.

QUEST ★★★★

RATING: Above Average. A stylish and powerful multi-tasker. **Strong points:** Powerful V6; uses regular fuel; smooth-riding suspension absorbs bumps quite well, without wallow (16-inch wheels); nice handling due to responsive steering, well-tuned suspension, and efficient brakes; plenty of passenger and cargo room, with lots of seating choices; adjustable pedals keep short drivers out of harm's way; easy entry/exit; most seats are easily folded and adjusted to maximize interior room; and many small storage compartments. **Weak points:** SE's 17-inch wheels are hard-riding; some body lean in turns, excessive wind noise; distracting dash reflection into the windshield and unusual layout of gauges in the middle of the dashboard; more front-seat legroom needed for tall drivers; second-row seats can't be folded flat and head restraints must be removed first; and Altima components and redesign may degrade overall reliability. **New for 2004:** Everything; Quest returns completely redesigned.

OVERVIEW: A totally different minivan than its predecessor, the $32,900–$50,000 2004 Quest is one of most stylish, smoothest-running, and (sigh) priciest minivans on the road. Now based on the Altima/Murano platform, it offers all the standard high-tech safety, performance, and convenience features one could want. Its long wheelbase allows for the widest opening sliding doors among front-drive minivans, rear-seating access is a breeze, and a capacious interior allows for flexible cargo and passenger configurations that can easily accommodate 4' x 8' objects with the liftgate closed. Standard fold-flat third-row seats and fold-to-the-floor centre-row seats allow owners to increase storage space, without worrying about where to store the extra seats; head restraints must be removed before the seats can be folded away.

Although it feels a bit heavy in the city, the Quest is very carlike when driven on the highway. Ride and handling are enhanced by a new four-wheel independent suspension, along with front and rear stabilizer bars, and upgraded antilock brakes.

Cost analysis/best alternatives: Forget about the 2003s; the 2004 model costs about $2,000 more this year, but represents the better choice, due to its many upgrades. Just try to get one made in March or later to keep first-year glitches to a minimum. **Best alternatives:** Other minivans worth considering are the GM Venture and Montana, Honda Odyssey, Mazda MPV, or Toyota Sienna. **Options:** Nothing really worthwhile. **Rebates:** Not likely. **Delivery/PDI:** $995. **Depreciation:** Predicted to be average. **Insurance cost:** Should be higher than average. **Parts supply/cost:** Good supply and reasonably priced, although the 2004 model will likely have some body part delays. **Annual maintenance cost: Should be less than** average. **Warranty:** Bumper-to-bumper 3 years/80,000 km; powertrain 5 years/100,000 km; rust perforation 5 years/unlimited km. **Supplementary warranty:** An extended warranty isn't needed. **Maintenance/repair costs:** Higher than average. **Highway/city fuel economy:** 8.2–12.4L/100 km.

Quality/Reliability/Safety

Pro: Quality control: Good quality control. **Reliability:** Good fit and finish, though early models have a few body rattles. A large dealer network means that servicing and parts are readily available. Plenty of glass provides excellent front and rear visibility. The Quest offers a large array of standard safety features that include front and side airbags that reduce deployment force as passenger weight decreases, head airbag protection for all occupants, side-impact beams, reinforced centre pillars, front and rear impact-absorbing zones, rear outboard three-point safety belts, standard traction control, ABS brakes with brake assist, and a childproof door lock in the sliding door. An integrated child seat is optional.

Con: Warranty performance: Average. **Owner-reported problems:** Because this is a totally new design, there hasn't been much owner feedback. Nevertheless, early reports signal some problems with electrical malfunctions, brake noise, vibration, binding, or overheating, and premature wear of the front discs, rotors, and pads. Fit and finish quality has been uneven on early production models, resulting in numerous squeaks and rattles. **Service bulletin problems:** N/A. **NHTSA safety complaints/safety:** N/A.

Road Performance

Pro: Acceleration/torque: V6 has plenty of power for all driving needs. **Transmission:** The 4-speed automatic transmission is competent, however, the SE's 5-speed delivers more mid-range grunt for passing and merging. **Routine handling:** Much improved, but still handles and manoeuvres like a large station wagon. Agile (the revised suspension helps in this area) and easy to drive on the highway; less so around town. The smooth, quiet ride on the highway isn't compromised by a full load. **Emergency handling:** Impressive highway stability even after sudden steering corrections or braking. **Steering:** Precise and predictable steering. **Braking:** Disc/disc; braking performance is excellent.

Con: 4-speed automatic lacks mid-range grunt found in the 5-speed automatic. Don't make too much of the Quest's carlike handling pretensions. Sure, it rides and handles much better than truck-based minivans like the GM Astro, but it can't match the quality, ride, or handling of the Honda Odyssey and Toyota Sienna. Quest requires a rather wide turning circle. Towing capacity of 1,590 kg (3,500 lb.) is possible only with an optional towing package.

Comfort/Convenience

Pro: Standard equipment: The interior is generally well appointed and comfortable. The SkyView longitudinal sunroof looks appropriate for a minivan, but out of place on the Maxima. Standard radio and DVD player gives great sound. **Driving position:** Carlike driving position and comfortable seats, which include a power lumbar support in the driver's seat. **Controls and displays:** Most controls and instruments are generally easy to use and read, except for the dimly lit front console (difficult to read at night). **Climate control:** Efficient, powerful system; you can turn on the AC and control the air outlet location separately. Side windows can be opened for maximum ventilation. **Entry/exit:** Easy. Helpful assist grab handles over the front and sliding doors. **Cargo space:** Quite good and easily accessed. **Trunk/liftover:** Low liftover height makes for convenient loading.

Con: Very limited colour selection. Some interior trim looks cheap, for example, covers for the SkyView sunroof don't appear to be top-quality. **Interior space/comfort F/R:** Not as big as some of its competitors. Sunroof trims front seat headroom by almost 4 cm. Seats could be longer with more lateral support. Can't get a second-row bench seat and the third-row seat isn't split-fold. Sliding door window should roll down. Difficult to change DVDs while driving (unlike the Odyssey). **Quietness:** Some wind noise.

Quest

List Price (negotiable)	**Residual Values** (months)			
	24	**36**	**48**	**60**
S: $33,995 (21%)	$23,000	$19,000	$16,000	$13,000
SL: $36,600 (23%)	$25,000	$21,000	$18,000	$15,000
SE: $43,400 (25%)	$29,000	$24,000	$21,000	$18,000

Technical Data

Powertrain (front-drive)
Engine: 3.5L V6 (240 hp)
Transmissions: 4-speed auto.
• 5-speed auto.
Dimension/Capacity
Height/length/width:
70/204.1/77.6 in.
Headroom F/R1/R2: 42/41.7/37.7 in.
Legroom F/R1/R2: 41.7/41.3/41.2 in.
Wheelbase: 124 in.
Turning circle: 40 ft.
Passengers: 2/2/3
Cargo volume: 148.7 cu. ft.
GVWR: 5,732 lb.
Tow limit: 3,500 lb.
Ground clearance: 5.8 in.
Fuel tank: 76/reg.
Weight: 4,000 lb.

Safety Features

	Std.	Opt.
Anti-lock brakes (4W)	■	—
Seat belt pretensioners F/R	■	—
Side airbags	■	—
Traction control	■	—

Note: The new Quest hasn't been crash-tested.

Subaru

In 1995, Subaru realized it was in a losing battle with Honda and Toyota for buyers of its front-drive compact cars and bet the farm on all-wheel-drive, versatile Outback and Forester models, and "Crocodile Dundee," an Australian actor cum Subaru pitchman. Since then, sales have soared, with most cars selling close to MSRP, and keeping much of their value come trade-in time.

Subaru's overall product lineup this year reinforces its AWD capabilities, as it moves upscale in adding more features and boosting prices. For 2004, the Forester and Baja sport-utility (the return of the Brat?) get turbocharged power plants, giving both engines a respectable 210 horses.

All Subarus provide reliable AWD capability, but most owners don't need their Subaru's off-road prowess; only 5 percent will ever use their Subaru for that purpose. The other 95 percent just like knowing they have the option of going wherever they please, whenever they please.

FORESTER, IMPREZA, WRX

RATING: *Forester:* Recommended; *Impreza, WRX:* Average. Now that the entry-level versions are dropped, these overpriced little Impreza sedans and wagons have priced themselves out of the market; if you don't need the AWD capability, you're wasting your money. **Strong points:** One of the most refined AWD drivetrains you'll find. Powerful WRX engine and impressive acceleration with the base 2.5L; excellent handling, without any torque steer; good braking; well-appointed base models; lots of storage space with the wagons; nice control layout; and better-than-average quality control. **Weak points:** Problematic entry/exit; the coupe's narrow rear window and large rear pillars hinder rear visibility; heater is insufficient and air distribution is inadequate; comfort compromised by WRX's short wheelbase, suspension, and 16-inch tires; front- and rear-seat legroom may be insufficient for tall drivers; small doors and entryways restrict rear access; WRX requires premium fuel and has a history of ABS failures, chronic surging, and stalling. **New for 2004:** *Impreza:* Fewer entry-level models and refreshed styling. *WRX:* An STi 300-hp turbocharged 2.5L sedan with upgraded suspension, steering, and brakes. *Forester:* A turbocharged 210-hp 2.5L 4-banger that includes a Variable Valve Control System, and a racier appearance to make the 2.5XT stand out.

OVERVIEW: The Impreza is essentially a shorter Legacy with additional convenience features. It comes as a four-door sedan, a wagon, and an Outback Sport wagon, all powered by a 165-hp 2.5L flat four engine. The rally-inspired WRX models have a more powerful 227-hp 2.0L turbocharged engine, lots of standard performance features, a sport suspension, aluminum hood with functional scoop, and higher-quality instruments, controls, trim, and seats.

Forester: Forester is a cross between a wagon and a sport-utility. Based on the shorter Impreza, it uses the Legacy Outback's 2.5L 165-hp engine coupled to a 5-speed manual transmission, or an optional 4-speed automatic. Its road manners are more subdued and its engine provides more power and torque for off-roading.

Cost analysis/best alternatives: Go for a 2003 Forester for the upgrades and all-around superior highway performance. WRX versions are expensive, problematic Imprezas, but when they run right, they'll equal the sporty performance of the Audi A4 and BMW 3 Series—cars costing $10,000 more. The Toyota Matrix/Pontiac Vibe front-drive and AWD models are excellent alternatives. If you really don't need a 4X4, here are some front-drives also worth considering: Honda Civic, Hyundai Elantra, Mazda6, and Toyota Corolla. **Rebates:** $2,000 rebates and low-interest financing. **Options:** Larger tires to smooth out the ride. **Delivery/PDI:** Be wary of delivery/PDI "creep"; fee is $1,195 this year, double what it once was. **Depreciation:** Slower than average. Foresters hold their value best of all. **Insurance cost:** Higher than average. **Parts supply/cost:** Parts aren't easy to find and can be costly; delayed recall repairs. **Annual maintenance cost:** Higher than average. Mediocre, expensive servicing is hard to overcome because independent garages can't service key AWD components. **Warranty:** Bumper-to-bumper 3 years/ 60,000 km; powertrain 5 years/100,000 km; rust perforation 5 years/ unlimited km. **Supplementary warranty:** Budget an extra $500 or more for an extended powertrain warranty to protect you from premature and repeated clutch failures. **Highway/city fuel economy:** *2.5L:* 7.7–10.5L/100 km.

Quality/Reliability/Safety

Pro: Quality control: Better than average. Above-average quality mechanical components. **Reliability:** Powertrain components should be durable, and there are few mechanical problems that take these Subarus out of service. However, servicing quality is spotty. **Warranty performance:** Base warranty is fairly applied and after-warranty assistance has been generous (especially with clutch failure claims). Servicing quality is spotty. **Safety:** Huge, fold-away side mirrors.

Con: A history of poor engine performance, premature clutch failures, frequent brake servicing; and some body panel and trim fit and finish deficiencies. Numerous reports of airbags failing to deploy or injuring passengers after deployment. **Owner-reported problems:** Cold weather stalling; poor engine idling; automatic and manual transmission malfunctions; early brake wear, rear wheel bearing failures; alloy wheels cause excessive vibration; premature exhaust system rust-out; minor electrical short circuits; catalytic converter failures; fit and finish deficiencies; water leaks and condensation problems from the top of the windshield or sunroof; windshield scratches too easily; and paint peeling. **Service bulletin problems:** Defective 4EAT transmission parking pawl rod; and clutch pedal sticking. **NHTSA safety complaints:** Driver burned from airbag deployment; clunking noise when brakes are applied; soft pedal and late brake engagement—pedal goes almost to the floor. Forester: Intermittent stalling; when backing into a parking space the "Hill Holder" feature activates, forcing the driver to use excessive throttle in Reverse; heater, defroster failure; front seats move fore and aft. *WRX:* Windshield cracking; chronic ABS brake failures.

Road Performance

Pro: Acceleration/torque: The 2.5L performs much better with the Impreza and Forester than with the Legacy Outback. It is smooth and powerful, with lots of low-end torque for serious off-road use. **Transmission:** The automatic transmission shifts smoothly. The all-wheel-drive system is a boon for people who often need extra traction, and it works well with either a manual or an automatic transmission. The manual transmission's "Hill Holder" clutch prevents the car from rolling backwards when starting out. **Routine handling:** Smooth and nimble. Hurtles through corners effortlessly with a flat, solid stance and plenty of grip. **Emergency handling:** Better than most small cars and sport-utilities. Tight cornering at highway speeds is done with minimal body lean and no loss of control. **Steering:** Precise and predictable. **Braking:** Disc/drum and disc/disc; smooth, effective braking with no brake fade after successive stops.

Con: Uncomfortable ride with a full load. WRX produces a choppy ride when passing over uneven pavement. Larger tires would improve handling.

Comfort/Convenience

Pro: Standard equipment: Well-appointed base models have a nice array of standard safety and convenience features. **Driving position:** Very good. Comfortable front seat and plenty of headroom and legroom. **Controls and displays:** Clear and simple dashboard and gauges. Very firm and supportive front seats. Versatile hatchback design. **Trunk/liftover:** Trunk has a low liftover. **Quietness:** Little noise intrudes into the cabin.

Con: The STi's narrow rear window and large rear pillars hinder rear visibility. Radio has awkward-to-access, poorly marked, tiny buttons. **Climate control:** The heater is insufficient and air distribution is inadequate. **Entry/exit:** Small doors and entryways restrict rear access. **Interior space/comfort F/R:** Front shoulder belts are uncomfortable and rear seat belts are hard to buckle up. Front- and rear-seat legroom may be insufficient for tall drivers. Rear seating is uncomfortable and limited to two small passengers. **Cargo space:** Not exceptional. Even though the wagon has extra storage capacity, overall capacity is a bit limited.

Forester, Impreza, WRX

List Price (negotiable)	**Residual Values** (months)			
	24	**36**	**48**	**60**
Wagon: $22,995 (18%)	$17,000	$13,000	$10,000	$8,000
Impreza 2.5 RS: $26,995 (18%)	$18,500	$14,500	$11,000	$9,000
Forester X: $27,995 (20%)	$19,500	$15,500	$12,000	$10,000
WRX Sedan: $35,495 (20%)	$25,000	$21,000	$17,000	$13,000
WRX STi: $46,995 (20%)	$32,000	$28,000	$24,000	$18,000

Technical Data

Powertrain (AWD)
Engines: 2.5L 4-cyl. (165 hp)
• 2.5L turbo (210 hp)
• 2.0L turbo (227 hp)
• 2.5 turbo (300 hp)
Transmissions: 5-speed man.

Headroom F/R: 39.2/37.4 in.
Legroom F/R: 43.1/32.4 in.
Wheelbase: 99.2 in.
Turning circle: 36 ft.
Cargo volume: 19.5 cu. ft.
GVWR: 4,120 lb.

• 4-speed auto.

Dimension/Capacity

Passengers: 2/3

Height/length/width:
60/172.2/67.1 in.

Tow limit: 2,000 lb.

Ground clearance: 7.5 in.

Fuel tank: 50L/reg.

Weight: 2,900 lb.

Safety Features/Crashworthiness

	Std.	Opt.
Anti-lock brakes	■	—
Seat belt pretensioners F/R	■	—
Side airbags	—	■
Traction control	—	—
Head restraints F/R		
Forester	*****	*****
Visibility F/R	*****	****
WRX STi	*****	**
Crash protection (front) D/P	****	*****
Forester	*****	*****
Crash protection (side) F/R	****	—
Forester	*****	****
Crash protection (offset)	*****	—
Forester	*****	—

LEGACY, OUTBACK 

RATING: Above Average. The AWD is what this car is all about. It handles difficult terrain without the fuel penalty or clumsiness of many truck-based SUVs. Without it, the Outback is just a well-equipped, middle-of-the-road vehicle, outclassed by most of the import competition. **Strong points:** A refined and reliable AWD system; a well-balanced, 6-cylinder engine; a comfortable ride; exceptional handling (if Vehicle Dynamic Control stability system works as it should); nice-handling GT version; lots of cargo room; and better-than-average crashworthiness. **Weak points:** The Vehicle Dynamic Control (VDC) feature adds exponential complexity to an already complicated-to-repair vehicle; serious stalling problems; and 6-cylinder-equipped vehicles are plagued by slow downshifts. The 2.5L is a sluggish performer, due undoubtedly to the car's heft; problematic automatic transmission performance when hooked to the smaller engine; mediocre handling on base models, without VDC; excessive 4-cylinder engine noise; cramped back seat; limited rear headroom for tall passengers; seat belts may be too short for large occupants; 6-cylinder requires premium fuel; and very dealer-dependent for parts and servicing. **New for 2004:** Nothing significant; a redesign is scheduled for the 2005 model.

OVERVIEW: A competent, full-time 4X4 performer for drivers who want to move up in size, comfort, and features. Available as a four-door sedan or five-door wagon, the Legacy is cleanly and conventionally styled, with even a hint of the Acura Legend in the rear end.

The Outback is a marketing coup that adds a V6 and customizes the AWD Legacy to give it more of an outdoorsy flair. American Motors tried the same marketing approach with the Eagle in the '70s and failed miserably, due to poor quality control, lousy marketing, and a passive public whose concept of off-road or high-performance thrills was driving through Stanley Park or on Toronto's 401.

Cost analysis/best alternatives: Get a cheaper 2003; it's almost identical to this year's model and it was considerably upgraded last year. Front-drives worth considering: Honda Accord and Toyota Camry. Worthwhile 4X4s: the Honda CRV, Hyundai Santa Fe, and Toyota RAV4. **Options:** Base models are severely performance-challenged. Optional integrated rear child seat is worthwhile. **Rebates:** $2,500 and low-interest financing. **Delivery/PDI:** $1,195. **Depreciation:** Slower than average. **Insurance cost:** Higher than average. **Parts supply/cost:** Parts aren't easily found and can be costly. **Annual maintenance cost:** Average. **Warranty:** Bumper-to-bumper 3 years/60,000 km; powertrain 5 years/100,000 km; rust perforation 5 years/unlimited km. **Supplementary warranty:** A good idea. **Highway/city fuel economy:** *2.5L:* 7.8–11L/100 km; *3.0L:* 8.4–12.4L/100 km.

Quality/Reliability/Safety

Pro: Quality control: Average, though powertrain defects have begun cropping up. Above-average body assembly and finish. **Warranty performance:** Although servicing quality is spotty, Subaru has been quite reasonable in processing warranty claims under the base warranty and through special, after-warranty assistance programs (particularly for engine and clutch repairs).

Con: Reliability: Powertrain defects can sideline the car for days. There are several reports of the transmission jumping out of First gear when using First to slow down or to descend a steep grade. **Owner-reported problems:** Cold-weather hard starts and chronic surging and stalling; snow gets inside of wheelwell, binding steering; minor electrical, fuel system, and automatic transmission problems. Owners report that the front brakes require more attention than average. Premature surface rust and exhaust system rust-out are common and misadjusted door strikers make for hard closing/opening. Servicing can be awkward because of the crowded engine compartment, particularly on turbocharged versions. Small horn buttons may be hard to find in an emergency. **Service bulletin problems:** Defective engine water pump; improved Sport Shift cold-weather operation; defective transmission parking pawl rod; roof rack wind noise; and premature suspension corrosion (see bulletin excerpt below):

> Certain rear suspension subframe components were produced with poor paint quality, which, after continued exposure to corrosive road salts for a period of several years, could result in rust-out of the component and possible breakage of the subframe. If such breakage

occurs while the vehicle is being operated, control of the vehicle could be affected, increasing the risk of a crash. Remedy: Dealers will clean and rustproof the rear suspension subframe.

NHTSA safety complaints/safety: Chronic stalling in forward gear and in Reverse, particularly with 6-cylinder-equipped models (see below):

> On multiple occasions, and with multiple drivers, I have had two 2003 Outbacks with frequent stalling when moving from Neutral to First, or to Reverse. This is potentially dangerous when moving out into traffic. Subaru denies receiving prior complaints (despite the material on your site) and in general denies that there is any problem. They also compelled me to sign a "gag" clause as part of a deal whereby they exchanged car #1 for car #2. I have arbitrarily put in a specific month, day, and year but it has happened on multiple occasions. Also car #1 had two incidents of the engine running too rough to be driven, progressively losing power and then quitting; when restarted, ran fine. Also car # 1 had one episode of turning over well but refusing to start.

Surging after coming to a stop; ABS brake failures; brakes feel spongy and take too much time to stop vehicle; braking pedal goes to the floor before braking effect, resulting in extended stopping distances; vehicle suddenly veers to the right when accelerating or braking; cruise control failed to disengage when brake pedal was depressed; fuel sloshes in fuel tank due to the absence of baffles; during a collision, airbags deployed but failed to inflate; the suspension's design causes severe pulling to one side; excessive steering and vehicle vibration when passing over uneven pavement; steering lock up while driving; knocking and clunking noise heard when turning; vehicle's rear end bounces about when passing over bumps; engine failure due to a cracked #2 piston; frequent surging from a stop; hard to shift from Park to Reverse; cracked seat belt buckle; seat belts are too short for large occupants; rear centre seat belt prevents the secure attachment of child safety seats; and Firestone tread failures.

Road Performance

Pro: Acceleration/torque: The 2.5L engine is a competent performer only with a manual gearbox. The 6-cylinder is adequate, but doesn't feel like it has much in reserve. **Routine handling:** Fairly soft and comfortable ride on smooth pavement. The GT's firmer suspension exhibits above-average handling. **Steering:** Acceptable steering response. Higher-end models handle well, although there is some excessive lean when cornering. Outback VDC has standard anti-skid/traction control. **Braking:** Disc/disc; a bit better than average.

Con: The 4-cylinder engine is noisy and rough running. It's tuned more for low-end torque than speedy acceleration. **Transmission:** The automatic transmission shifts into too high a gear to adequately exploit the engine's power and is reluctant to downshift into the proper gear. Manual transmission's shift linkage isn't suitable for rapid gear changes. Base models don't handle well. **Emergency handling:** Base vehicles bounce around on uneven pavement, rear end tends to swing out during high-speed cornering, and there's too much body lean in turns at lesser speeds.

Comfort/Convenience

Pro: Standard equipment: Well appointed. **Driving position:** Low but comfortable driving position, with plenty of headroom and legroom for most drivers. Good fore and aft visibility. The left footrest prevents cramping. Adjustable driver's seat on the deluxe model. **Controls and displays:** Well-designed, sweeping dashboard (similar to the Accord's) and control layout. Easy-to-read analogue gauges. **Climate control:** Excellent climate control system that's both efficient and quiet. **Entry/exit:** Tall doors make for easy front and rear access. **Interior space/comfort F/R:** Seating for four average-sized adults offers good front and rear headroom and legroom. Comfortable cloth-covered seats with plenty of side and shoulder support. **Cargo space:** Better-than-average cargo capacity. **Trunk/liftover:** The wagon's cavernous trunk is expandable with split rear seatbacks, and is more useful than the sedan's rear-seat pass-through.

Con: Tall drivers could use additional rearward seat travel and six-footers will find rear headroom barely adequate. It's also a tight fit for the middle rear-seat passenger. Power window and lock switches aren't easily accessible. Limited-service spare tire. Sedan has a small trunk, which is expandable only through a rear-seat pass-through. Trunk hinges can damage cargo and cut into storage space. **Quietness:** Excessive engine noise.

Legacy, Outback

List Price (firm)	**Residual Values** (months)			
	24	**36**	**48**	**60**
L Sedan: $27,795 (20%)	$19,000	$16,000	$13,000	$10,000
H6 3.0: $39,995 (20%)	$30,000	$25,000	$20,000	$15,000

Technical Data

Powertrain (AWD)
Engines: 2.5L 4-cyl. (165 hp)
• 3.0L 6-cyl. (212 hp)
Transmissions: 5-speed man.
• 4-speed auto.

Dimension/Capacity
Passengers: 2/3
Height/length/width: 65/175/68 in.
Headroom F/R: 40.2/39.5 in.
Legroom F/R: 43/33.4 in.
Wheelbase: 99.4 in.
Turning circle: 39 ft.
Cargo volume: 13 cu. ft.
Tow limit: 1,500 lb.
Fuel tank: 60L/reg.
Weight: 3,140 lb.

Safety Features/Crashworthiness

	Std.	Opt.
Anti-lock brakes (4W)	■	—
Seat belt pretensioners F/R	■	—
Side airbags	—	■
Traction control	■	—
Head restraints F/R	***	***
Visibility F/R	*****	*****
Crash protection (front) D/P	****	*****
Crash protection (side) F/R	***	****
Wagon	****	*****
Crash protection (offset)	*****	—

Suzuki

Suzuki has been making very good entry-level small cars and sport-utility vehicles for over a decade, but most buyers aren't familiar with the company's products because they were mostly sold under GM's name. In fact, the company makes only two mainstream vehicles in its own name: the all-new Aerio small sedan and the Vitara sport-utility—a vehicle sold by General Motors as the Tracker and rated "Above Average" in *Lemon-Aid SUVs, Vans, and Trucks 2004.*

No Suzuki Firenza for Canada

Last year, GM bought a controlling interest in the assets of bankrupt Korean automaker Daewoo, and renamed the company GM Daewoo Auto and Technology Company, or GMDAT. The first three cars from this new venture are the Chevrolet Aveo, Optra, and Epica.

Similar in size to the Geo Metro, the Swift returns after a two-year absence.

In Canada, the Aveo will be offered as the Suzuki Swift and the Epica will be sold as the Suzuki Verona. U.S. Chevrolet dealers will get the Aveo, but the Optra and Epica will be sold by Suzuki as the Firenza and Verona. None of these vehicles can be recommended during their first year on the Canadian market.

Hmmm…I wonder why GM and Suzuki Canada don't want to bring back the Firenza nameplate?

VERONA ★

RATING: Not Recommended during its first year on the market. A Verona is an updated Daewoo Leganza, given a sexy Italian moniker and a bit of European styling. It carries a 155-hp 2.5L inline 6-cylinder engine to power its mid-sized frame. This put it in the same league as Honda's Accord, Nissan's Altima, and the Toyota models, which use more efficient 4-cylinders to do the same tasks with less fuel penalty. Its ride and handling are acceptable, but lack the refinement of its Japanese competitors. A 2004 Verona GL sells for $22,995, plus $995 for freight and PDI.

AERIO ★★★★★

RATING: Recommended. This little car is a winner due to its better-than-average overall performance and low price. Suzuki's front-drive 2003 Aerio replaced the Esteem with a much-improved compact sedan and wagon that offer real competition to other Asian automakers. About the same size as the Esteem, the Aerio uses a 155-hp 2.3L 4-cylinder engine that's one of the most powerful engines found in this small car class and much more powerful than its predecessor's 122 horses. It has been re-styled with cleaner, simpler lines, and offers more standard features than the Esteem. And, with the addition of all-wheel drive, the $20,495 (plus $995 delivery/PDI) Aerio AWD S competes head-to-head with Subaru AWDs, the Pontiac Vibe, and Toyota Matrix. Some of the car's standard features include AC, power windows, power mirrors, tilt steering wheel, CD player, a tachometer, and split-fold rear seats. **New for**

2004: A small horsepower boost (10 horses). These cars return without a price increase, making them a steal. The base GL sedan sells for $15,995, but substantial rebates and low-financing programs should kick in during the summer and fall of 2004.

OVERVIEW: On the road, the Aerio performs fairly well. Its light weight and relatively powerful engine gets it quickly up to cruising speed. The automatic transmission shifts smoothly, although fuel economy is sharply reduced. Ride quality is good; brakes are adequate, though a bit soft; road and engine noise are omnipresent; and radio and climate controls aren't easily reached. Good forward visibility, but the rear view is restricted by wide rear pillars. Unlike the sedan's low roofline, the wagon's tall roof assures that there's plenty of headroom for tall passengers. There's a surprising amount of passenger and cargo room; legroom is on par with or beats most of its competition; the same is true for cargo space.

The Aerio hasn't been crash-tested by NHTSA, but IIHS gives the car a "Good" rating for head-restraint protection and offset crashworthiness.The only quality problem seen so far is paint chipping. Service bulletins deal with minor deficiencies like suspension popping and windshield wiper fluid leaking onto the cowl.

Aerio

List Price (very negotiable)	**Residual Values** (months)			
	24	**36**	**48**	**60**
GL Sedan: $15,995 (13%)	$11,000	$9,000	$7,000	$6,500
AWD GL: $22,895 (14%)	$16,000	$13,000	$11,000	$9,000

Technical Data

Powertrain (front-drive)
Engine: 2.3L 4-cyl. (155 hp)
Transmissions: 5-speed man.
• 4-speed auto.
Dimension/Capacity (sedan)
Passengers: 2/3
Height/length/width:
60.8/171.3/67.7 in.
Headroom F/R: 40.6/37.6 in.
Legroom F/R: 41.4/36 in.
Wheelbase: 97.6 in.
Turning circle: 32.8 ft.
Cargo volume: 14.6 cu. ft.
Tow limit: N/A
Fuel tank: 50L/reg.
Weight: 2,875 lb.

Safety Features/Crashworthiness

	Std.	Opt.
Anti-lock brakes	—	■
Seat belt pretensioners F/R	■	—
Side airbags	—	—
Traction control	—	—
Head restraints F/R	*****	*****
Visibility F/R	*****	***
Crash protection (offset)	*****	—

Toyota

Toyota has been a quality leader for three decades, despite a few bumps in the road caused by excessive front-brake vibrations and powertrain deficiencies since 1997. However, one does get the impression the automaker has been skating on its reputation for the past few years, while Honda, Nissan, and Mazda vehicles got better. As its redesigned 2004 Sienna minivan puts on the kilometres, we'll see if Toyota's recent quality decline is temporary, or an early sign the company will, henceforth, put profits ahead of quality.

Toyotas do hold up very well over the years; are especially forgiving of owner neglect; and cost very little to service at independent garages. Also, warranty performance is far better than what you'll experience from Chrysler, Ford, and GM. But the kicker for most buyers is how little their vehicles depreciate; it's not unusual to see a 5-year-old sedan selling for almost half its original selling price—a value reached by most Detroit Big Three vehicles a bit after their second year of ownership.

On the downside, a perusal of *Lemon-Aid* readers' letters and emails as well as NHTSA reports shows that some models have been plagued by engineering mistakes, much like Honda, that result in serious safety-related failures, powertrain breakdowns, and electrical system glitches affecting power doors and accessories. Consequently, I have lowered some ratings to reflect this deficiency and warn buyers of the potential for harm. Another disturbing development is Toyota's increasing reliance on sky-high freight/PDI charges of $1,000 or more. It makes no sense for the company to pat itself on the back for keeping prices low, while at the same time charging these exorbitant fees. Toyota customers are not stupid; they can add.

In addition to launching an all-new Sienna, an upgraded Prius hybrid, and Solara, Toyota continues to roll out its new Scion brand (which is only available in the States) to target young drivers.

PRIUS ★★★

RATING: Average. **Strong points:** Good fuel economy, five-passenger capacity, comprehensive powertrain warranty. **Weak points:** An incredibly high $1,110 delivery/PDI fee; underpowered; rear seating is cramped for three adults; sales and servicing may not be available outside of Montreal, Toronto, Ottawa, and Vancouver; and there's no long-term reliability data. **New for 2004:** Approaches the Camry in size, with 15 cm added to the wheelbase and overall length increased by 12 cm; and the engine gets two more horses. Toyota's new Hybrid Synergy Drive is more sophisticated, powerful, and efficient than last year's system. The engine is rated at 76 hp, six more than before; the drive motor can produce 67 hp, up from 44; and this year's battery is smaller, lighter, and more powerful.

OVERVIEW: This year's Prius is longer, wider, and more accommodating for both passengers and cargo. Although slightly smaller than a Camry, it is now classed as a mid-sized sedan. It employs both a gasoline engine and an electric motor for maximum fuel economy and power—reaching 100 km/h within 10 seconds—2 seconds less than its predecessor.

An electric motor, not the gasoline engine, is its main source of power; and it uses an innovative continuously variable transmission for smooth and efficient shifting. The motor is used mainly for acceleration, with the gasoline engine kicking in when needed. Once underway, the 1.5L 4-cylinder gasoline engine takes over to provide power and recharge the battery pack. Braking automatically shuts off the engine, as the electric motor acts as a generator to replenish the nickel-metal-hydride battery pack.

Clearly, Toyota's hybrid is engineered to be as versatile and hassle-free as possible, based as it is on the Echo's dimensions and designed to never need recharging. Interestingly, due to its primary reliance upon electrical energy, fuel economy is higher in the city than on the highway—the opposite of what one finds with gasoline-powered vehicles. Other drawbacks: handling is marginal, due mainly to the low rolling resistance tires; and inadequate brake feedback.

Some shortcomings: interior room is limited for tall occupants; the steering wheel position takes some getting used to; and the seats could use better bolstering and lumbar support.

Going hybrid doesn't mean that drivers have to forsake modern conveniences and safety features. Each Prius comes with standard ABS (Hydraulic Anti-lock Braking System with Generative Braking Assist), front airbags, power windows, locks, and mirrors, and air conditioning. **Owner-reported problems:** Steering and tire failures. **NHTSA safety complaints/safety:** Driver tapped the brakes to disengage the cruise control and vehicle suddenly veered to the left, out of control; sudden steering failure; complete power loss on the highway; vehicle suddenly shuts down due to a faulty battery; vehicle runs out of gas, despite gauge showing plenty of fuel in tank; tail lights don't work; and Bridgestone tire blowout.

Cost analysis/best alternatives: Get a 2004 Prius: it costs no more than last year's version and the extra power and room will come in handy. Other small cars that represent good investments are the Honda Civic, Hyundai Accent, and Toyota Echo. The Honda Hybrid isn't as refined as the Prius. **Options:** Head-protecting side airbags. **Rebates:** These cars aren't selling very well: only 1,100 Canadians have purchased a Prius so far. Nevertheless, Toyota is betting that the Prius' popularity will rise along with fuel prices; so rebates aren't likely. Keep in mind the credits that some provincial governments will give you if you buy a hybrid: $2,000 from the Ontario government, and $1,000 from the B.C. government. **Delivery/PDI:** $1,110. **Depreciation:** Average. **Insurance cost:** Average. **Parts supply/cost:** Parts aren't easily found and can be costly. **Annual maintenance cost:** Too early to tell. **Warranty:** Bumper-to-bumper 3 years/60,000 km; powertrain 5 years/100,000 km; hybrid-related components (HV battery, battery control module, inverter with converter) 8 years/160,000 km; major emission components 8 years/130,000 km; rust perforation 5 years/unlimited km. **Supplementary warranty:** Not needed. **Highway/city fuel economy:** Fuel economy is nearly twice as good as a 4-cylinder Camry:

4.3L/100 km (65 mpg) in combined city and highway driving compared to 8.6L/100 km (34.5 mpg).

Prius

List Price (very negotiable)	**Residual Values** (months)			
	24	**36**	**48**	**60**
Prius: $29,990 (15%)	$21,000	$18,000	$14,000	$10,000

Technical Data

Powertrain (front-drive)
Engine: 1.5L 4-cyl. (76 hp)
Transmission: CVT
Dimension/Capacity
Passengers: 2/2
Height/length/width: 58.1/175/67.9 in.
Headroom F/R: 39.1/37.1 in.
Legroom F/R: 41.9/38.6 in.
Wheelbase: 106.3 in.
Turning circle: 34.1 ft.
Cargo volume: 16.1 cu. ft.
Tow limit: N/A
Fuel tank: 45L/reg.
Weight: 2,890 lb.

Safety Features/Crashworthiness

	Std.	**Opt.**
Anti-lock brakes	■	—
Seat belt pretensioners F/R	■	■
Side airbags	—	■
Traction control	—	■
Head restraints F/R	*****	*****
Visibility F/R	*****	*****
Crash protection (front) D/P	***	****
Crash protection (side) F/R	***	***

ECHO ★★★★★

The Echo hatchback: A Yaris in Europe and Scion xA in the United States.

RATING: Recommended. An incredibly practical small car, if you can get by the tall, function-over-form styling. **Strong points:** Plenty of usable power; excellent fuel economy; and lots of interior space for passengers along with an incredible array of storage areas, including a huge trunk and standard 60/40 split-fold rear seats. Reasonably well equipped with good-quality materials, well-designed instruments and controls, comfortable seating, easy rear access, and excellent visibility fore and aft. Quite nimble when cornering, and very stable on the highway. Surprisingly quiet for an economy car. **Weak points:** Tall profile and light weight makes the Echo vulnerable to side wind buffeting; base tires provide poor wet traction; excessive torque steer (sudden pulling to one side when accelerating); narrow body width limits the rear bench seat to two adult passengers; and unorthodox styling "cute" to some but ugly to others. **New for 2004:** A new hatchback, which isn't new at all. It's sold as the Yaris in Europe and the Scion xA in the States.

OVERVIEW: Echo is Toyota's entry-level five-passenger econocar that gives great fuel economy without sacrificing performance. Both two- and five-door sedans and hatchbacks are available, and the car costs substantially less than the Corolla. It offers about the same amount of passenger space as the Corolla, thanks to a high roof and low floor height.

Cockpit controls and instrumentation are particularly user-friendly—located high on the dash and more toward the centre of the vehicle, rather than directly in front of the driver, where many gauges and controls are hidden by the steering column. Toyota says the repositioning of the instrument cluster farther away from the driver requires less refocusing of the eyes from far to near. This produces less driving fatigue and eye strain.

Echo is powered by a 108-hp 1.5L DOHC 4-cylinder engine featuring Variable Valve Timing cylinder head technology. It's the same design used in Lexus to combine power and fuel economy in a low-emissions vehicle. Normally, an engine this small would provide wimpy acceleration for most cars, but, thanks to the Echo's light weight, acceleration is more than adequate with a manual gearbox, and acceptable with the automatic.

Standard safety features: five three-point seat belts (front seat belts have pretensioners and force limiters), two front airbags (side airbags are optional), four height-adjustable head restraints, rear child seat tether anchors, and rear child door locks.

Cost analysis/best alternatives: Get the 2004, since discounted 2003s aren't easily found. The hatchback is a bargain for light commuting duties. Other cars worth considering are the Honda Civic, Hyundai Accent, Mazda Protegé, and the Nissan Sentra. **Options:** Consider snow tires and better-quality 14-inch tires for improved traction in inclement weather. Forget the silly-looking rear spoiler. Beware of option loading, where you have to buy a host of overpriced, non-essential, impractical features in order to get one or two amenities you require. **Rebates:** $1,020 rebates and low-interest financing. **Delivery/PDI:** $960. **Depreciation:** Much slower than average. **Insurance cost:** Below average. **Parts supply/cost:** Easily found and reasonably priced. **Annual maintenance cost:** Costs over the long term are predicted to be low. **Warranty:** Bumper-to-bumper 3 years/60,000 km; powertrain 5 years/ 100,000 km; rust perforation 5 years/unlimited km. **Supplementary warranty:** Not needed. **Highway/city fuel economy:** 5.5–7.1L/100 km with an automatic transmission.

Quality/Reliability/Safety

Pro: Quality control: Remarkably well built for an entry-level subcompact. **Reliability:** Nothing has turned up yet. **Warranty performance:** Toyota has few written warranty extensions; nevertheless, the company has been fair and generous in interpreting warranty obligations, even after the base warranty expires.

Con: Owner complaints: Electrical malfunctions. **Service bulletin problems:** N/A. **NHTSA safety complaints/safety:** Sudden, unintended acceleration; airbags failed to deploy; complete brake loss; brake and accelerator are mounted too close together; vehicle drifted off the highway at 110 km/h.

Echo

List Price (very negotiable)	**Residual Values** (months)			
	24	**36**	**48**	**60**
Hatchback: $12,995 (8%)	$9,000	$7,000	$6,000	$5,000
Sedan: $14,600 (9%)	$11,000	$9,000	$7,500	$6,500

Technical Data

Powertrain (front-drive)
Engine: 1.5L 4-cyl. (108 hp)
Transmissions: 5-speed man.
• 4-speed auto.
Dimension/Capacity
Passengers: 2/3
Height/length/width:
59.1/164.5/65.4 in.
Headroom F/R: 39.9/37.6 in.
Legroom F/R: 41.1/35.2 in.
Wheelbase: 93.3 in.
Turning circle: 32.8 ft.
Cargo volume: 13.6 cu. ft.
Tow limit: 1,000 lb.
Fuel tank: 45L/reg.
Weight: 2,095 lb.

Safety Features/Crashworthiness

	Std.	Opt.
Anti-lock brakes	—	■
Seat belt pretensioners F/R	■	—
Side airbags	—	■
Traction control	—	—
Head restraints F/R	*****	*****
Visibility F/R	*****	*****
Crash protection (front) D/P	****	****
Crash protection (side) F/R	***	****

COROLLA ★★★★

RATING: Above Average. **Strong points:** Pleasant ride; good braking; better quality control than Detroit Big Three; improved interior ergonomics (lots more space); improved crashworthiness scores; and a high resale value. **Weak points:** Average acceleration requires constant shifting to keep in the pack; automatic transmission-equipped versions are slower still. Clumsy emergency handling; lots of high-speed wind and road noise; limited front legroom; head restraints block rear visibility; many report that these cars literally stink, due to a "rotten-egg" exhaust smell entering the interior; some reports of airbags deploying inadvertently or failing to deploy; unimpressive crashworthiness scores on the 2002 model; overall quality has deteriorated; and high freight/PDI charges. **New for 2004:** Nothing significant.

OVERVIEW: A step up from the Echo, the Corolla has long been Toyota's conservative standard-bearer in the compact sedan class. Over the years, however, the car has grown in size, price, and refinement to the point where it can now be considered a small family sedan. All Corollas ride on a front-drive platform with independent suspension on all wheels. There are three variants: the value-leader CE and the more upscale S and LE. Power is supplied by a torquey 130-hp 1.8L twin-cam four, teamed with either a standard 5-speed manual gearbox or an optional 4-speed.

Standard equipment includes AC, tilt steering wheel, split-fold rear seat, power mirrors, and a CD player. ABS and front side airbags are optional.

The 2003 model gained fresh styling, a larger interior, a much-needed 4-speed automatic transmission, five more horses, and a Matrix wagon spin-off.

Cost analysis/best alternatives: The 2004 costs about $1,200 more than a similarly equipped 2003. Get a discounted 2003, if you can find one, and make it a second-series 2003 built after March to ensure you get all the latest assembly-line improvements. Other small cars that represent good investments are the Honda Civic, Hyundai Elantra and Tiburon, Mazda6, and VW Jetta. **Options:** Be wary of the S model. You'll pay big bucks for fancy trim, a leather-covered steering wheel, revised shift knob, foglights, and tachometer. Consider the built-in rear child seat. For better steering response and additional high-speed stability, order the optional 185/65R14 tires that come with the LE. **Rebates:** $1,000 rebates, low-interest financing, and modest discounting. **Delivery/PDI:** $1,010. **Depreciation:** Much slower than average. **Insurance cost:** Higher than average. **Parts supply/cost:** Parts are easily found and reasonably priced. **Annual maintenance cost:** Lower than average. **Warranty:** Bumper-to-bumper 3 years/60,000 km; powertrain 5 years/100,000 km; rust perforation 5 years/unlimited km. **Supplementary warranty:** Not needed. **Highway/city fuel economy:** *1.8L and manual:* 5.3–7.1L/100 km.

Quality/Reliability/Safety

Pro: Quality control: Good component and assembly quality, but trending downward. Toyota can do better, judging by the few performance- and safety-related complaints elicited by the Avalon, Echo, and Celica. **Reliability:** Quite reliable in the past; one hopes that this year's redesign will eliminate many of the powertrain deficiencies we've seen lately, like chronically malfunctioning airbags and powertrains. Unfortunately, the Camry's redesign didn't improve quality much. Keep your fingers crossed. **Warranty performance:** Toyota is usually fair and generous in interpreting warranty obligations, even after the base warranty expires. Nonetheless, the company has tried to blame owners for its own inadequacies.

Con: The Corolla has fallen from its "Recommended" status, due mainly to the flurry of safety complaints reported to government safety officials. This may mean that Toyota has gone too far in de-contenting its cars since the 1997 model year. **Owner-reported problems:** A stinking exhaust; engine and transmission failures; a litany of airbag defects; premature front- and rear-brake pad wear and vibrations; suspension and steering malfunctions; and electrical glitches. **Service bulletin problems:** Special service campaign to tighten the rear hub bolts; no airflow from centre vents; door glass window run, instrument panel finish, and upper and lower door improvements; headlights come on when switched off; trunk difficult to open; low air volume from AC; front suspension area tapping. **NHTSA safety complaints/safety:** Fire ignited due to loose fuel line; sudden, unintended acceleration; vehicle suddenly shut down while underway at 100 km/h; in another case, vehicle suddenly accelerated while cruise control was engaged; several owners report that a hole in the oil pan caused vehicle to stall; when the vehicle was parked, the parking brake was released and both airbags deployed; both airbags deployed right after driver turned on the ignition switch; airbags deployed after car passed over a bump in the road; in a rear-end collision, front airbag came out, but failed to deploy; side airbag failed to deploy in a side impact; both airbags failed to deploy in a frontal collision:

> [Our] [t]hirteen-year old daughter sustained [a] severe traumatic brain injury after [a] head-on collision. Her airbag deployed, [and the] top of passenger-side airbag was completely blown apart from one side to the other, allowing her head to strike [the] dash.
>
> We both had seat belts on. I sustained bruising of [my] hips and right rib. She was in [a] coma and [is] just now beginning to move [her] left side. She still cannot sit up, stand, walk, talk coherently, eat, or do anything for herself.

Faulty seat belt wiring could cause a fire; floormat catches accelerator pedal; brake failure when decelerating; when brakes are applied, pedal goes soft, resulting in extended stopping distances; when coming to a gradual stop, brakes locked up, causing extended stopping distance; rear welding broke away from the frame, resulting in complete loss of control; sudden collapse of the rear axle; rear control arm broke while driving at 110 km/h; left rear tire fell off; defective Uniroyal and Goodyear Integrity tires; warped wheels; shifter refused to go into gear while driving; transmission sometimes goes from Drive to Neutral while driving; refuses to shift out of Park; vehicle wanders all over the road and pulls to the left or right at moderate speeds; in windy conditions, vehicle becomes hard to steer, veering left or right; suspension bottoms out when passing over small bumps in the road; centre console has insufficient padding; persistent strong "rotten-egg" sulfur smell from the exhaust; a weak climate control; high beams shoot up into the sky, blind driver, and don't adequately light the highway; glow-in-the-dark inside trunk release doesn't glow in the dark because it is rarely exposed to light (owner actually crawled inside trunk to test it out); poor quality wiper blades; poorly designed headlights: "Both headlights have stray beams on high beam that project upward at a 45-degree angle. Very distracting when driving in light mist conditions. Dealer has attempted to adjust headlights with no effect on condition; low-beam headlights are too dim; sunlight washes out dash readings.

Road Performance

Pro: Acceleration/torque: Adequate, though not impressive, acceleration times. **Transmission:** Both manual and automatic transmissions shift smoothly. The manual transmission gives the Corolla extra pep and uses a light clutch for effortless shifting. **Routine handling:** Average handling under normal driving conditions; the ride is busy, but comfortable. This is characteristic of Toyota products, where handling takes second place to comfort. **Braking:** Disc/drum; better-than-average performance without ABS.

Con: AC and an automatic transmission can seriously reduce engine horsepower. **Emergency handling:** A bit clumsy, with some body roll, and sometimes the car plows straight ahead in hard cornering. Ride quality deteriorates as the load increases. **Steering:** Not much road feel. The sedan's ABS braking isn't impressive—too much weaving and veering to one side.

Comfort/Convenience

Pro: Standard equipment: Pretty well equipped, though some features that were standard last year are now optional. **Controls and displays:** The instrument panel and control layout are exceptionally user-friendly. **Climate control:** First-class heating, defrosting, and ventilation system. **Entry/exit:** Wide rear doors facilitate access to the rear. **Interior space/comfort F/R:** Comfortable front seating; the wagon's rear seat will hold two comfortably. **Cargo space:** Spacious cargo area on the wagon. The rear seatback can be folded down. **Trunk/liftover:** Large trunk on the sedan has a low sill for easy loading.

Con: Driving position: Short-statured drivers may have trouble seeing over the wagon's hood and rear head restraints block rearward visibility. Driver-side sun visor strikes rear-view mirror. **Interior space/comfort F/R:** Tight rear seating and limited rear footroom. **Quietness:** Lots of engine and road noise at highway speed.

Corolla

List Price (very negotiable)	**Residual Values** (months)			
	24	**36**	**48**	**60**
CE: $16,410 (9%)	$12,000	$10,000	$8,500	$6,500
LE: $20,965 (12%)	$16,000	$13,000	$11,000	$9,000

Technical Data

Powertrain (front-drive)
Engine: 1.8L 4-cyl. (130 hp)
Transmissions: 5-speed man.
• 4-speed auto.
Dimension/Capacity
Passengers: 2/3
Height/length/width:
58.8/178.3/66.9 in.
Headroom F/R: 39.1/37.1 in.
Legroom F/R: 41.3/35.4 in.
Wheelbase: 102.4 in.
Turning circle: 35.2 ft.
Cargo volume: 14 cu. ft.
Tow limit: 1,000 lb.
Fuel tank: 50L/reg.
Weight: 2,590 lb.

Safety Features/Crashworthiness

	Std.	Opt.
Anti-lock brakes	—	■
Seat belt pretensioners F/R	■	—
Side airbags	—	■
Traction control	—	—
Head restraints F/R	****	***
Visibility F/R	****	**
Crash protection (front) D/P	*****	*****
Crash protection (side) F/R	****	****
Crash protection (offset)	*****	—

CAMRY, SOLARA ★★★★

RATING: Above Average, but is the Camry coasting on its past high-quality reputation? Compare its sorry quality record with the Avalon, Echo, and Celica. **Strong points:** *Camry:* Good powertrain set-up (V6); upgraded 4-cylinder is also a competent performer; pleasant ride; quiet interior; well-laid-out instrumentation and controls; nicely padded dash and door panels; lots of interior passenger and storage space; and a high resale value. *Solara:* A better-performing 4-cylinder engine; more firmly damped suspension; and more attractively styled than the sedan. **Weak points:** *Camry:* Engine horsepower isn't top-in-class; V6 engines require premium fuel; suspension may be a bit too soft for some; little steering feedback; no rear disc brakes on 4-cylinder-equipped models (except for the high-end XLE); annoying windshield reflections at night; and complicated navigation system controls. The $1,110 transport and PDI charges are unacceptably high. Also, an unusually large number of performance- and safety-related complaints registered by government safety investigators points the finger at Toyota's cost-cutting, de-contenting strategy adopted since 1997. *Solara:* Tricky entry/exit; trunk has a small opening and high sill; limited rear visibility. **New for 2004:** *Camry:* SE version gets a 225-hp 3.3L V6, hooked to a revised 5-speed automatic, improved fuel injection, and additional soundproofing. Standard foglights and an in-dash 6-CD changer are new to the XLE and XLE V6. *Solara:* This year's redesign makes it much larger than its predecessor, but interior space remains practically the same. It also gets the new 3.3L V6 coupled to a 5-speed sequential-shift automatic transmission.

OVERVIEW: The Camry is available only as a four-door sedan—gone is the station wagon, and the coupe and convertibles have been given the Solara moniker (see below). Power is supplied by a peppy 2.4L 157-hp 4-cylinder engine, a carried-over 210-hp 3.0L V6, and the new 225-hp 3.3L V6. These engines can be coupled to a 5-speed manual, an electronically controlled 4-speed automatic, or a 5-speed automatic. The suspension features MacPherson struts that reduce suspension noise through the use of revised springs, shocks, and an anti-sway bar. The rear suspension is similarly constructed with the addition of dual lower control arms to increase stability and reduce highway wander. Control is further maintained through speed-sensing variable power steering. V6-equipped SE and XLE models get standard four-wheel disc brakes, while the LE and 4-cylinder-equipped Camrys get discs up front and rear drums (shame!).

There is also a nice array of safety features: traction control is standard on all V6-equipped models; rear seats have shoulder belts for the middle passenger; low beam lights are quite bright; headlights switch on and off automatically as conditions change; and the front airbags are a third-generation de-powered design.

Camrys have fallen behind in the standard equipment offered with all of its models, obviously a result of product de-contenting to keep prices down, or, as I suspect, keep profits high. For example, the MacPherson strut suspension is no match for the Honda Accord's double-wishbone suspension at all four corners.

Solara

The Solara's small, but it's not cheap. Built in Cambridge, Ontario, a base model Solara costs $26,800, but put in the Sienna and Lexus RX330's V6 power plant and you can expect to pay thousands more.

Introduced in the summer of 1998, the Solara is essentially a longer, lower, bare-bones, two-door coupe or convertible Camry with a sportier powertrain and suspension and a more stylish exterior. But don't let this put you off. Most new Toyota model offerings, like the Sienna, Avalon, and RAV4, are Camry derivatives.

You have a choice of either a 4- or 6-cylinder power plant. Unfortunately, if you choose the V6, you also get a gimmicky rear spoiler and a headroom-robbing moon roof. Convertibles come with a standard automatic transmission. The stiff body structure and suspension, as well as tight steering, make for easy sports car-like handling, with lots of road feel and few surprises.

Cost analysis/best alternatives: With Toyota keeping prices at last year's level (Okay, the Solara SE does cost $150 more this year) the 2004s are the better buy—particularly in view of the fact that the SE gets a larger engine and both the SE and XLE will use an upgraded 5-speed automatic transmission. Other cars worth considering are the Honda Accord, Lexus ES 300, and Mazda6. **Options:** The built-in child safety seat ($150) is a nice feature. Stay away from the optional moon roof; it robs you of much-needed headroom. **Rebates:** $2,000 rebates, plus low-interest financing. **Delivery/PDI:** $1,110. **Depreciation:** Slower than average. **Insurance cost:** Higher than average. **Parts supply/cost:** Parts are easily found and moderately priced. **Annual maintenance cost:** Less than average. **Warranty:** Bumper-to-bumper 3 years/ 60,000 km; powertrain 5 years/100,000 km; rust perforation 5 years/ unlimited km. **Supplementary warranty:** Not needed. **Highway/city fuel economy:** *3.0L and auto.:* 7.8–11.9L/100 km.

Quality/Reliability/Safety

Pro: Quality control: Better than average for most components, but owner complaints are increasing. **Reliability:** Once one of the most reliable cars on the market; however, latest redesign shows quality control has lapsed. **Warranty performance:** Usually good. With only a few exceptions, customers usually get a fair shake, even if the warranty has expired. Safety is enhanced with a driver- and passenger-side front and rear curtain airbag system, which is combined with individual seat-mounted side inflators. The airbags are safer, de-powered, "smart." Improved crash protection via pre-cut steel panels that create multiple crush-protection zones. Also, anti-intrusion brake pedal absorbs and deflects impact forces away from the driver. Child restraint system has user-friendly anchors for easy installation and three-point seat belts are provided in all seating positions.

Con: Owner-reported problems: Chronic braking problems—squealing, squeaking, and grinding noises even though pads and calipers are replaced repeatedly; rear disc brakes cut grooves into rotors; suspension failures; heat shield rattling; squealing side window operation; and body/accessories glitches. The Solara convertible produces excessive body flexing (common to most convertibles), resulting in a creaky top. The unusually large number of safety-related complaints raises concerns that Toyota's last Camry redesign has cheapened the product. **Service bulletin problems:** Engine surging; front-brake vibration (see following).

Front Brake Vibration

BR006-02 December 24, 2002

Front brake vibration (replace pads and rotors)

^ 2002–03 model year Camry V6 XLE and V6 SE

This repair is covered under the Toyota Comprehensive Warranty. This warranty is in effect for 36 months or 36,000 miles [57,600 km], whichever occurs first, from the vehicle's in-service date.

* Warranty application is limited to correction of a problem based upon a customer's specific complaint.

Transmission shuddering during 2–3 upshift; steering column noise when turning (TSB # ST002-03, issued June 2003); windshield creaking, ticking; AC performance and durability improvement (see below).

AC - Blower Volume Gradually Diminishes

AC001-03 April 15, 2003

AC performance and durability improvement

2002–03 Camry

Some 2002–03 model year Camry owners may experience a condition where the blower volume gradually decreases after about 1-1/2 hours of driving. It has been determined that, in hot, high humidity conditions, the AC system evaporator is freezing over, blocking the airflow path. There may be some instances where this condition may affect the AC compressor and clutch assembly. An in-line thermistor resistor harness is now available to correct this concern.

Front-door wind noise and water leaks and sliding roof repair tips. **NHTSA safety complaints/safety:** Starter caught on fire in the parking lot; underhood fire (left side) while vehicle was parked overnight; fire ignited from underneath vehicle while driving; airbags failed to deploy:

> My brother was involved in a fatal crash in a new 2002 Toyota Camry XLE. The accident occurred in North Carolina and is being investigated by the highway patrol. The car left the highway and rolled over, eventually impacting a tree. The car was equipped with front, side, and side curtain airbags. No airbag deployed in this accident. I believe the lack of airbag deployment was contributory to my brother's death.

Inadvertent airbag deployment; faulty cruise control caused vehicle to suddenly accelerate; sudden acceleration without braking effect; excessive grinding noise and long stopping distances associated with ABS braking; pedal went to floor but no braking effect; ABS brakes suddenly locked up when coming to a gradual stop; defective rear brake drum; many owners complain of poor brake pedal design: "Arm that holds up brake pedal is interfering with the driver's foot. Driver stated if consumer had a large-sized foot, it could easily get wedged and stuck on brake pedal. Foot gets caught between the floormat and the brake arm, needs to be redesigned."; front right axle broke six months after car was purchased; vehicle tends to drift to the right at highway speeds; excessive steering wheel vibrations at speeds over 100 km/h; entire vehicle shakes excessively when cruising; vehicle's weight is poorly distributed, causing the front end to lift up when the vehicle speed exceeds 90 km/h; suspension bottoms out too easily, damaging the undercarriage; too-compliant shock absorbers make for a rough ride over uneven terrain; rear suspension noise at low speeds; automatic transmission slippage, or suddenly shifts to a lower gear; particularly poor shifting when in Overdrive; engine was running with transmission in Park position and the car rolled down a hill—the two small girls inside the car jumped out, but one was run over. Car rolled backwards after driver put it into Park and removed ignition key; vehicle parked overnight had its rear window suddenly blow out; windshield distortion is a strain on the eyes; floor-mounted gear shift indicator is hard to read; seat belts are too tight on either side and tighten up uncomfortably with the slightest movement; shoulder belt twists and won't lie straight; leaking suspension struts and strut rod failure; trunk lid may suddenly collapse; faulty driver-window track; driver-side door latch sticks; "rotten-egg" exhaust smell; fuel tank makes a sloshing noise when three-quarters full; clunking noise heard from rear of vehicle when gas tank is half full; tire jack collapsed while changing tire; high rear end cuts rear visibility; turn signal volume is too low.

Road Performance

Pro: Acceleration/torque: Better than average, with sufficient reserve torque for passing. Both the 4- and 6-cylinder power plants are quieter this year and the 4-banger is a better, more powerful performer. This car shines on long drives, and for that you must have the better-performing 3.0L, or torquier 3.3L V6. **Transmission:** Smooth-shifting automatic gearbox with a dual-mode feature that allows the driver to choose either a power or economy setting. **Routine handling:** Nimble and predictable handling. Supple but steady ride on all but the worst roads. Noise, vibration, and harshness have been reduced considerably. **Emergency handling:** Very good. Minimal body roll and front-end plow. Responds well to sudden steering corrections. **Steering:** Quick to respond and predictable. **Braking:** Disc/drum/disc/disc; better-than-average braking.

Con: Neither the 4-banger nor the V6 are high-performance engines, but the Camry doesn't pretend to be a high-performance car. The heavier Solara convertible is more sluggish than coupes and sedans and will be redesigned in late 2004. The overly compliant suspension makes for a busy ride when passing over bumps, even for the more tightly sprung coupes. Little road feel with power steering.

Comfort/Convenience

Pro: Standard equipment: Fully equipped. **Controls and displays:** Much improved dash and instrument layout is both practical and attractive. Heater and climate controls are well situated and easy to use, with large buttons and logical placement. **Climate control:** Efficient heating, defrosting, and ventilation. **Entry/exit:** Front and rear seats are easily accessible. **Interior space/comfort F/R:** Firm and supportive seats. Roomy interior includes comfortable front seats, and large rear seats will easily accommodate three adults, though rear headroom is a bit tight for six-footers. **Cargo space:** Better than average. Huge glove box and lots of little storage bins and door map pockets. Huge but narrow cargo area on the wagon. **Trunk/liftover:** Reasonably sized trunk has a low liftover sill and carries a full-sized spare tire. All models have split folding rear seatbacks with in-trunk security locks. **Quietness:** Lots of improvement here. Little engine or road noise intrudes into the cabin.

Con: Standard equipment: Cumbersome front seat latches and seat belts. Front seat cushions should be deeper. Tilt steering wheel has limited range of motion. Climate controls need larger LEDs. Sound system controls aren't user-friendly on GPS-equipped vehicles. **Driving position:** Limited over-the-shoulder visibility with the Solara. **Entry/exit:** Difficult with the Solara's narrow rear passageway and the lack of a driver's seat slide-forward mechanism. **Cargo space:** Solara's trunk isn't user-friendly with its small opening and high liftover.

Camry, Solara

List Price (negotiable)	**Residual Values** (months)			
	24	**36**	**48**	**60**
LE: $24,800 (12%)	$20,000	$18,000	$15,000	$13,000
LE V6: $27,070 (13%)	$22,500	$20,500	$17,500	$15,500
XLE V6: $32,615 (15%)	$25,500	$23,500	$21,500	$19,500
Solara SE: $26,800 (13%)	$21,000	$18,500	$15,500	$13,500

Technical Data

Powertrain (front-drive)
Engines: 2.4L 4-cyl. (157 hp)
• 3.0L V6 (210 hp)
• 3.3L V6 (225 hp)
Transmissions: 5-speed man.
• 4-speed auto.
• 5-speed auto.
Dimension/Capacity (Camry)
Passengers: 2/3
Height/length/width:
58.7/189.2/70.7 in.
Headroom F/R: 39.2/38.3 in.
Legroom F/R: 41.6/37.8 in.
Wheelbase: 107.1 in.
Turning circle: 36.7/35 ft.
Cargo volume: 14.1 cu. ft.
Tow limit: 2,000 lb.
Fuel tank: 70L/reg.
Weight: 3,208 lb.

Safety Features/Crashworthiness

	Std.	Opt.
Anti-lock brakes	■	■
Seat belt pretensioners F/R	■	—
Side airbags	—	■
Traction control	—	■

Head restraints F/R	****	***
Solara	***	***
Visibility F/R	*****	*
Crash protection (front) D/P	*****	****
Crash protection (side) F/R	***	*****
Solara	***	*****
Crash protection (offset)	*****	—

AVALON ★★★★★

RATING: Recommended. **Strong points:** Excellent powertrain performance; a roomy interior with plenty of storage space; easy access to comfortable rear seats; exceptional reliability when compared with some of Toyota's other models; quiet interior; a high resale value; and sportier handling than the Camry. **Weak points:** High freight/PDI charges; rear corner blind spots; mushy brake pedal; and a bit fuel-thirsty. **New for 2004:** Optional Vehicle Skid Control.

OVERVIEW: Toyota's largest model, this six-passenger near-luxury four-door offers more value and reliability than do other cars in its class that cost thousands of dollars more. A front-engine, front-drive, mid-sized sedan, based on a stretched Camry platform, the Avalon is similar in size to the Ford Taurus and bigger than the rear-drive Cressida it replaced.

The Avalon comes with a 210-hp version of the Camry's 3.0L V6 power plant, coupled to a 4-speed electronically controlled automatic transaxle. Base models offer a nice array of standard comfort and convenience features, including AC, power windows, power door locks, cruise control, and an AM/FM cassette sound system. Safety features include standard ABS and a three-point shoulder belt for the rear centre-seat passenger.

Cost analysis/best alternatives: Insignificant 2004 changes; look for a discounted 2003 model. If you want a more driver-involved experience in a Toyota/Lexus, consider a Lexus ES 300 or GS 300. Other cars you may wish to look at: the Nissan Maxima and Toyota Camry V6. **Options:** The engine-immobilizing anti-theft system is worthwhile. **Rebates:** $3,000+ on the 2003s, plus low-interest financing. **Delivery/PDI:** $1,110. **Depreciation:** Slower than average. **Insurance cost:** Higher than average. **Parts supply/cost:** Parts are relatively inexpensive and easily found. **Annual maintenance cost:** Less than average. **Warranty:** Bumper-to-bumper 3 years/60,000 km; powertrain 5 years/100,000 km; rust perforation 5 years/unlimited km. **Supplementary warranty:** Not needed. **Highway/city fuel economy:** 7.4–11.2L/100 km.

Quality/Reliability/Safety

Pro: Quality control: Good-quality powertrain and body components. Much-improved brake durability. **Reliability:** Interestingly, there are fewer reliability or durability problems reported than with Toyota's other offerings.

Warranty performance: Better than average, except for engine sludge fudging.

Con: Owner-reported problems: Owners report that brakes perform erratically, appear to be automatically applied, dragging the vehicle; excessive wind noise intruding into the passenger compartment, and some fragile trim items. Sunroof glass flutters at cruising speed due to weak springs in the sunroof assembly. **Service bulletin problems:** N/A. **NHTSA safety complaints/ safety:** Sudden acceleration with no brakes; airbag deployed for no reason; front airbags failed to deploy; side airbags failed to deploy in a side impact; airbag warning light comes on constantly; transmission failure; transmission failed to hold vehicle on a hill while waiting for a red light; fuel dampener and fuel pump failure caused fuel leak and fumes to enter interior; Bridgestone tire failure; upper steering knuckle broke; steering wheel turned to the right, yet car fails to respond; dashboard lighting reflects into the windshield; and the dashboard gear display is hard to read at night.

Road Performance

Pro: Acceleration/torque: Brisk acceleration with the smooth, powerful, and quiet V6 engine; and there's plenty of torque for passing and traversing hilly terrain (0–100 km/h: 7.8 seconds). **Transmission:** The electronically controlled 4-speed automatic transmission is well matched to its power without sacrificing performance. In fact, the powertrain set-up is one of the smoothest, best-integrated combinations available. **Emergency handling:** Slow, but sure-footed. **Routine handling:** Better-than-average handling, thanks to the stiffened suspension. Ride quality is flawless, providing living-room comfort on virtually any kind of road. **Braking:** Disc/disc; better than average (100–0 km/h: 40 m).

Con: Steering: Power steering is over-assisted at all speeds, and the car has a tendency to oversteer. Tends to plow ahead when cornering at high speed.

Comfort/Convenience

Pro: Standard equipment: Loaded with standard safety and convenience features. **Driving position:** Plenty of room and comfortable, supportive seating. **Controls and displays:** Well-designed instrument layout is both practical and attractive, with controls that are easy to see and well within reach (instruments are similar to the Camry's, with controls placed in all the familiar places). **Climate control:** Powerful climate control system—efficient heating, defrosting, and ventilation. User-friendly controls. **Entry/exit:** Easy access to both the front and rear seating areas. **Interior space/comfort F/R:** Roomier than the Camry; front and rear seats are exceptionally comfortable, with plenty of thigh support and more room as a result of this year's lengthened wheelbase. Large rear seat will easily seat three people. **Trunk/liftover:** Enormous trunk has a low liftover sill. **Quietness:** Lots of sound insulation makes for an interior that's quieter, generally speaking, than the Camry's.

Con: Bland styling, with a hint of Lexus' GS 300. Limited rear corner visibility. The cupholders are flimsy, and the fuzzy headliner has a cheap appearance. **Cargo space:** The Avalon doesn't have fold-down rear seats for trunk access and hauling long objects.

Avalon

List Price (negotiable)	**Residual Values** (months)			
	24	**36**	**48**	**60**
Avalon XLS: $45,830 (21%)	$33,000	$29,000	$24,000	$20,000

Technical Data

Powertrain (front-drive)
Engine: 3.0L V6 (210 hp)
Transmission: 4-speed auto.
Dimension/Capacity
Passengers: 2/3
Height/length/width:
57.7/191.9/71.7 in.
Headroom F/R: 38.7/37.9 in.
Legroom F/R: 41.7/40.1 in.
Wheelbase: 107.1 in.
Turning circle: 40 ft.
Cargo volume: 15.9 cu. ft.
Tow limit: 2,000 lb.
Fuel tank: 70L/reg.
Weight: 3,461 lb.

Safety Features/Crashworthiness

	Std.	**Opt.**
Anti-lock brakes	■	—
Seat belt pretensioners F/R	■	—
Side airbags	■	—
Traction control	—	■
Head restraints F/R	***	***
Visibility F/R	*****	**
Crash protection (front) D/P	****	*****
Crash protection (side) F/R	****	*****
Crash protection (offset)	*****	—

CELICA ★★★★★

RATING: Recommended. **Strong points:** Exceptionally well-matched engine and transmission; great road holding and handling; comfortable front seating; impressive reliability; few performance- or safety-related owner complaints; reasonably priced GT; and a high resale value. **Weak points:** High freight/PDI charges; automatic transmission compromises GT engine performance; difficult rear-seat access, rear seating not for three adults, and rear bucket seats aren't suitable for long drives. Pricey GT-S; noisy 180-hp engine at high rpms; no crashworthiness data; and a higher-than-average rate of accident injury claims. **New for 2004:** The availability of HID headlights.

OVERVIEW: The front-drive Celica is a four-passenger, two-door hatchback that offers benchmark reliability, good handling, and great fuel economy in an attractive sports car package. The GT and GT-S have a firm suspension, well-equipped interior, ABS brakes, and a more sporting feel than do other versions. All handle competently and provide the kind of sporting performance expected from a car of this class. The extra performance in the higher-line versions does come at a price, but this isn't a problem given the high resale value and excellent reliability for which Celicas are known. Overall, it's one of the best choices in the sporty car field.

Cost analysis/best alternatives: Buy an identical 2003 model, if you can find one and the price is right. Keep in mind that the 2004s cost only $1,000 more than last year's model, so look for a sizeable rebate on the 2003. For driving pleasure and performance, the GT is your best value. There's a generous array of standard equipment, performance is excellent, and it looks much like the more expensive GT-S. If you miss the GT version, try the Ford Mustang, Mazda Miata, and Subaru Impreza RS Coupe. GT-S rivals include the Acura Integra GS-R and BMW 323Ci. **Options:** Not much worth buying. The Celica comes fairly well equipped. GT models have standard front disc/rear drum brakes, which offer excellent braking performance; anti-lock brakes are not available on the GT, and judging by their high failure rate on other cars, this doesn't seem to be much of a loss. **Rebates:** Look for $3,000 rebates by mid-2004, in addition to low-interest financing. **Delivery/PDI:** $1,110. **Depreciation:** Slower than average. **Insurance cost:** Higher than average. **Parts supply/cost:** Parts are relatively inexpensive and easily found. **Annual maintenance cost:** Less than average. **Warranty:** Bumper-to-bumper 3 years/ 60,000 km; powertrain 5 years/100,000 km; rust perforation 5 years/ unlimited km. **Supplementary warranty:** Not necessary. **Highway/city fuel economy:** 7.3–10.2L/100 km with an automatic transmission.

Quality/Reliability/Safety

Pro: Quality Control: Once the best among automakers, but has declined of late. **Reliability:** The Celica uses the same mechanical components that are employed on Toyota's other models, and this explains why reliability or durability problems are similarly worrisome. **Warranty performance:** Warranty claims are handled fairly, with only a few exceptions. Safety features include dual front airbags, height-adjustable front head restraints, impact-absorbing materials on the roof and doors, three-point seat belts with pretensioners and force limiters for front passengers, three-point seat belts for rear passengers, and rear child seat anchors. When the airbags are deployed, a sensor disconnects the fuel pump to minimize the risk of fire.

Con: Owner-reported problems: The front brakes are troublesome; some audio systems and trim items have also been failure-prone. **Service bulletin problems:** Two Special Service Campaigns: Campaign #301 says Toyota will inspect or replace the throttle body; #30B allows for the inspection or repair of the fuel tank check valve, which may become separated from the inlet pipe and fall into the fuel tank. Troubleshooting tips for sulfur "rotten-egg" exhaust odour.

NHTSA safety complaints/safety: Airbags failed to deploy; passenger wheel suddenly locked up, causing a collision; cruise control may suddenly disengage, causing vehicle to lose power; transmission is too easy to accidentally shift from Fifth to Second gear; automatic transmission won't shift into Overdrive; fuel leak from faulty hose; engine clicks and stalls.

Rear head restraints aren't available and the audible reverse alarm isn't Toyota's brightest idea. Audible only inside the vehicle, it adds a forklift cachet to your Celica. For info on disabling this feature, log onto *www.celica.net*, an excellent site for technical advice and owners' unbiased reviews.

Road Performance

Pro: Emergency handling: Better than average. **Steering:** Responsive, predictable power steering also transmits plenty of road feel. **Acceleration/torque:** Both the 140- and 180-hp variants of the 1.8L provide plenty of high-performance thrills. GT engine has more torque at lower rpms than the GT-S, which only delivers sporty performance after hitting 6000 rpm. **Transmission:** Standard 5-speed manual transmission has easy throws and light clutch action. The GT-S' E-shift automatic transmission is very user-friendly, and shifting times are reasonably quick. **Routine handling:** Nimble and predictable handling in all conditions. Sportier handling on GT and GT-S models makes for a firm but not uncomfortable ride. **Braking:** Excellent performance (100–0 km/h: 37 m).

Con: Automatic transmission saps GT's engine performance. Suspension may be too firm for some.

Comfort/Convenience

Pro: Standard equipment: Long on standard, innovative features. For example, the 10-speaker 200-watt "System 10" radio is one of the most advanced systems currently on the market. Re-styled exterior resembles a Lexus coupe from the front and a Supra from the rear. **Driving position:** Very good. The driver's seat has a manual adjustment that's easy to use. **Controls and displays:** Complete and well-designed controls. **Climate control:** Works well and is easy to calibrate while driving. **Interior space/comfort F/R:** Basically a two-seater; front seats are particularly comfortable, especially on the GT and GT-S. Long objects can be accommodated by folding down the 50/50 split rear seatbacks. **Quietness:** Fairly quiet, well-insulated interior.

Con: Tacky-looking plastic cover for the storage bin. **Entry/exit:** Very poor rear access forces you to practically crawl into the cramped back seat. **Interior space/comfort F/R:** Models equipped with the sunroof offer minimal headroom for tall front-seat passengers. Limited outward vision. The high beltline and low seating position induce claustrophobia. **Cargo space:** Rather limited.

Trunk/liftover: Small trunk has a high liftover. Some road noise and transmission whine may intrude into the interior.

Celica

List Price (firm)	**Residual Values** (months)			
	24	**36**	**48**	**60**
Celica GT: $25,640 (14%)	$20,000	$18,000	$15,000	$12,000
GT-S: $34,400 (17%)	$26,000	$23,000	$20,000	$16,000

Technical Data

Powertrain (front-drive)
Engines: 1.8L 4-cyl. (140 hp)
• 1.8L 4-cyl. (180 hp)
Transmissions: 5-speed man.
• 6-speed man. (4-speed auto. E-shift)

Dimension/Capacity
Passengers: 2/2
Height/length/width: 51.4/170.4/68.3 in.
Headroom F/R: 38.3/35 in.
Legroom F/R: 44.1/27 in.
Wheelbase: 102.3 in.
Turning circle: 38 ft.
Cargo volume: 12.9 cu. ft.
Tow limit: N/A
Fuel tank: 55L/prem.
Weight: 2,460 lb.

Safety Features/Crashworthiness

	Std.	**Opt.**
Anti-lock brakes	—	■
Seat belt pretensioners F/R	■	—
Side airbags	—	■
Traction control	—	—
Head restraints F/R	*****	*****
Visibility F/R	*****	*****
Convertible	*****	**
Crash protection (front) D/P	****	****
Crash protection (side) F/R	***	—

SIENNA ★★★★★

RATING: Recommended (2004). Taller, wider, longer, larger and much more refined, for little more than last year's price. The best all-around minivan choices are the Toyota Sienna, Honda Odyssey, and Nissan Quest. The Mazda MPV is a small minivan that's more reliable than the top-rated models, but it lacks power, interior room, and ride comfort. The GM minivans are good performers, but—like the Ford Windstar/Freestar and Chrysler minivans—are plagued by serious reliability problems. **Strong points:** Takes the "mini" out of minivan; reasonable base price; acceleration second only to the Odyssey; an AWD option; a better-performing, economical 5-speed automatic transmission; a tighter turning circle; enhanced seating versatility and storage capacity; quiet interior; and uses regular fuel. **Weak points:** Unacceptably high freight and PDI charges ($1,245); automatic transmission power drain; lacks the trailer-towing brawn of rear-drive minivans and towing capability may be much less than advertised, says this *Lemon-Aid* reader:

> We have just leased a 2004 Toyota Sienna CE. As we mentioned many times to the salesperson (and sales manager), it was our intention to use this vehicle for towing our Coleman tent trailer. The trailer weighs about 1,700 lb. [770 kg] empty and we were assured by the dealership that since the Sienna '04 is rated for 3,500 lb. [1,590 kg], it would be ideal for our needs. They even arranged for the installation of a class III hitch prior to our taking possession.
>
> Imagine our surprise when we discovered within the owner's manual a "Caution" stating that one must not exceed 72 km/h (45 mph) while towing a trailer. This limit is not stated in the promotional literature we were provided, or on the Toyota.ca website, or in any trailer-towing rating guide. This limit was also not mentioned at any time during our purchase negotiations. Alarmingly, Toyota defines a "Caution" as a "warning against anything which may cause injury to people if the warning is ignored." As it turns out the dealer was not aware of this speed limit....

Mediocre braking; third-row head restraints are mounted too low; and options are way overpriced. **New for 2004:** Almost everything.

OVERVIEW: This year's Sienna is quite different from earlier models. It provides lots more interior room that accommodates up to eight passengers, handles much better, rides more comfortably, and uses a more powerful fuel-efficient engine. Standard 4-wheel-disc brakes and all-wheel drive is offered for the first time.

A new 3.3L 230-hp V6 turns in respectable acceleration times under 9 seconds, almost as good as the Odyssey. Handling is completely carlike, there's less vulnerability to wind buffeting, and minimal road noise.

Sienna's interior and exterior have been gently re-styled. The third-row seats split and fold away, head restraints don't have to be removed when the seats are stored, and second-row bucket seats are easily converted to bench seats.

The Sienna can seat seven or eight and offers dual power-sliding doors with optional remote controls, as well as a V6 power plant. It's built in the same Kentucky assembly plant as the Camry and comes with lots of safety and convenience features that include side airbags, anti-lock brakes, and a low-tire-pressure warning system. As with most minivans and vans, you can save money by buying the Sienna's cargo version, but you won't get as many features.

Cost analysis/best alternatives: 2004 Siennas are so improved and reasonably priced that, except for the Honda Odyssey, they eclipse all the competition.

The 2004 costs only about $700 more this year, but extras are pricey. Bargain hunters take note: Cargo Siennas usually cost $3,000–$4,000 less than the base CE. Other minivans worth considering are the Honda Odyssey, Mazda MPV, and Nissan Quest. **Options:** Power windows and door locks, rear heater, and AC unit. Be wary of the power-sliding door. As with the Odyssey and GM minivans, these doors can injure children and pose unnecessary risks to other occupants. Go for Michelin or Pirelli original equipment tires. **Rebates:** Not likely. **Delivery/PDI:** $1,260. **Depreciation:** Much slower than average. **Insurance cost:** A bit higher than average. **Parts supply/cost:** Excellent supply of reasonably priced parts taken from the Camry parts bin. Automatic transmission torque converters are frequently back ordered due to their poor reliability. **Annual maintenance cost:** Like the Camry, much lower than average. **Warranty:** Bumper-to-bumper 3 years/60,000 km; powertrain 5 years/ 100,000 km; rust perforation 5 years/unlimited mileage. **Supplementary warranty:** An extended warranty isn't really necessary. **Highway/city fuel economy:** *3.3L V6:* 8.1–12.2L/100 km; *3.0L V6:* 8.8–12.4L/100 km.

Quality/Reliability/Safety

Pro: Reliability: Better than the Detroit average, and fewer complaints than with some of Toyota's other models, like the Corolla and Camry. **Warranty performance:** Acceptable, but could be much better.

Con: The most serious reliability problems concern engine sludge buildup and defective transmission torque converters that cause the Check Engine light to remain lit, which can lead to sudden transmission failure (Honda Odysseys have a similar tranny problem that is covered by a 7-year goodwill warranty). Fortunately, Toyota has extended its warranties to cover these problems on all the model years affected, whereas Honda hasn't included all afflicted models. Other reliability concerns include stalling when the AC engages; electrical shorts; excessive brake noise; sliding door defects; and various other body glitches, including excessive creaks and rattles and paint that's easily chipped. **Owner-reported problems:** Premature brake wear and noisy brakes; engine replacement due to sludge buildup; automatic transmission failures; distracting windshield reflections and distorted windshields; power sliding door malfunctions; and interior squeaks and rattles. **Service bulletin problems:** Steering angle sensor calibration; loose sun visor remedy; and service tips to reduce front door wind noise. **NHTSA safety complaints/safety:** Neither front nor side airbag deployed in a collision; loss of steering; automatic transmission suddenly went into Neutral while on the highway; unsafe transmission overdrive design; defective transmission torque converter causes the engine warning light to come on; sudden unintended acceleration; sudden stalling when the AC is turned on; vehicle rolled away with shifter in Park on a hill, and ignition shut off; vehicle jerks to one side when accelerating or stopping; power sliding door unlocks when vehicle is underway or door can close on children because it requires too much pressure to stop when closing. Next to the power sliding door there's a power door switch that continues to operate even if the child door lock is on—a child can easily activate the switch; rear brakes make a loud clunking noise when vehicle is backing up or accelerating; reflection of the dashboard on the windshield impairs visibility; when turning the steering wheel to the left, the steering wheel doesn't return smoothly; child safety seat second-row tethering is poorly designed; centre-rear seat belt doesn't tighten sufficiently when children are restrained; rear seat belts can't be adjusted; slope of the windshield makes it hard to gauge where the front end stops; excessive door rattling; rear window exploded as front door was closed; sunroof flew off when opened while Sienna was underway.

Road Performance

Pro: Emergency handling: No problem. Sienna can change course abruptly without wallowing or losing directional stability. **Steering:** Excellent steering feel and quick response. **Acceleration/torque:** Toyota's V6 handles most driving chores effortlessly without noise or vibration (0–100 km/h: 10.2 seconds). **Transmission:** Quiet and smooth shifting. **Routine handling:** A pleasure. Handling is crisp and effortless. **Braking:** Good, on pre-2004 models; a bit longer stopping distance with the all-disc 2004s.

Con: Base engine struggles a bit when carrying a full load uphill.

Comfort/Convenience

Pro: Standard equipment: Very well appointed. **Driving position:** Comfortable seating with excellent forward visibility. **Controls and displays:** Easy-to-read displays and user-friendly controls. **Climate control:** Easy to adjust and operates efficiently. **Entry/exit:** Easy entry/exit up front. **Interior space/comfort F/R:** First-class seating with plenty of headroom and legroom. **Cargo space:** Adequate. Seats can be removed for additional cargo room. **Trunk/liftover:** Easy to load or unload. **Quietness:** No engine or road noise.

Con: There's a confusing array of stalks sprouting out from the steering column. Rear visibility obstructed by middle roof pillars and rear head restraints. Radio speakers are set too low for acceptable acoustics. **Interior space/comfort F/R:** The wide centre pillars make for difficult access to the middle seats. Removing middle- and third-row seats is a two-person chore. **Cargo space:** Third-row seats lack a fore/aft adjustment to increase cargo space. **Trunk/liftover:** The rear hatch is inordinately heavy. **Quietness:** A proliferation of body rattles undermines Toyota quality claims.

Sienna

List Price (negotiable)	**Residual Values** (months)			
	24	**36**	**48**	**60**
CE: $30,000 (16%)	$23,000	$20,000	$16,000	$13,000
LE: $34,750 (17%)	$25,000	$23,000	$18,000	$15,000
LE AWD: $39,160 (17%)	$28,000	$25,000	$21,000	$18,000

Technical Data

Powertrain (front)
Engine: 3.3L V6 (230 hp)
Transmission: 5-speed auto.
Dimension/Capacity (LE)
Height/length/width:
69/200/77 in.
Headroom F/R1: 42/40.2 in.
Legroom F/R1: 42.9/39.6 in.
Wheelbase: 119 in.
Turning circle: 38 ft.
Passengers: 2/2/3; 2/3/3
Cargo volume: 70.1 cu. ft.
GVWR: 5,690 lb.
Tow limit: 3,500 lb.
Ground clear.: 6.9 in.
Fuel tank: 79L/reg.
Weight: 4,123 lb.

Safety Features/Crashworthiness (2003)

	Std.	**Opt.**
Anti-lock brakes (4W)	■	—
Seat belt pretensioners F/R	■	—
Side airbags	—	■
Traction control	■	—
Head restraints F/R	****	***
Visibility F/R	*****	***
Crash protection (front) D/P	*****	*****
Crash protection (side) F/R	****	*****
Crash protection (offset)	*****	—

EUROPEAN VEHICLES

More bucks than brains

European vehicles are generally driver's cars, noted for a high level of performance combined with a full array of standard comfort and convenience features. They are fun to drive, well appointed, and attractively styled.

But, there's a downside.

European cars, SUVs, and minivans are also way overpriced, poorly serviced, and not all that reliable. And, as consumers balk at outrageously high prices, automakers try to justify their price-gouging by adding non-essential problem-prone gadgets that drive up costs as dealers sell the sizzle because the steak is bad.

Writes British independent automotive journalist Robert Farago, on The Truth About Cars website (*www.robertfarago.com/tac*):

> Once upon a time, a company called Mercedes-Benz built luxury cars. Not Elk aversive city runabouts. Not German taxis. Not teeny tiny hairdressers' playthings. And definitely not off-roaders.... In the process, the Mercedes brand lost its reputation for quality and exclusivity. In fact, the brand has become so devalued that Mercedes themselves abandoned it, reviving the Nazi-friendly Maybach marque for its top-of-the-range limo. Now that Mercedes has morphed with Chrysler, the company is busy proving that the average of something good and something bad is something mediocre.

You won't read this kind of straight reporting from the cowering North American motoring press as they fawn over any new techno gadget-laden vehicle hailing from England, Germany, or Sweden. It's easy for them; they get their cars and press junket for free.

Some European automakers, unable to hide the lousy quality of their vehicles, abandoned the U.S. and Canadian markets altogether, while others are barely hanging on. Vauxhall, Fiat (hoping to be acquired by dummy GM), Renault, and Peugeot were the first to turn tail, while Jaguar (more Taurus bull than Jaguar cat), Saab (soon sitting on GM's truck chassis), and Ford's Land Rover land mine continue to struggle with losses that have drained their cash reserves for years. Even Ford's Volvo division is having one of its worst years in over a decade.

Buyers aren't falling for the car ads' appeal to snobbery. They are still wary of European automakers' reputation for poor quality, high parts and servicing costs, and weak dealer networks. And they have independent data to support their misgivings. One MIT study concluded over a decade ago that European automakers built poor-quality vehicles and then, much like the Detroit Big Three, attempted to fix their mistakes at the end of the assembly line. As one reviewer of *The Machine That Changed the World* by James P. Womack, Daniel T. Jones, and Daniel Roos (HarperCollins, November 1991) wrote:

> This study of the world automotive industry by a group of MIT academics reaches the radical conclusion that the much-vaunted Mercedes technicians are actually a throwback to the pre-industrial age, while Toyota is far ahead in costs and quality by building the automobiles correctly the first time.

Canadian drivers have their own examples of spotty European-built quality. Take, for example, this recent email:

> I have a 2000 Jetta TDI. It has 125,000 km on it and has a long list of problems with it. I have printed the words "LEMON @ 125k" on the back window with the brightest yellow paint I could find. I have had about six people with Jettas or had friends with Jettas stop and talk to me. I also found a website with a chat group about VWs. There seems to be a problem with either the airflow metre or some sort of sensor in the airflow. It seems to happen at the 100,000-km mark, give or take 20,000 km. Anyways, the airflow metre is a $500 part, plus installation. This problem results in a huge loss of power. On long inclines, my car cannot reach 100 km/h. If I happen to be going faster than that, my car will lose speed in a hurry.
>
> I sat outside the dealership a few Saturdays ago. With my car painted up, it took all of 10 minutes for the sales and service managers to come out. They asked me what was wrong with the car and I started my list: power-steering rack was gone, ABS system gone, car loses power going up hills. The service manager interrupted at that point, saying it was an airflow sensor. (How about that? Didn't even look at the car and he knew what it was.) I didn't get to finish my list, mind you.
>
> He then went on to say if I had my vehicle serviced there all the time, they would have done something for me. (Sure they would have, at $85/h). Of course, I am just trying to save money like Volkswagen did when it moved the Jetta plant to Mexico to save labour. Why would I pay $65 for an oil change when I can do it myself for less than half that? I get all major servicing done at the dealer (timing belt, etc.), but anything I can do myself, I do it. I am a plumber by trade and I am pretty good with my hands. I have also fiddled around with cars since I was a teenager. Oil changes, filter changes, etc. are easy. Basically, if I feel I can't do it I will have a mechanic do it.
>
> Anyway, here is a list of things wrong with my vehicle:
>
> 1. ABS system doesn't work.
> 2. Lack of power (airflow metre).
> 3. Power steering gone.

4. Passenger-side seat belt sometimes doesn't lock up under hard braking.
5. Glove box hinge snapped off.
6. Cupholder broken (this is the second one—$185 to replace).
7. Leather seats in rear tore when folded down.
8. Hood severely chipped.
9. Three of the locking wheel studs had to be drilled out after they stripped (cheap alloy). Wheels will not come off if they can't be taken off. Good thing I didn't get a flat in the middle of nowhere to find this out.
10. Driver-side heated mirror doesn't work.
11. Inside panel on driver-side door dislodges each time door is opened.

I am in the process of sending letters to the dealership and VW Canada outlining my gripe. Until then I am going to sit in front of the dealership on Saturdays for a while.

J.L.

Lemon-Aid readers who own pricey European imports invariably tell me of nightmarish electrical glitches that run the gamut from the annoying to the life-threatening. Other problems noted by owners include premature brake wear and excessive brake noise, AC malfunctions, faulty computer modules leading to erratic shifting, poor driveability, hard starts, and frequent stalling. Interestingly, although Ladas were low-quality Soviet imports, their deficiencies pale in comparison to what I've seen coming out of the European luxury corral. Maybe that's the logic behind GM's recently announced co-venture with Lada to market a Cavalier-Lada 4X4 hybrid to European consumers.

Although poor servicing is usually more acute with vehicles that are new on the market, it has long been the Achilles heel of European importers. Owners give them low ratings for mishandling complaints, inadequately training service representatives, and hiring an insufficient number of mechanics—not to mention the abrasive, arrogant attitude typified by some automakers and dealers who bully customers because they have a virtual monopoly on servicing in their region. Look at their dealer networks and you'll see that most European automakers are crowded into Ontario and the West Coast, leaving their Eastern and Western Canada customers to fend for themselves. This makes their chances of finding someone to do competent repairs about as likely as getting Sheila Copps to "kill" the GST.

In light of all their shortcomings, why are some European vehicles so popular? Basically, because their well-heeled buyers have more bucks than brains. As I said before, these cars look so good and make driving so much fun and so comfortable that you quickly forget about Franz, Ingmar, and Luigi waiting for your return to top off their Canada Pension Plan.

Audi

Saddled in the early '80s with a reputation for poor-quality cars that would suddenly accelerate out of control, Audi fought back and staged a spectacular comeback with well-built, moderately priced front-drive and AWD Quattro sedans and wagons. Through an expanded lineup of sedans, coupes, and a Cabriolet, Audi has gained a reputation for making sure-footed, all-wheel-drive luxury cars that are loaded with lots of high-tech bells and whistles.

As with most European makes, Audis excel in comfort and performance. But servicing and warranty support remains problematic, especially now that VW/Audi has closed down its Canadian headquarters and runs its Canadian operations from the U.S. and Germany. Also, the fact that Audi only warrants the powertrain up to 4 years or 80,000 km is far from reassuring. Nevertheless, the company has elicited few customer complaints over the past few years and initial build quality is first-class.

This year, Audi returns with an A4 convertible and a brand new S4, carrying a potent 340-hp V8 ready to take on the high-performance competition. The S4 is essentially an all-dressed A4, but its 6-speed manual transmission and soon-to-arrive 6-speed semi-manual gearbox add considerably to its performance credentials.

Audi is my choice for the top European luxury and sports car brand, providing a fun driving experience without the fear that my wallet will be emptied by outrageous service charges and recurring product failings.

A4 ★★★★★

RATING: Recommended. **Strong points:** Loaded with safety, performance, and convenience features; optional AWD; powerful and smooth-running base engine; comfortable ride; exceptional handling; cargo room (wagon); front and rear Seat belt pretensioners F/R; and impressive build quality. **Weak points:** Not as fast as its rivals; plain, functional interior; an obtrusive centre console limits leg movements; quite limited rear-seat room; climate controls mounted too low; some tire drumming and engine noise; and servicing remains a problem. **New for 2004:** The S4 high-performance brother to the A4.

OVERVIEW: This is Audi's most popular and oldest model. It bills itself as Audi's family sport sedan and targets the BMW 3 Series and Volvo S40/S60 customer by featuring a roomy interior; independent suspension; an efficient automatic climate control; a light body; low-speed traction enhancement; an optional 5-speed automatic/manual transmission; and all-wheel drive. Equipped with standard dual and side airbags, ABS, and a 170-hp 1.8L turbo or a 220-hp 3.0L V6 engine coupled to an electronic 5-speed Tiptronic automatic transmission, this entry-level four-door is reasonably powerful, easy to handle, and holds its value very well. The convertible, called a Cabriolet, arrived late last year and returns this year as a 2004 model.

Cost analysis/best alternatives: Go for a cheaper second-series 2003 model, unless you must have the S4 tire burner, and then I suggest you buy something from the second six-month production. Other vehicles worth taking a look at: the Audi's TT Coupe, the BMW 3 Series, Infiniti I35, Lexus IS 300, Mazda Millenia, Mercedes C-Class, Toyota Avalon, and Volvo S60 or V70. **Options:** An automatic transmission and all-wheel drive. Think twice about getting the power moon roof if you're a tall driver. **Rebates:** $3,000 rebates and a variety of dealer incentive plans and low-interest financing programs. **Delivery/PDI:** $1,205 (double what it should be). **Depreciation:** Very slow. Audi values no longer nosedive when the base warranty expires and repair costs become the owner's responsibility. **Insurance cost:** Higher than average. **Parts supply/cost:** Often back ordered and expensive. Forget about saving money by getting parts from an independent supplier; they carry few Audi parts. **Annual maintenance cost:** Higher than average, but not exorbitant. **Warranty:** Bumper-to-bumper 4 years/80,000 km; powertrain 4 years/ 80,000 km; rust perforation 10 years/unlimited km. **Supplementary warranty:** Don't leave the dealership without it. **Highway/city fuel economy:** *1.8L:* 7.5–11.4L/100 km; *3.0L V6:* N/A.

Quality/Reliability/Safety

Pro: Quality control: Better than average. Overall quality control has improved over the past several years, with fewer body, trim, accessory, brake, and electrical glitches than exhibited by previous models. **Reliability:** No major headaches. Electrical problems have taken these cars out of service for extended periods in the past. The 1.8L engine has performed very well. **Warranty performance:** Better than average. "Good" head restraint performance, says IIHS.

Con: Owner-reported problems: The electrical system is the car's weakest link and it has plagued Audi's entire lineup for the past decade. Normally, this wouldn't be catastrophic; however, as the cars become more electronically complex, with more functions handled by computer modules, you're looking at some annoying glitches to say the least. Owners also report long servicing delays, some premature brake wear and grinding when in Reverse, transmission suddenly downshifting or jerking into forward gear, sudden stalling, and other powertrain glitches. Ignition coil packs fail frequently, causing the engine to shut down:

> I have been in contact with close to 100 other 2002 Audi owners through the Audiworld website and the coil pack problem is a serious issue. The coil packs have been failing on at least 10 percent of the 2002 vehicles and Audi says they are all isolated cases. When they fail, the car barely runs and can create a lot of personal safety issues. I want them to recall the vehicles and replace the remaining coil packs.

Steering grinds when turned; mirror memory setting doesn't work. **Service bulletin problems:** Faulty ignition coils (see below).

Ignition System - MIL ON/Engine Runs Rough/Low Power

Bulletin No: Group: 01 **Date:** Feb. 27, 2003

Number: 03-01

Ignition coil, misfire diagnosis

Models: All with Eng. Code (AMB, AMU, ATC, ATW, AWM, AWP AVK) 2001 —> 2003

Condition:

^MIL light ON/flashing
^ Engine running rough
^ Vehicle exhibits loss of power
^ Diagnostic Trouble Codes (DTCs) P0300 and any of the following DTCs - P0301, P0302, P0303, P0304, P0305 or P0306 are stored in DTC memory indicating cylinder misfire

Inoperative keyless entry transmitter and fresh air control lever light. **NHTSA safety complaints/safety:** Car fails to start (not coil-related, they say); sudden acceleration; airbags failed to deploy; chronic complaints of delayed braking; no brakes in rainy weather; parking brake failure; premature replacement of the front brake rotors; sudden headlight failure; and blue-white headlights blind oncoming drivers.

Road Performance

Pro: Acceleration/torque: Surprisingly, the base 1.8L engine provides gobs of low-end torque and accelerates better with the automatic transmission than with the manual gearbox. The turbocharger works well, with no turbo delay or torque steer. **Transmission:** Both the manual and Tiptronic automatic transmission perform flawlessly. AWD is extended to entry-level models at a time when most automakers are dropping the option on passenger cars. **Routine handling:** Handling is exceptional, with no passenger discomfort. **Emergency handling:** Better than average. **Steering:** Crisp and predictable, with lots of road feedback. **Braking:** Disc/disc; impressive braking performance (100–0 km/h: 36 m).

Con: The ride is a bit firm and the car still exhibits some body roll and brake dive under extreme conditions. Braking is a bit twitchy at times.

Comfort/Convenience

Pro: Standard equipment: A nice array of functional, though not lavish, features. **Driving position:** Acceptable. Comfortable, firm seating and a telescopic steering column help you easily find the right driving position. Excellent visibility fore and aft. **Controls and displays:** Most major controls can be easily reached and the instrument layout is both practical and complete. **Climate control:** Efficient climate control system, although some of the controls take getting used to. **Entry/exit:** Easy front and rear access. **Interior space/comfort F/R:** Plenty of headroom and interior space for passengers up front. **Cargo space:** Average cargo space for a vehicle this size. **Trunk/ liftover:** Large trunk is easy to load and unload. **Quietness:** Exceptionally quiet interior.

Con: The wide centre console robs the driver of much-needed legroom and kneeroom, and the optional power moon roof cuts headroom by almost 5 cm (2 in.). Rear passenger room is very limited. **Controls and displays:** Radio and climate controls aren't intuitive; keep the owner's manual handy.

A4

List Price (negotiable)	**Residual Values** (months)			
	24	**36**	**48**	**60**
A4 1.8L: $38,145 (25%)	$26,000	$22,000	$18,000	$14,000
A4 3.0L Sedan: $45,490 (25%)	$30,000	$25,000	$20,000	$16,000
Cabriolet: $64,600 (25%)	$42,000	$36,000	$31,000	$24,000

Technical Data

Powertrain (front/AWD)
Engines: 1.8L 4-cyl. (170 hp)
• 3.0L V6 (220 hp)
• 2.7L V6 (250 hp)
Transmissions: 5-speed man.
• 6-speed man.
• 5-speed auto.

Dimension/Capacity
Passengers: 2/3
Height/length/width:
56.2/179/76.3 in.
Legroom F/R: 41.3/33.5 in.
Headroom F/R: 38.2/36.8 in.
Wheelbase: 104.3 in.
Turning circle: 36 ft.
Cargo volume: 13.4 cu. ft.
Tow limit: 2,000 lb.
Fuel tank: 70L/prem.
Weight: 3,252 lb.

Safety Features/Crashworthiness

	Std.	**Opt.**
Anti-lock brakes	■	—
Seat belt pretensioners F/R	■	—
Side airbags	■	—
Traction control	—	■
Head restraints F/R	****	***
Visibility F/R	*****	*****
Crash protection (front) D/P	****	****
Crash protection (side) F/R	*****	****
Crash protection (offset)	*****	—

A6/S6 ★★★★★

RATING: Recommended. **Strong points:** A more powerful, torquier base engine; superb handling; comfortable seating; plenty of passenger and cargo room (beats out both BMW and Mercedes in this area); easy front and rear access; and very good build quality. The Avant wagon performs like a sporty sport-utility, with side airbags and intense HID xenon headlights. IIHS considers offset crash protection to be "Acceptable" and head restraint protection to be "Good." **Weak points:** Limited availability; firm suspension can make for a jittery ride; some tire thumping; the wagon's two-place rear seat is rather small; no crashworthiness rating; a chintzy powertrain warranty; and servicing can be problematic. **New for 2004:** Nothing significant. The RS6, newly arrived as a 2003 model, has been dropped.

OVERVIEW: The $51,950 (plus a grossly inflated $1,205 freight/PDI) A6 FrontTrak 3.0 is essentially a larger, fully equipped A4. It's a comfortable, spacious, front-drive drive luxury sedan or wagon that offers standard dual front side airbags and ABS. It comes with a base 220-hp 3.0L or for $3,000 more you can get the Quattro version. Other Quattros feature a biturbo, 250-hp 2.7L V6, a 300-hp 4.2L V8, or a 340-hp variant of the same engine. Its Tiptronic automatic transmission also has a manual gearshift capability. The car's base price jumps to over $69,370 if you choose the higher-end 4.2L Quattro version.

The 2003 $88,500 S6 Avant wagon, launched several years ago, carries a powerful 340-hp 4.2L V8 engine, and sport suspension and tires. The 3.0L sedan offers either front-drive or a Quattro AWD configuration.

Cost analysis/best alternatives: Choose the 2003 models, which are practically identical to this year's offerings. Other vehicles worth taking a look at: the Acura TL, BMW 5 Series, Infiniti I35, and Volvo V70. **Options:** All-wheel drive, if needed. Think twice about getting the power moon roof if you're a tall driver. **Rebates:** $3,000 rebates and low-interest financing. **Delivery/PDI:** $1,205. **Depreciation:** Slower than average. Audi values no longer plummet when the base warranty expires, and high repair costs have been brought under control. **Insurance cost:** Higher than average. **Parts supply/cost:** Very dealer-dependent and expensive. Independent suppliers carry few Audi parts. **Annual maintenance cost:** Low during the warranty period, then it climbs steadily. **Warranty:** Bumper-to-bumper 4 years/80,000 km; powertrain 4 years/80,000 km; rust perforation 10 years/unlimited km. **Supplementary warranty:** A prerequisite to Audi ownership, and it guarantees a good resale price. **Highway/city fuel economy:** 7.8–13.5L/100 km; *Quattro:* 8.4–13.9L/100 km.

Road Performance

Pro: Acceleration/torque: Better-than-average acceleration with the upgraded base engine and automatic transmission. Impressive performance is delivered by the 2.7 T, however, with 0–100 km/h times under 7 seconds. **Braking:** Better-than-average braking performance (100–0 km/h: 38 m). Average offset crash protection and acceptable head restraint performance.

Con: Vehicle shakes when put into Reverse and the front bumper catches on parking bumpers. **Service bulletin problems:** Ignition coil packs fail frequently, causing the engine to shut down (see bulletin in A4 section); inoperative keyless entry transmitter, and faulty fresh air control lever light. **NHTSA safety complaints/safety:** Airbags failed to deploy; front seat failed to stay anchored in a rear-ender; severe hesitation upon acceleration; vehicle loses power or stalls when making right turns; premature transmission failure; transmission slips from automatic to manual mode without warning; chronic hesitation at low speed; faulty gas tank sensors transmit wrong indication of remaining fuel; distorted front windshield.

TT COUPE ★★★★★

RATING: Recommended. **Strong points:** Impressive acceleration with the optional 225-hp power plant; all-wheel drive available; good handling and road holding; very well appointed and tastefully designed interior; comfortable, supportive seats; plenty of passenger and cargo space (especially with rear seatbacks folded); standard ABS; "smart" dual front airbags; "Good" front head restraint protection; "Average" rear head restraint protection; optional Bose sound system; and a predicted high resale value. **Weak points:** Base engine lacks low-end torque; excess weight limits handling; poor rear and side visibility; ride may be too firm for some; confusing interior controls; a useless back seat; difficult rear-seat access; lots of engine and road noise; no crashworthiness test results; and limited availability. **New for 2004:** An automatic transmission sometime in 2004.

OVERVIEW: Costing an estimated $55,475 (plus $1,205 for freight/PDI), the TT Coupe and Roadster debuted in the spring of 1999 as a sporty front-drive hatchback with 2+2 seating and set on the same platform used by the A4, Golf, Jetta, and New Beetle. A two-seat convertible version, the Roadster, was launched in the spring of 2000. The base 180-hp 1.8L engine (lifted from the A4) is coupled to a manual 5-speed, while the optional 250-hp 3.2L V6 uses a 6-speed manual transaxle. Shorter and more firmly sprung than the A4, the TT's engines are turbocharged, though only the optional engine comes with standard AWD. A spoiler and anti-skid system are also standard.

More beautifully styled and with better handling than the Prowler, the TT comes with lots of high-tech standard features that include four-wheel disc brakes, airbags everywhere, traction control (FWD models), a power top (Quattro), a heated-glass rear window, and a power-retractable glass windbreak between the roll bars (convertible). An alarm system employs a pulse radar system to catch prying hands invading the cockpit area.

Cost analysis/best alternatives: The 2004's base price has risen by about $4,000, even though there is little that is new this year. Go for a 2003. Other vehicles worth taking a look at: the BMW Z Series, Honda S2000, and Mazda Miata. **Options:** All-wheel drive is recommended only if it's essential for your driving needs. **Service bulletin problems:** Ignition coil packs fail frequently, causing the engine to shut down (see bulletin in A4 section); AC control knobs fall off; and faulty fresh air control lever light. Troubleshooting tips on eliminating AM radio static. **NHTSA safety complaints/safety:** Incredibly few consumer complaints. Xenon headlights make it difficult to see the sides of the road; central computer causes door locks to jam; windshield wipers often stop working; faulty gas tank sensors transmit wrong indication of remaining fuel; and Michelin tire tread separation.

BMW

BMW has fully recovered from its 1994 purchase of the Rover Group—a major mistake that drained cash reserves, delayed the upgrading of its models, and weakened management through dismissals and resignations. The canny Germans have bounced back, by unloading the Land Rover Group upon unsuspecting Ford, keeping the successful Mini Cooper franchise, and through upgrading their best-selling 3 Series, 5 Series, flagship 7 Series, and profit-laden X5 sport-utility, which will soon spin off a smaller X3 variant. The Z8, BMW's slow-selling exotic version, is axed this year.

BMW continues to build well-appointed cars that excel at handling and driving comfort. Its vehicles have excellent road manners, depreciate slowly, and have an "I got mine!" cachet that buyers find hard to resist. Unfortunately, they also have an incredibly complicated centre console-mounted "iDrive" feature that some have renamed the "iDie" controller. An article in *Popular Science,* titled "uDrive Me Crazy," describes it this way:

> With a push, turn, or shove, this automotive supermouse controls 700 functions, which are displayed in menus on the screen above. Most features that 7 Series drivers had previously selected and modulated via switches, knobs, sliders, pushbuttons, and stalks are now operated by the knob.... It's somewhere between silly and wonderful. It's more a step sideways than a step forward—a few good ideas, a few good features, and a whole bunch of bad implementation.

Pity that BMW has just installed iDrive on its redesigned 2004 5 Series.

Other BMW minuses: limited interior room (except for the high-end models), some quality control deficiencies (notably the powertrain, electrical system, and notoriously soft brake pads), and its vehicles can be difficult and expensive to service. Incidentally, a good website listing the more common problems with BMWs is *www.roadfly.org.*

The 5 Series is the cream of the crop; however, there are many good reasons for buying other Bimmer models: good overall reliability (with just a few exceptions, like the X5 SUV); impressive, high-performance road handling; prestige value; and a low rate of depreciation. Keep in mind, though, that there are plenty of other cars that cost less, offer more interior room (Audi comes to mind), and are more reliable and better performing. So if you're buying a BMW, remember that the entry-level versions of these little status symbols are more show than go, and that just a few options can blow your budget. The larger, better-performing, high-end models are more expensive and don't give you the same standard features as do many Japanese imports. Also, be prepared to endure long servicing waits, body and trim glitches, and brake, electrical, and accessory problems.

BMW Mini Cooper

Arriving late to the nostalgia niche dominated by the Dodge PT Cruiser, Ford T-Bird, and VW New Beetle, BMW has finally let loose with its own flashback from the past—the Mini Cooper, last marketed by Austin in 1967. The front-drive, $25,200 (S version: $29,950 plus $1,250 freight/PDI) Mini Cooper returns with more power (200 km/h top speed) and safety features than its predecessor. Dealer markup is 9 percent and freight/PDI often gets boosted to $1,550. $2,000 discounts are out there. Plus, unlike the T-Bird, the Mini still has the distinctive looks of the original. Fortunately, it has ditched the original's 10-inch wheels for 16- and 17-inchers.

Although not as small as the original Austin Cooper, the Mini is still less than 4 m long, with a 239-cm wheelbase, a width of 1.9 m, and a height of only 1.4 m (yes, smaller than a Chevy Metro). The 112-hp 1.6L 4-banger can be teamed to a 5-speed manual or automatic transmission, and turns in a 0–100 km/h time of a leisurely 9.2 seconds. Four-wheel ABS disc brakes are standard, along with a host of safety features. For 2004, we'll see fewer cracked windshields now that the A-pillars have been reinforced.

Okay, so the Mini's neither cheap nor fast, but it is stylish, with its uniquely hunkered-down, *cute*, macho look. It has more interior space than the Austin (but, weren't we all smaller then?), there are 50/50 split folding seats for additional storage space, and the fully independent suspension carries a body we are told is much more rigid than what's offered by the competition. Plus, fuel economy is said to be in the 9.5L per 100 km range (30 mpg).

So what's there not to like?

First, take away the retro styling and you get an undersized, underperforming, and untried British import thrust into a market where many proven competitors do more for less money. Second, it's built in England, which guarantees a plethora of quality bugs. Says the *Christian Science Monitor* in an early review, "Quality is suspect in both [models]. Thrumming wheel bearings, whining steering, loose and missing interior parts marred two weeks of driving."

IIHS offset crash tests give the Mini an "Above Average" rating. Finally, only 20,000 Minis are to be sent to North America, with Canada allotted less than 2,000—not enough for an adequate parts supply and competent mechanics to be trained to service these little beasties.

Guess some things never change.

NHTSA safety complaints/safety: Inaccurate speedometer and jammed passenger-side seat belt shoulder retractor. Other deficiencies mostly concern poor ergonomics, a surprising oversight for a German-engineered car. For example: inside door handles are located too far back on the doors; getting the spare tire from under the vehicle is a chore; and shoulder belts are uncomfortable. **Service bulletin problems:** N/A; see *www.mini.ca* or *www.mini2.com/forum.* **Best alternatives:** Mazda Miata, Mercedes-Benz SLK230, and Porsche Boxster.

3 SERIES ★★★★★

RATING: Recommended. One of the best-handling sport sedans on the market. Still, many competitors deliver more interior room and standard features for less money. **Strong points:** Good acceleration (M3 acceleration is incredibly fast); excellent handling, impressive braking, and top-notch quality control. **Weak points:** Many complaints of engine overheating, sudden shutdown, and engine compartment fires; a somewhat harsh ride (M3 harsher than most); insufficient front headroom and seat lumbar support for tall occupants; limited rear-seat room and cargo area; tricky entry/exit, even on sedans; confusing navigation system controls; excessive tire noise, especially with the M3; and requires premium fuel. **New for 2004:** Rain-sensing wipers, re-styled wheels, new headlight and grilles on the 330i and 330xi, and the SMG transmission option is extended to all rear-drives.

OVERVIEW: With BMW's recent mechanical upgrades, styling changes, and increased exterior and interior dimensions, the 3 Series has come to resemble its more expensive big brothers with super-smooth powertrain performance and enhanced handling. You will find three powertrains, rear- or all-wheel-drive, and four body styles among 10 different models. The base coupes now come with the 2.5L and 3.0L 6-cylinder engines used in BMW's entry-level sedans and standard Sport suspension. These new four-seaters are also lower, longer, and wider than their four-door equivalents, use different body panels, and present a more aero appearance. The 323 convertible and wagon versions (launched last spring) have been made over in a similar fashion. The 315-hp 3.2L 6-cylinder engine is reserved for the unbelievably fast and nimble M3 Coupe and Convertible.

Recently upgraded cockpit amenities include a standard tilt/telescope steering wheel and optional power memory seats, power lumbar adjustments, an in-dash CD player, and steering wheel audio and cruise controls. Traction and stability control and rear side-impact airbags are available options.

Cost analysis/best alternatives: The 2004s return practically unchanged. Therefore, a discounted 2003 would be the better choice, if you can find one. Other cars worth considering are the Audi A4 and Lexus IS 300. **Options:** If you buy a convertible, invest $1,500 in the rollover protection system that pops up from behind the rear seat. The optional Sport suspension does enhance handling and steering, but it also produces an overly harsh, jiggly ride on rough pavement. The Sport Package comes with a harsh-shifting SMG gearbox (the one you loved to hate on the M3); it isn't worth the extra money. Wider tires compromise traction in snow. **Rebates:** The Japanese competition and a sluggish economy have produced $3,000 rebates, low-interest financing incentives, and more generous leasing terms. **Delivery/PDI:** $1,495 (way too much). **Depreciation:** Slower than average. **Insurance cost:** Higher than average. **Parts supply/cost:** Parts are less expensive than for other cars in this class. Unfortunately, they aren't easily found outside of the dealer network, where they're often back ordered. **Annual maintenance cost:** Average, until the warranty runs out; then your mechanic starts sharing your paycheque. **Warranty:** Bumper-to-bumper 4 years/ 80,000 km; powertrain 4 years/ 80,000 km; rust perforation 6 years/ unlimited km. **Supplementary warranty:** A prerequisite to Bimmer ownership. **Highway/city fuel economy:** *2.5L:* 7.3–11.5L/100 km.

Quality/Reliability/Safety

Pro: Quality control: Better than average. Body assembly and workmanship are quite good. **Reliability:** Average. **Safety:** All 3 Series models are pre-wired for an alarm system and use a Coded Driveaway Protection system that won't allow the car to start unless the ignition key matches the ignition switch code, which changes each time the car is started.

Con: Warranty performance: Average. BMW usually resolves disputes through individual "goodwill" settlements; however, their handling of the engine-overheating problem has been deplorably stupid. Parts are scarce outside major metropolitan areas, and independent mechanics are rare. **Owner-reported problems:** Brakes, electrical system, and some body trim and accessories are the most failure-prone components. Engine overheating is a serious and common failure. Steering column is kinked to the left; steering wheel creaking; and CD player skips. *M3:* Loud clunking from the rear end when shifting or decelerating said to be caused by a faulty driveshaft attachment at the differential. **Service bulletin problems:** Erratic shifts (see following).

A/T - Hard/No Shift Condition/DTCs Stored in TCM

No.: 24 02 00 **Date:** April 2002

SUBJECT: E36/7, E39, E46, E53 with GM5; "hard" or "no-shift" with FC 50, 52, 53, or 55 in TCM

MODEL: E36/7 (Z3) with M54 and GM5 from 06/00
E39 with M52TU & M54 and GM5 from 09/99 up to 03/01
E46, 323i (sedan/coupe/convertible/sport wagon) with M52TU & GM5 from 05/98 up to 03/00
E46, 328i (sedan/coupe) with M52TU & GM5 from 05/98 up to 06/00
E46, 325iTA (sport wagon) with M54 & GM5 from 08/00 up to 03/01
E46, xi (all-wheel-drive, sedan/sport wagon) with M54 & GM5 from 08/00
E53 with M54 & GM5 from 04/00

SITUATION: Customer may complain of "hard" or "no-shift", and one of the following "Gear Error" fault codes will be stored in the TCM: FC 50, 52, 53, 55. No other fault codes are present in the transmission control module.

CAUSE: Valve body malfunction/sticking valves.

Engine wiring harness replacement for ignition coils. *M3:* Service action to improve mirror tilt when in Reverse gear. **NHTSA safety complaints:** *All models:* Many incidents where cooling fan failure caused engine to overheat, or a fire to ignite (see *www.roadfly.org*); airbag failed to deploy. *325i:* Transmission failure within five days of purchase; transmission replaced and still downshifts harshly; RPM idle fluctuation; vehicle jerked forward when decelerating; sudden loss of power due to computer failure; electrical system fire; right door airbag deployed even though vehicle was hit on the left; sunroof glass suddenly exploded (several incidents reported); door locks without prior warning; faulty fuel pump; and Bridgestone tire tread separation. *328:* Premature tire failure (bubbles in the tread); sudden acceleration; when accelerating, engine cuts out, then surges forward (suspected failure of the throttle assembly); severe engine vibrations after a cold start as Check Engine light comes on; if driver wears a size 12 shoe or larger, when foot is flush against the accelerator pedal, the top of the shoe rubs up against the panel above the pedal, preventing full pedal access. Rear quarter blind spot with the convertibles. *330i:* Side airbag deployed when vehicle hit a pothole; vehicle overheats in low gear; tires lose air; vehicle slips out of Second gear when accelerating.

Road Performance

Pro: Emergency handling: No-surprise suspension and steering makes for crisp high-speed and emergency handling. Much better rear-end stability. **Steering:** Exceptionally accurate and sensitive. Lots of road feedback. **Acceleration/torque:** The inline 6-cylinder engines and transmissions are the essence of harmonious cooperation, even when coupled to an automatic transmission; there's not actually that much difference between the two from a performance perspective. **Transmission:** Light and precise gear shifting with easy clutch and shift action. **Routine handling:** Competent and predictable handling on dry surfaces. **Braking:** Smooth, efficient braking produces short stopping distances with the 328i (100–0 km/h: 37 m).

Con: The ride is firm and occasionally uncomfortable on rough roads. Mediocre acceleration with the heavier ragtop coupled to an automatic transmission. Some body flexing and shake when the convertible is run over rough pavement.

Comfort/Convenience

Pro: Driving position: Comfortable driving position enhanced by a tilt/telescope steering wheel, nice ergonomics, plenty of front headroom and passenger legroom for average-sized adults, and good all-around visibility. **Controls and displays:** Excellent control layout and design, though it could use additional gauges. Wiper action automatically drops from constant to intermittent when you slow down or stop. All power windows have an express down and up. **Climate control:** The 325i's additional room and separate driver and passenger climate controls make for a more hospitable interior. **Interior space/comfort F/R:** Sufficient space and comfort for four occupants—not five, as BMW would like you to believe. Barely enough rear-seat room in the coupes. **Trunk/liftover:** Trunk has a low liftover for easy loading. **Quietness:** A rigid body design keeps rattles and clunks to a minimum, although the convertibles have more than their share of wind noise.

Con: Standard equipment: Competitors offer more (can you believe the convertible's power softtop is a $2,000 option?). Bland styling and very austere, narrow interior. Limited front headroom for tall occupants. As well, front seats could use additional lumbar support for tall occupants. Interior plastics look tacky. Convertible provides limited rear visibility. **Controls and displays:** Confusing switches on the dual climate-control system. **Entry/exit:** Tight rear access with coupes, despite the front seats' automatic forward positioning feature. The four-door model offers marginally improved access to the rear seat. **Interior space/comfort F/R:** Seats are too firm for some. Limited headroom when equipped with a sunroof. Centre console cuts into driver's legroom and kneeroom. Not much kneeroom or legroom for rear-seat passengers with the front seats pushed rearward (particularly bad with the convertible). Rear seating is limited to two adults. **Cargo space:** Less than average with the sedans; worse with the coupes. **Quietness:** Noisy high-performance tires, wind rush with convertible.

3 Series

List Price (firm)	**Residual Values** (months)			
	24	**36**	**48**	**60**
325i: $39,450 (18%)	$28,000	$21,000	$17,000	$13,000
325Ci: $42,250 (18%)	$34,000	$27,000	$22,000	$17,000
330Ci Coupe: $49,550 (20%)	$38,000	$33,000	$27,000	$24,000

Technical Data

Powertrain (rear-drive)
Engines: 2.5L 6-cyl. (184 hp)
• 3.0L 6-cyl. (225 hp)
• 3.2L 6-cyl. (333 hp)
Transmissions: 5-speed man.
• 4-speed auto.
Dimension/Capacity (328i)
Passengers: 2/3
Height/length/width:
55.7/176/68.5 in.

Headroom F/R: 37.2/36.7 in.
Legroom F/R: 41/34 in.
Wheelbase: 107.3 in.
Turning circle: 36 ft.
Cargo volume: 15 cu. ft.
Tow limit: 1,000 lb.
Fuel tank: 62L/prem.
Weight: 3,100 lb.

Safety Features/Crashworthiness

	Std.	**Opt.**
Anti-lock brakes	■	—
Seat belt pretensioners F/R	■	—
Side airbags	■	—
Traction control	■	—
Head restraints F/R	***	***
Visibility F/R	*****	*****
Crash protection (front) D/P	****	*****
Crash protection (side) F/R	***	*****
Crash protection (offset)	*****	—

Jaguar

Jaguar's name conjures up images of posh sedans and fast sports cars rolling through the English countryside. They also reek of elegance, taste, and money. The car's styling, ride, handling, and comfort still entice motorists full of nostalgia for British cars of the 1960s, but what they are given is a hodgepodge collection of Ford Taurus parts masquerading as Jaguar original equipment (OE). No wonder that Taurus-sourced mechanical, electronic, and body problems persist—like sudden, unintended acceleration, automatic transmission and brake failures, shimmying, and excessive noise when driving with the rear windows open. Nonetheless, these cars are better built than they were a decade ago.

But the Jaguar's poor-quality image has been tough to shake, which may explain why these luxury cars have such a high rate of depreciation, and why most buyers prefer to lease rather than purchase their "Jag."

When Ford bought Jaguar it had three goals: improve quality, make the cars more affordable, and make the division profitable. It took over five years, but Jaguar quality is no longer as bad as it once was and Ford has introduced two cheaper models, the S-Type and X-Type series, joined by the more upscale XJ8 and XK8. As far as making Jaguar profitable—forget it. Jaguar continues to post a $500+ million operating loss that it blames on others.

Caught in a dilemma similar to that of the Corvette, high-end Jaguars have excess weight that makes it necessary to install lots of complicated and difficult-to-troubleshoot devices, as well as larger engines, in order to make these cars decent highway performers. Entry-level Jags have a different problem: convincing buyers their X-Types are more than gussied-up $41,195 Contours (with a larger trunk and less headroom) and that S-Types are worth $59,950, even though they have fewer luxury features than many competitors.

Confident that its poor-quality cars are ancient history, Jaguar offers a comprehensive 4-year/80,000 km base mechanical warranty. The automaker also provides a 6-year/unlimited km rust perforation warranty, and Jaguar Club benefits that include no-cost maintenance, roadside assistance, and trip interruption services. Spotty servicing quality is still a problem. Unfortunately, there aren't many Jaguar dealers to choose from, so if you don't find a competent and conscientious one, it's doubtful that you'll be able to go elsewhere for a second opinion.

X-TYPE ★

RATING: Not Recommended. Aimed at the Acura TL, Audi A4, BMW 3 Series, Mercedes C-Class, and Lexus crowd, the X-Type's $41,195 base price and $44,995 MSRP for the V6 model is a lot of money to pay for what is essentially a *faux* Jaguar competing in a very real crowd. **Strong points:** Head-protecting side airbags; comfortable ride; excellent steering and handling, particularly with the AWD; lots of trunk space. **Weak points:** This isn't a luxury car, as advertised; equipped with the failure-prone Contour's 2.5L V6 power plant and underpinnings; Lincoln LS 3.0L is barely adequate; limited dealer network; Ford mechanics will be less familiar with this model; and Contour parts aren't easily found. Navigation system isn't easy to program or understand. Transmission set-up is a bit confusing and imprecise. **New for 2004:** Nothing significant; the 2005 X-Type will get a 2.7L V6 diesel engine. Does this make sense?

OVERVIEW: This all-wheel drive is a derivative of the Ford Mondeo and Contour, and has the same interior dimensions, except for a larger trunk, and a bit less headroom. Power is supplied by two engines: the Contour's 194-hp 2.5L V6, and an optional, Lincoln LS-derived, 231-hp 3.0L V6, hooked to either a manual or automatic 5-speed transmission. The optional Sport Package offers a firmer suspension, and dynamic stability control, 17-inch tires, and exclusive trim features. Safety complaints registered by NHTSA: ignition switch can be hit by driver's knee, causing vehicle to shut down; and engine surging.

S-TYPE ★★

RATING: Below Average. On sale since the summer of 1999, the S-Type shares its rear-drive platform with Lincoln's LS sedan. **Strong points:** Impressive engine performance; superb handling and road holding; quiet running; improved reliability; and a well-appointed interior. **Weak points:** Erratic-shifting manual/automatic transmission; not much room for cargo; a cramped interior; limited dealer network—Ford mechanics unfamiliar with this new model; some parts delay; and accelerated depreciation. **New for 2004:** Nothing significant. Last year's base V8 got 20 additional horses, a new suspension, and revised interior. A 5-speed manual transmission and optional 6-speed automatic were also offered for the first time.

OVERVIEW: Finally, Ford comes out with a mid-priced Jaguar, the first product built upon Ford's shared platform concept that was also used with the 2000 Lincoln LS sedan. So, what's not to like? How 'bout a transmission that's more suitable to John Deere? An interior that won't hold three rear occupants in comfort? Plastic trim instead of real wood?

The S-Type has a longer wheelbase than the XJ sedans and is wider and taller. It's just a bit shorter in length, however. Retro styling continues at Ford with the oval grille vertical bars, four round headlights, and bland, pinched-looking tail lights. From the side, this Jag looks like an Infiniti J30.

Aimed at the BMW 5 Series, Mercedes E-Class, and Lexus GS crowd, the S-Type is pricier than its Lincoln LS twin at $59,950 for the V6 (the same as last year's price) and $13,000 more for the V8 ($72,950).

A rear-drive, $89,950 S-Type R, equipped with a supercharged 4.2L 400-hp V8, arrived as a late 2003 model.

Quality control, although improved of late, is still likely to be a long-term problem, particularly in view of this being a new model churned out by Jaguar's old manufacturing complex near Birmingham, England. The car does have four-wheel ABS, traction control, front head/chest side airbags, Ford's Duratec-based 240-hp 3.0L V6, and an optional Jaguar 300-hp 4.0L V8. The 5-speed automatic is a Ford-Jaguar joint venture. It has a history of being slow to react to a throttle-induced downshift. Stomp all you want on the gas pedal, the transmission still takes its own sweet time to engage. It may also hunt for the proper gear (more pronounced with the V6) or change gears for no apparent reason, after hesitating a few seconds.

Other deficiencies: steering is a bit over-assisted for high speeds (the optional Sport Package will help), limited passenger (rear passengers sit knees-to-chin with scrunched toes) and cargo space, restricted rear visibility, and a confusing array of audio and climate controls.

Owner-reported problems include: poor headlight illumination, AC failure, passenger-side front wheel fell off; Continental tire blowouts; and sliding roof may not close properly or bounces open.

Cost analysis/best alternatives: Jaguar is a low volume seller, and discounting is quite common. There isn't any compelling reason to buy or lease a 2003 or 2004 S-Type when the car is compared with less expensive convertibles/roadsters like the BMW Z3, Mercedes SLK, and Porsche Boxster. Other, more reliable, luxury cars you may wish to consider, with as much or more cachet, are the Lexus models, the Lincoln Town Car, Infiniti models, Mercedes S-Class, and the Toyota Camry or Avalon. **Options:** The voice-activation system is a pricey gadget that adds complexity without providing the convenience promised. **Rebates:** Zero percent financing, $3,000+ rebates on the 2003s, and up to $5,000 for the S-Type and high-end models. **Delivery/PDI:** $670 (about half the fee charged for a Toyota Sienna). **Depreciation:** Faster than average. **Insurance cost:** Higher than average, almost usurious. **Parts supply/cost:** Parts are often back ordered because not a lot of Ford dealer inventory goes into stocking Jaguar parts. The S and X versions will be able to benefit from Ford's generic Lincoln parts, but Contour components will still likely be back ordered. Parts are moderately expensive as well, and there are few independent suppliers to inject price competition into the equation. **Annual maintenance cost:** Predicted to be higher than average. **Warranty:** Bumper-to-bumper 4 years/80,000 km; rust perforation 6 years/unlimited km. **Supplementary warranty:** Don't leave home without it. **Highway/city fuel economy:** 8.9–13.8L/100 km with the 3.0L V6 and 9.1–14.2L/100 km with the V8.

S-Type

Safety Features/Crashworthiness

	Std.	Opt.
Anti-lock brakes	■	—
Seat belt pretensioners F/R	■	—
Side airbags	■	—
Traction control	■	—
Head restraints F/R		
S-Type	***	**
X-Type	*****	*****
Visibility F/R	*****	**
Crash protection (front) D/P		
X-Type	****	****

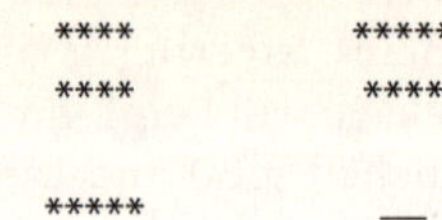

Crash protection (side) F/R		
S-Type	****	*****
X-Type	****	****
Crash protection (offset)		
X-Type	*****	—

Mercedes-Benz

Mercedes-Benz has long made it a point to design and engineer cars at the forefront of technology and safety, and to clothe them in conservative, though attractive, garb. So what has it done for Chrysler since buying the company? Not much, except for sending hundreds of Detroit executives running for the doors and their pensions.

Mercedes' quality myth

Mercedes says it wants to instill a new quality control ethic into Chrysler's executive ranks, but I'm not sure if the German automaker's quality perceptions are that much better than Chrysler's.

For example, Mercedes' M-Class sport-utilities launched a few years back were abysmally bad. You couldn't have made a worse vehicle, judging by the unending stream of desperate-sounding service bulletins sent to dealers during the first few years following the official launch. Two bulletins stand out in my mind: one was an authorization for dry cleaning payouts to dealers whose customers' clothing was stained by the dye from the leather burgundy-coloured seats. The other was a long explanation as to why drivers were jolted by static electricity when entering or leaving the vehicle.

Car columnists have always known that Mercedes has made some bad cars and SUVs, but it took business reporters (not their auto beat writers) from the gutsy *Wall Street Journal* to spill the beans.

In a February 2, 2002, article headlined "An Engineering Icon Slips," the *WSJ* cited several confidential industry-initiated surveys that showed Mercedes' quality and customer satisfaction had fallen dramatically since 1999—to a level below GM's Opel, a brand with one of the worst reputations for poor quality in Europe.

An earlier J.D. Power and Associates study of vehicle durability in the States put Mercedes 12th in its ranking—behind Lincoln, Cadillac, and Jaguar. Its chief failings were in the transmission and features/controls categories. Following this study, J.D. Power and Associates announced to a National Automobile Dealers Association convention that Mercedes' rating for overall quality had been lowered to "Fair" from "Good."

Industry insiders tell me that Mercedes-Benz quality has been diluted by the more than doubling of its product lineup since 1997 and by the free ride it has gotten from a fawning press. Helpful, too, have been the company's aggressive PR campaigns and the Teutonic mindset that tends to blame the driver rather than the product—both spectacularly successful in keeping the quality myth alive.

For a small taste of how well Mercedes manages the press when it circles its wagons, look on the Internet for stories relating to the A-Class "baby Benz" flipping over during its press launching, allegations relating to M-B's Holocaust involvement, and its tight-lipped reaction to the above-cited quality surveys, and reports that Mercedes residual values have fallen dramatically.

2003–04 product lineup

In an attempt to respond to critics' charges that its cars were priced out of most buyers' reach and that its styling was antiquated, Mercedes tried out its own new aero look in the beginning of this decade with the entry-level 190 series (another so-called "baby Benz"). It was a flop.

The aero look has been retained and refined, however, and is now what makes these luxury imports distinctive, along with a revamped model lineup full of high-tech safety and convenience features. This has worked well in attracting younger buyers, while keeping Mercedes' older, more conservative clientele.

For 2004, a new E-Class wagon and CLK convertible are scheduled to arrive and the company will bring out a new 7-speed automatic transmission.

C-CLASS ★★★★

RATING: Above Average; *C230 Kompressor:* Average. Once you take away the three-pointed star and add the 230's cloth upholstery, the Emperor's new clothes look shopworn. Save up for an E-Class. **Strong points:** Plenty of high-tech safety features like standard "smart" airbags, side airbags, traction control, and brake assist; good powertrain matchup enhanced with AWD and a supercharger; comfortable ride; easy handling; good braking; innovative anti-theft system; and good overall crash protection. Resale value remains high. **Weak points:** C230 is bereft of many standard performance and convenience features one would expect to see in a luxury car; limited rear seat and cargo room; tight entry/exit; a choppy ride; some tire thumping, engine, and wind noise. **New for 2004:** Not much, except larger 17-inch wheels and a sport package of dubious value.

OVERVIEW: These cars are Mercedes' entry-level offerings, spun off of its base four-door sedan. They include a C240 and C320 sedan, the C230 Coupe, and the C320 wagon, along with AMG variants.

The $34,950 (plus $1,495 freight/PDI charges) C230 two-door hatchback coupe is the automaker's cheapest car—and it looks it. Gone are the standard amenities one finds with most Mercedes models. The base engine is a modest 189-hp supercharged 1.8L 4-cylinder hooked to either a 5-speed automatic, or a 6-speed manual transmission. More powerful engines, including a V6 version and AWD are available as you move up the model range.

Mercedes press relations flacks call the C230 a sporty coupe and wince whenever the hatchback term is used by less respectful, but more accurate, journalists. The company knows that North American buyers, until recently, were disdainful of luxury hatchbacks (more money for less car?), and would much prefer to call the car something else (Kompressor sounds good, yes, Helmut?).

Based on the 2001 C-Class sedan's 271.5 cm (106.9 in.) wheelbase, the C230 Kompressor's sport pretensions are supported somewhat by its supercharged engine, rakish styling, 6-speed manual transmission (a 5-speed automatic is optional), and standard sport suspension. Still, drivers say the coupe feels no more "sporty" than the sedan, although it may be a bit more nimble.

The C320 wagon is another C-Class variant costing about $52,850, pricing it above the Audi A4. 2.8 Avant Quattro, the A6 Avant, and the BMW 325 Sport Wagon. It's a nicely styled wagon that carries a 215-hp 3.2L V6 and provides storage space comparable to the Audi and BMW competition.

Cost analysis/best alternatives: Mercedes' three-point star is sinking fast, forcing the automaker to cut prices and add incentives to move its slow-selling entry-level models. Bottom line? Take a pass on this year's entry-level offering and shop the deals on the practically identical 2003 lineup. Also take a look at the Audi TT or a BMW 325CI Coupe—not great performers either, but they look the part. Other cars worth considering are the Audi A4, BMW 3 Series, and Volvo S40, V40, V70 wagon, or V70XC. Rivals to the CLK and the SLK230 would be the BMW Z4, Chevrolet Corvette, Porsche Boxster, Volvo C70 and S80, and other models in the Mercedes-Benz SL-Class stable. **Options:** The Bose sound system is a sound investment. **Rebates:** $3,500–$4,000 rebates, generous leasing deals, and low-interest financing. **Delivery/PDI:** $1,495 (cut it in half through smart bargaining). **Depreciation:** Slower than average. **Insurance cost:** Higher than average. **Parts supply/cost:** Limited availability, but parts aren't that expensive. **Annual maintenance cost:** Less than average. **Warranty:** Bumper-to-bumper 4 years/80,000 km; powertrain 4 years/80,000 km; rust perforation 5 years/120,000 km. **Supplementary warranty:** Not needed. **Highway/city fuel economy:** *1.8L:* 7–11.2L/100 km; *2.6L:* 8.4–13.7L/100 km; *3.2L V6:* 8.6–12.3L/100 km.

C-Class

List Price (soft)	**Residual Values** (months)			
	24	**36**	**48**	**60**
C230 Coupe: $34,950 (18%)	$23,000	$17,000	$15,000	$13,000
C240 wagon: $43,200 (20%)	$27,000	$21,000	$18,000	$15,000
C320 Sports Coupe: $40,100 (22%)	$25,000	$18,000	$17,000	$14,000

Technical Data

Powertrain (front-drive/AWD)
Engines: 1.8L 4-cyl. (189 hp)
• 2.6L V6 (168 hp)

Height/length/width:
54.3/171/68 in.
Headroom F/R: 38.5/36.3 in.

• 3.2L V6 (215 hp)
• 3.2L V6 (349 hp)
Transmissions: 5-speed auto.
• 6-speed man.

Dimension/Capacity

Passengers: 2/2
Legroom F/R: 42/33 in.
Wheelbase: 106.9 in.
Turning circle: 35.3 ft.
Cargo volume: 9.9 cu. ft.
Tow limit: N/A
Fuel tank: 62L/prem.
Weight: 3,252 lb.

Safety Features/Crashworthiness

	Std.	Opt.
Anti-lock brakes	■	—
Seat belt pretensioners F/R	■	■
Side airbags	■	—
Traction control	■	—
Head restraints F/R	*****	*****
Visibility F/R	****	**
Crash protection (front) D/P	****	*****
Crash protection (side) F/R	*****	*****
Crash protection (offset)	*****	—

Quality/Reliability/Safety

Pro: Quality control: Average. **Reliability:** Except for intermittent electrical system shutdowns, components that would affect reliability are relatively trouble-free. Vehicle starts and then stalls (fuel sensors suspected). **Safety:** Standard three-point seat belts, side airbags, and high-tech anti-theft alarm.

Con: Service bulletin problems: *C230:* Engine oil leaks; piston slapping, ticking; buzzing, droning noises. **NHTSA safety complaints:** *C230:* Airbags failed to deploy. Many owner complaints that the vehicle suddenly loses power and then shuts down:

> I have taken my vehicle in for repeated repairs and the dealer has not fixed the problem with the car shutting off on me. There is an electrical problem, because the car will shut off on the highway, in traffic, going up a hill, and then the dashboard will light up, mentioning system failure. I have used the SOS button in my car repeatedly to get help, but that is still not good enough. I have also contacted Mercedes-Benz USA and Mercedes in Stuttgart, Germany. I have put them on notice to purchase the car back from me, and they have a week and a half left to respond. I try not to drive the car a lot, which defeats the purpose of purchasing a new vehicle, but frankly, I am scared that something might happen, because it has shut off on me on I75 going 60+ mph [100+ km/h]. It is certainly a safety hazard.

No throttle response; vehicle shakes violently when put into gear; transmission slips in cold weather; or jerks into gear when accelerating; excessive vibration until vehicle warms up; dash gauges wash out in sunlight; driver-side seat belt buckle attachment located between the seat and console is set too far down, making it impossible for driver to latch or release buckle; blind spot between the rear window and the two back side windows; and "rotten-egg" exhaust smell. *240:* Airbags failed to deploy; chronic stalling; several incidents of sudden unintended acceleration; accelerating from a stoplight vehicle appears to stall, then surges forward, other cases where car hesitates when accelerating; and total engine failure. *320:* Dash vent reflects onto the windshield.

E-CLASS ★★★★★

RATING: Recommended. This is the Mercedes to buy, not some hatchback-cum-sporty coupe (see previous entry). **Strong points:** Strong acceleration in the higher gear ranges; well appointed with many safety, performance, and convenience features; good engine and transmission combo. 4Matic all-wheel drive operates flawlessly; easy handling; good braking; comfortable ride; roomy interior; cargo room (wagon); innovative anti-theft system; excellent quality control; reasonable freight and PDI fees; a high trade-in value; four-star driver and passenger frontal crash protection, side protection rated five stars for both occupants; and a "Better than Average" rating by IIHS for offset crashworthiness. **Weak points:** Sluggish acceleration in First gear; handling not quite as crisp as the BMW 5 Series; suspension is a bit soft; complicated, difficult-to-use COMAND electronic control centre; navigation system controls are a pain to use; a surprisingly small trunk; and tall drivers may be bothered by the knee bolsters. The failure-prone electronics can screw up dozens of safety and performance features. **New for 2004:** A new rear-drive and AWD wagon with V6 and V8 engines arrives with the 2004s and another diesel engine returns next year. The E500 gets a new 7-speed automatic transmission.

OVERVIEW: Mercedes' best-selling sedans offer inline 6-cylinder engines, but a V8 power plant is offered on the rear-drive E320 ($71,350); the AWD will cost an extra $5,000 with a 217-hp 3.2L. The E430 is ($79,950) with a 275-hp 4.3L V8. An Autobahn pocket rocket E55 AMG sedan carries a 349-hp 5.5L V8 and a price tag over $100,000.

Redesigned twice in the past nine years, the E-Class is considered to be state of the art in German auto technology. These cars do everything well, and manage to hold five people in relative comfort. They're first class in combining performance, road manners, and comfort. True, they're not the best riding, handling, or accelerating cars available, but they're able to perform each of these tasks almost as well as the best cars in each specific area, without sacrificing some other important element in the driving equation.

E-Class cars continue to offer standard traction control that prevents wheelspin upon acceleration. Another interesting feature is the remarkably smooth and quiet base 24-valve 217-hp high-performance version of the inline 6-cylinder engine that powers the 300 series.

Mercedes E-Class models have front and side airbags for both the driver and right front passenger, plus dual-locking shoulder belts. The front airbags are designed to deploy at higher crash speeds when occupants are belted than when they're unbelted. Belts in the front seat have tensioners that activate in a crash to reduce belt slack. Sensors in the seat and belt deactivate the airbags and belt pretensioner on the passenger side if no occupant is riding in this seat. The middle back seat has a lap/shoulder belt. Energy-absorbing padding between the footwell and floor carpet is designed to reduce the forces on drivers' legs in serious frontal crashes.

Cost analysis/best alternatives: The 2003s are the better buy, since most of the 2004 models simply carry over last year's improvements at a higher cost. Other cars worth considering are the Audi A6/allroad Quattro, BMW 5 Series, Lexus GS 300 or 430, or Volvo S80, V70, and V70 XC. **Options:** Stay away from the COMAND option. **Rebates:** Look for deep discounting as leftovers are moved out to make way for the revised sedan. **Delivery/PDI:** $1,495 (double what's acceptable). **Depreciation:** Slower than average. **Insurance cost:** Higher than average. **Parts supply/cost:** Hard to find outside of the dealer network, and can be expensive at times (body parts especially). **Annual maintenance cost:** Less than average. **Warranty:** Bumper-to-bumper 4 years/ 80,000 km; powertrain 4 years/80,000 km; rust perforation 5 years/ 120,000 km. **Supplementary warranty:** Not needed. **Highway/city fuel economy:** *3.2L V6:* 7.5–11.4L/100 km; *3.2L V6 diesel:* 6.1–8.9L/100 km.

Quality/Reliability/Safety

Con: Although these cars are built for safety, their declining quality and complicated engineering can make them a nightmare to own and service. **Service bulletin problems:** Engine ticking; piston slap. **NHTSA safety complaints/ safety:** *300E:* Passenger-side curtain and rear side door airbag deployed for no reason; steering suddenly locked up while turning. *320:* When accelerating, vehicle hesitates, then surges forward; intermittent stalling when decelerating; if water enters the automatic transmission control module, the transmission won't shift; stalling, hesitation, leakage, and noise due to the failure of the transmission's electronic circuit board; malfunctioning gateway module and software causes electrical system to run haywire and causes the SRS light to come on; electrical shorts cause brake failures (E-Class and SL-Class); ABS light comes on for no reason; and faulty electrical wires.

Road Performance

Acceleration/torque: Better-than-average acceleration with the base engine, but the 4.2L V8's performance is dazzling (0–100 km/h: 8 seconds). **Transmission:** Flawless. **Braking:** It'll be hard to find better braking with any other car in this class (100–0 km/h: 35 m).

Volkswagen

Like most European cars, Volkswagens are practical and offer excellent handling and great fuel economy without sacrificing interior comfort. But overall reliability isn't very good (particularly after the fifth year of ownership), and servicing is often better and much cheaper at independent garages, which have grown increasingly popular as owners flee more expensive VW dealerships. Although parts are fairly expensive, independent repair agencies usually have no trouble finding them.

This year, Volkswagen introduces a new Passat W8, a New Beetle Convertible, and a 2004 Eurovan. VW also plans to cut prices in the new year—this will be done through low-interest financing programs, dealer incentives, and clearance sales. In other words, VW will offer rebates without having to use the term.

NEW BEETLE ★★★★

RATING: Above Average. Like the Mini Cooper, the New Beetle is an expensive ($23,690) trip down memory lane. Personally, I don't think it's worth it—with or without its speed-activated spoiler and dash-mounted bud vase. **Strong points:** Powerful 1.8L turbocharged engine; easy handling; sure-footed and comfortable, though firm, ride; impressive braking; most instruments and controls are user friendly; comfortable and supportive front seats with plenty of headroom and legroom; cargo area can be expanded by folding down the front seats; reasonable freight and PDI charges; good crash protection scores; and top-quality mechanical components and workmanship. **Weak points:** Serious safety defects reported by owners; base engine runs out of steam around 100 km/h; diesel engine lacks pep and produces lots of noise and vibration; delayed shifts from Park to Drive; easily buffeted by crosswinds; large head restraints and large front roof pillars obstruct front visibility; limited rear legroom and headroom; excessive engine noise; skimpy interior storage and trunk space. **New for 2004:** Upgraded head-protecting airbags and front head restraints; improved spoiler; and new wheels. The GLX model has been axed.

OVERVIEW: Why so much press coverage for the return of an ugly German import that never had a functioning heater, was declared "Small on Safety" by Ralph Nader and his Center for Auto Safety, and carried a puny 48-hp engine? The simple answer is that it was cheap, and it represented the first car most of us could afford as we went through school, got our first job, and dreamed of getting a better car. Time has taken the edge off the memories of the hardship the Beetle made us endure—like having to scrape the inside windshield with our nails as our breath froze—and left us with the cozy feeling that the car wasn't that bad, after all.

But it was.

Now VW has resurrected the Beetle and produced a competent front engine, front-drive, compact car—set on the chassis and running gear of the Golf hatchback—that's much safer than its predecessor, but oddly enough is still afflicted by many of the same deficiencies we learned to hate with the original.

Again, without the turbocharger, the 115-hp base engine is underwhelming (the 90-hp turbodiesel isn't much better) when you get it up to cruising speed. There's still not much room for rear passengers, engine noise is disconcerting, radio buttons and power accessory switches located on the door panels aren't user-friendly, front visibility is hindered by the car's quirky design, and storage capacity is at a premium.

On the other hand, the powerful, optional 1.8L turbocharged engine makes this Beetle an impressive 180-hp performer and when dressed in Turbo S garb, the 6-speed manual transmission makes this Beetle a sparkling performer. The heater works fine; steering, handling, and braking are quite good; and the interior is not as spartan or tacky as it once was.

Cost analysis/best alternatives: Go for the 2004 to get its safety upgrades. Other cars worth considering are the Hyundai Elantra, Mazda Protegé, Mini Cooper (if you must have cachet), Nissan Sentra, Toyota Corolla, and VW Cabrio. **Options:** The turbocharged 180-hp 1.8L engine is a must if you plan lots of highway use. **Rebates:** The novelty has worn off; look for price cuts in the new year. **Delivery/PDI:** $1,205. **Depreciation:** Much slower than average, especially during the first two years. **Insurance cost:** Higher than average. **Parts supply/cost:** Not hard to find since they're taken from the Golf/Jetta parts bin, but they may be more expensive than parts for most other cars in this class. **Annual maintenance cost:** Less than average during the first three years. After this, expect repair costs to start to climb dramatically. **Warranty:** Bumper-to-bumper 2 years/40,000 km; powertrain 5 years/80,000 km; rust perforation 12 years/unlimited km. **Supplementary warranty:** A good idea. **Highway/city fuel economy:** *1.9L turbodiesel:* 4.8–6.9L/100 km; *150-hp 1.8L:* 7.9–10.5L/100 km (no data on the 180-hp version); *2.0L:* 7.7–10.5L/100 km.

Quality/Reliability/Safety

Pro: Quality control: Average. **Reliability:** Average. **Warranty performance:** Slow but fair treatment of warranty claims by customer service staff located in the States. **Safety:** Standard three-point seat belts and anti-theft alarm (thieves just adore these cars). IIHS has given the Beetle a "Good" rating for offset crashworthiness and front-seat head restraint protection, and an "Acceptable" rating for the rear.

Con: Owner-reported problems: Frequent mass airflow sensor failures; intermittent automatic transmission leaks; won't shift from Reverse to Drive; "slams" into forward gear, or downshifts abruptly when coming to a stop; axle oil pan and oil pump failures; high-maintenance brakes due to prematurely worn pads and rotors; malfunctioning dashboard gauges; poor fit and finish accompanied by omnipresent squeaks, rattles and buzzing; inoperative power windows; driver door and trunk won't shut; faulty radios; window regulators are often broken; and front light covers retain water and short out. **Service bulletin problems:** Rough-running engine and chronic stalling caused by faulty ignition coils (see Audi A4 section); trunk lid hard to close; and loose armrests. **NHTSA safety complaints/safety:** Sudden acceleration; delayed acceleration; when vehicle is cruising on the highway, it suddenly loses most of its power for about 30 seconds, and then returns to normal; steering wheel locked up when accelerating; steering wheel shakes excessively and pulls vehicle to the right; back glass suddenly exploded; rear windshield glass hard to see through; front and rear windshield distortion; chronic stalling; Check Engine light comes on for no reason; temperature gauge warning light malfunctioned and reservoir tank sensor failed, causing vehicle to overheat; side airbag deployed for no reason, injuring occupant; side-impact airbags did not deploy, resulting in death; in another incident, driver-side airbag didn't deploy; airbag light won't go off; driver seatback failure following a rear-end collision; busted fuel tank leaked fuel; plastic fuel tank is easily punctured due to its vulnerable, low-mounted position; headrest is 15 cm (6 in.) too high to fit driver's head and obstructs rear visibility; open sunroof sucks exhaust fumes into the cabin; horn failure; the left front strut slipped down through the spindle, causing the spindle to hit the wheelwell; brake and turn signal lights cannot be seen by other drivers in bright sunlight due to the slope design.

New Beetle

List Price (firm)	**Residual Values** (months)			
	24	**36**	**48**	**60**
GLS: $23,690 (17%)	$17,000	$15,000	$12,000	$8,500
GLS 1.9 TDI: $25,470 (18%)	$18,000	$16,000	$13,000	$9,500
GLX: $30,570 (19%)	$22,000	$19,000	$16,000	$12,000
Cvt.: $29,610 (19%)	$23,000	$20,000	$17,000	$13,000
GLX: $36,190 (22%)	$27,000	$23,000	$21,000	$17,000

Technical Data

Powertrain (front-drive)
Engines: 2.0L 4-cyl. (115 hp)
• 1.9L TD 4-cyl. (100 hp)
• 1.8L T 4-cyl. (150 hp)
• 1.8L T 4-cyl. (180 hp)
Transmissions: 4-speed auto.
• 5-speed man.
• 6-speed man.
Dimension/Capacity
Passengers: 2/2
Height/length/width:
59.5/161.1/67.9 in.
Headroom F/R: 42/34 in.
Legroom F/R: 45.5/23 in.
Wheelbase: 98.9 in.
Turning circle: 36 ft.
Cargo volume: 12 cu. ft.
Tow limit: N/A
Fuel tank: 55L/reg.
Weight: 2,769 lb.

Safety Features/Crashworthiness

	Std.	Opt.
Anti-lock brakes	■	—
Seat belt pretensioners F/R	■	—
Side airbags	■	—
Traction control	—	■
Head restraints F/R	*****	***
Visibility F/R	**	*
Crash protection (front) D/P	****	****
Crash protection (side) F/R	*****	***
Crash protection (offset)	*****	—

GOLF, JETTA

RATING: Average. This year's rating has been downgraded due to the high incidence of inadvertent airbag deployments and other safety failures. **Strong points:** Superb all-around front-drive performers that offer power to spare with the manual shifter (GTI VR6, GLX), first-class handling, a comfortable ride, and good fuel economy. **Weak points:** Poor acceleration with the 4-cylinder engine and an automatic tranny; powerful V6 produces excessive torque steer; harsh automatic transmission downshifts; difficult entry/exit; restricted rear visibility; limited rear legroom; and a high number of safety-related complaints. Many incidents reported of airbags deploying for no reason and injuring or killing occupant. Keep in mind that the $1,205 freight and PDI fee is more than double what is acceptable; maintenance costs increase dramatically after the fifth year of ownership. **New for 2004:** *Golf, Jetta, GTI:* upgraded 100-hp 1.9L diesel engine (10 more horses) and new wheels. A high-performance Golf R32 model, equipped with a 237-hp 3.2L V6 and 6-speed manual transmission, will arrive in mid-2004. The Jetta gets a revised trunk lid and tail lights. *Cabrio:* Replaced by the convertible.

OVERVIEW: Practical and fun to drive. That pretty well sums up the main reasons why these VWs continue to be so popular. Yet they offer much more, including lots of front interior room, a competent engine (with a manual transmission), responsive handling, great fuel economy, and better-than-average reliability over the first three years.

The Jetta is a more expensive Golf with a trunk (probably why Jettas always outsell Golfs), and the convertible (formerly called a Cabrio) is a more expensive, roofless Golf. The Golf GTI V6 is a sporty performer that comes with lots of standard equipment, including air conditioning, an upgraded sound system, and a split folding rear seat. Jettas offer standard cruise control, power mirrors, and alloy wheels.

If we include the new sporty R32, there are six engines available: a 100-hp 1.9L turbo diesel, a 115-hp 2.0L base 4-cylinder, a 150-hp 1.8L 4-cylinder turbo, a 180-hp 1.8L 4-cylinder turbo variant, a 201-hp 2.8L V6 borrowed from the Corrado and Passat, and the 237-hp 3.2L V6. A manual 5-speed transmission is also offered as a standard feature, along with an optional 5-speed automatic and 6-speed manual. Power steering is optional on base-level models. Although the entry-level Golf's engine is no tire burner, it's well mated to the base manual transmission.

Cost analysis/best alternatives: This year's price increases are minimal inasmuch as most 2004s are carried over unchanged, making the 2004 improved diesels the best deal. If you wait until mid-2004, you'll benefit from price cuts and get a better-built Golf or Jetta, although it would be best to wait until the end of 2004 for the R32's first-year defects to be corrected. Other cars worth considering are the Honda Civic, Mazda6, Nissan Sentra and Altima, and Toyota Corolla and Matrix. Convertible shoppers might also want to test-drive a Chrysler Sebring, Chevrolet Cavalier or Sunfire, or a Ford Mustang. **Options:** Stay away from the electric sunroof; it costs a bundle to repair and offers not much more than the well-designed manual sunroof. Plus, you lose too much headroom. The diesel option isn't a good idea unless you travel more than 30,000 km a year. Granted, there are fewer things to go wrong with diesel engines, and fuel economy is high, but gasoline-powered VWs already have an excellent track record and good fuel efficiency. A lower fuel cost isn't a strong enough argument to weigh against the reduced performance of diesel engines. And diesel fuel prices have risen considerably over the past five years and are expected to go even higher—wiping out much of the fuel savings. **Delivery/PDI:** $1,205. **Rebates:** $3,000 rebates and low-interest financing. **Depreciation:** Slower than average, especially the Jetta and all Cabriolet versions. **Insurance cost:** Higher than average. **Parts supply/cost:** Not hard to find, but parts can be more expensive than most other cars in this class. **Annual maintenance cost:** Less than average while under warranty. After that, repair costs start to climb dramatically. **Warranty:** Bumper-to-bumper 2 years/40,000 km; powertrain 5 years/80,000 km; rust perforation 12 years/unlimited km. **Supplementary warranty:** A good idea. **Highway/city fuel economy:** *1.9L TDI:* 4.8–6.9L/100 km; *2.0L:* 7.7–10.5L/100 km; *2002 2.8L V6:* 8.4–12.6L/100 km.

Quality/Reliability/Safety

Pro: Quality control: Average. No rusting due to a galvanized body and stainless steel exhaust system. **Reliability:** Good overall reliability during the first few years (that's why so many are used as taxis). **Warranty performance:** Slow but fair treatment of warranty claims by customer service staff located in the States. **Safety:** Standard three-point seat belts and anti-theft alarm (thieves just adore these cars).

Con: Ownership costs rise dramatically after the fifth year of ownership. **Owner-reported problems:** The brakes and the electrical, fuel, and exhaust systems are especially troublesome. Chronic stalling. Many incidences where the airbag warning light stays lit or the airbag suddenly deploys. Engine timing belt failure and excessive oil consumption; automatic transmission grinding, howling, whining; erratic shifting; locking up; won't go into Fourth gear; when shifting gears, clutch goes out without warning; manual transmission slips out of gear and engages Reverse; steering moan when turning; tachometer registers 4700 rpm before downshifting. Owners report that doors are poorly hung causing rattling and water intrusion into the cabin; window regulators fail regularly, allowing glass to fall into door:

> Passenger's side front window ground to a halt, then collapsed into the door frame. Window would not come back up. Dealer reports this is a known issue: the Teflon guide sleeves break, especially in hot climates, and allow the window to fall into the door. Dealer will not replace other slides until they fail, even though the vehicle has less than 1,300 mi. [2,092 km] on it!
>
> Dealer claims replacement parts have been redesigned to include metal reinforcement. This is a huge issue, especially since there is no warning before failure. In inclement weather the car becomes totally undriveable and cannot be secured.
>
> Other model years, including 2001 Jettas, have the same problem. Please look into this. VW should repair all four windows on all Jettas with this type of window guide sleeve.

Door ajar light comes on for no reason; bent driver's door; instrument panel buzz; rattles are omnipresent, dashboard controls and interior and exterior trim items aren't very durable (rattles throughout the interior, dents beneath the doors, body paint scratches); and convertible top storage obstructs rear vision. **Service bulletin problems:** Inoperative air blower motor and prematurely worn rotors and rear brake pads. VW says they will replace these pads for free up to 12 months or 19,312 km (12,000 mi.), so use that as your benchmark for negotiating a refund. **NHTSA safety complaints/safety:** Coil failures (see Audi A4 section) and frequent failure of the brake light switch. *Golf:* Airbag sensors are too sensitive, cause deployment when car passes over a small pothole; in another incident airbags failed to deploy; intermittent stalling; vehicle suddenly veers to one side; uncomfortable driver's seat creates excessive fatigue on long trips; horn can't be located when wheel is turned; improper lug nuts allow wheel to separate from the car. *Jetta:* Shoulder and lapbelt disengaged during a collision; restraints failed, allowing driver to hit the windshield even though the airbag deployed; vehicle was hit from all sides and neither front nor side airbags deployed; car hit the curb and airbag deployed, hitting driver's head and killing him; many incidents where airbag deployed for no apparent reason:

> Driving on the turnpike, the driver-side airbag deployed without any sort of impact. There is no visible damage to the vehicle, which was only two months old at the time. Fortunately the only injury was a burn from the airbag on the side of my arm. My biggest fear is knowing relatives and friends who drive Jettas and who have young children in their car. This incident could easily have been fatal. Hopefully your establishment can encourage VW where I have failed, to do something about this defect.

Passenger-side airbag deployed seconds after impact; another time, it deployed six hours after a fender-bender; airbag warning light comes on for no apparent reason and shuts down the airbag system; bolts from drive shaft to transmission sheared off; transmission sticks in Reverse gear; fuel tank leaks; when the brakes are released, vehicle suddenly accelerates; cruise control won't shut off; intermittent stalling; when vehicle is driven at speeds over 57 km/h (35 mph), no-start caused by door wiring harness failure; Check Engine light comes on, vehicle shakes violently, followed by complete brake loss; brake master cylinder failure; ABS failure; brake pedal went to the floor with no braking effect, then slowly returned to normal; brake pedal could not be pushed down; vehicle rolled down incline with emergency brake engaged; while cruising at 80 km/h, brakes suddenly engaged; moaning noise heard when turning; faulty heated seats are fire-prone, or as one owner so succcinctly wrote government investigators, "My heated seats caught fire and burnt my ass...."; the exhaust pipe extends underneath the bumper, revealing raw edge of pipe; trunk stainless steel protection plate at the latch cutout area is razor sharp; the exhaust pipe extends underneath the bumper, revealing raw edge of pipe; adhesive that secures the brake light in the rear window can melt in sunlight; gear console gives an inaccurate gear reading; interior and instrument lights fail intermittently; headlight condensation; windshield stress crack; front seats rock front to back; windshield wipers shut off intermittently; window regulator failures:

> My car window failed only three days after I bought it. I was told by the dealers and a representative from the Volkswagen corporate offices that the reason they can't get it fixed in a reasonable amount of time is it is on back order from the factory because the driver's side or passenger's side window clips break in over 50 percent of new Volkswagen GTIs and Golfs within the first year.

Road Performance

Pro: Emergency handling: Acceptable despite the softer suspension, which creates additional body roll. **Steering:** Precise and predictable steering. **Acceleration/torque:** The standard 2.0L 4-cylinder engine is acceptable for city driving and leisurely highway cruising with the manual gearbox, thanks mainly to the car's light weight and improved fuel injection. The 180-hp 4-cylinder and V6 are the prerequisite engines for performance thrills. **Routine handling:** Excellent handling. Ride quality is less firm than with previous models. **Braking:** Acceptable (100–0 km/h: 40m).

Con: The base 4-cylinder engine provides sluggish acceleration when hooked to an automatic transmission. Excessive torque steer with the V6. Diesel engines equipped with cruise control can't handle small hills very well, and usually drop 10–15 km/h. **Transmission:** Manual 5-speed transmission is imprecise and requires long lever throws to change gear. Harsh automatic transmission downshifts at full throttle. The soft suspension produces lots of body roll when cornering under power.

Comfort/Convenience

Pro: TDI comes with impressive features like standard cruise control, side airbags, pretensioning seat belts, a full-sized spare, four-wheel disc brakes, ABS, tilt/telescoping steering column, heated mirrors, anti-theft alarm system, central power locking system, and air conditioning. **Driving position:** Excellent. Front seat headroom is adequate for most drivers, even with a sunroof. Ergonomics are impressive. **Controls and displays:** The dash is well laid-out and gauges are easy to read. Most controls are easy to find and use. **Climate control:** Generally works well and is easy to adjust. However, some owners have complained that heating is barely sufficient. Check this out on the dealer's lot before accepting delivery. **Interior space/comfort F/R:** Although the interior is austere, it's well finished and fairly spacious up front. The rear seats include headrests. **Cargo space:** The Golf's hatchback design is a bit more versatile for diverse cargo. **Trunk/liftover:** Spacious trunk has a low liftover.

Con: Standard equipment: Basic amenities are disappointing, such as the crank windows and cloth upholstery. Rear visibility is somewhat restricted. The front seats are too firm for some and the low-profile tires make small bumps unusually harsh. The rear seats are too narrow and there's not enough kneeroom. Radio and power window buttons are too small. The lack of a centre console cuts down on storage space. The Cabrio's trunk is too small. The odd way the trunk opens compromises loading. **Entry/exit:** A bit tricky. **Quietness:** Despite additional sound deadening, the TDI's engine is fairly noisy and tends to produce excessive vibrations.

Golf, Jetta

List Price (firm)	**Residual Values** (months) 24	36	48	60
Golf CL: $19,300 (14%)	$11,500	$9,500	$7,500	$6,000
Golf GL: $20,230 (14%)	$13,000	$11,000	$9,000	$7,500
Golf GL TDI: $22,290 (15%)	$14,500	$12,500	$10,500	$8,500
Golf GLS TDI: $24,720 (16%)	$16,500	$14,500	$12,500	$10,000
Golf GLS 2.0L: $23,010 (16%)	$15,500	$13,500	$11,500	$9,000
GTI 1.8T: $26,550 (18%)	$18,500	$16,500	$13,500	$12,000
VR6: $30,000 (19%)	$19,000	$17,500	$14,000	$12,500
Jetta GLS 1.8: $26,390 (18%)	$17,500	$15,500	$13,500	$11,000
Jetta GLS TDI: $26,080 (19%)	$19,500	$18,000	$14,500	$13,000
1.8T Wagon: $27,860 (19%)	$19,500	$18,000	$14,500	$13,000

Technical Data

Powertrain (front-drive)
Engines: 2.0L 4-cyl. (115 hp)
• 1.8LT 4-cyl. (150 hp)
• 1.8LT 4-cyl. (180 hp)
• 1.9L 4-cyl. diesel (100 hp)
• 2.8L V6 (201 hp)
• 3.2L V6 (237 hp)
Transmissions: 5-speed man.
• 6-speed man.
• 4-speed auto.
• 5-speed auto.
Dimension/Capacity (GTI VR6)
Passengers: 2/3
Height/length/width: 56.7/163.3/68.3 in.
Headroom F/R: 37.9/36.5 in.
Legroom F/R: 41.3/33.3 in.
Wheelbase: 98.9 in.
Turning circle: 36 ft.
Cargo volume: 18 cu. ft.
Tow limit: 1,000 lb.
Fuel tank: 55L/reg.
Weight: 2,723 lb.

Safety Features/Crashworthiness

	Std.	Opt.
Anti-lock brakes	■	—
Seat belt pretensioners F/R	■	—
Side airbags	■	—
Traction control	■	■
Head restraints F/R		
Golf 2d	*****	*****
4d	***	***
Jetta	***	**
Visibility F/R	*****	**
Crash protection (front) D/P	*****	*****
Crash protection (side) F/R		
Jetta	****	****
Crash protection (offset)	*****	—

PASSAT ★★★★

RATING: Above Average, mainly for performance, not quality control. **Strong points:** Lively acceleration with the manual transmission; refined road manners; sophisticated, user-friendly all-wheel drive; no turbo lag; quiet running; plenty of passenger and cargo room; and exceptional driving comfort. Well appointed and holds its value well. **Weak points:** Mediocre acceleration (automatic transmission-equipped GLS 1.8T and TDI), rear corner blind spots, many safety and performance complaints (faulty airbags and transmissions, distorted windshields, and chronic stalling); high-priced; and poor fuel economy. **New for 2004:** Passat gets a 134-hp 2.0L turbocharged diesel that can be coupled to either a front-drive, or an all-wheel-drive powertrain.

OVERVIEW: Volkswagen's largest front-drive compact, the Passat is an attractive mid-sized car that rides on the same platform as the Audi A4, giving it an 8-cm (3-in.) larger wheelbase and more passenger room. It has a more stylish design than the Golf or Jetta, but still provides a comfortable, roomy interior and gives good all-around performance for highway and city driving. The car's large wheelbase and squat appearance give it a massive, solid feeling, while its aerodynamic styling makes it look sleek and clean. Most Passats come fully loaded with air conditioning, tinted glass, power-assisted disc brakes on all four wheels, front and rear stabilizer bars, full instrumentation, and even a roof rack with the wagon. Engine offerings include a new 170-hp 1.8L turbocharged inline four, a 190-hp 2.8L V6, and a 270-hp 4.0L hooked to either a 5- or 6-speed manual or a 5-speed Tiptronic automatic transmission.

Last year's late arrival of the W8 sedan and wagon ratchets performance up another notch. It harnesses a powerful 8-cylinder engine to a 5- or 6-speed manual transmission and VW's 4Motion all-wheel drive.

Cost analysis/best alternatives: Get a practically identical, discounted 2003 model, unless you want the new turbodiesel. Other cars worth considering are the BMW 3 Series, Nissan Altima, and Toyota Camry. **Options:** The 2.8L V6 and a good anti-theft system. The AWD is an excellent investment, if you need the extra surefootedness and traction. **Rebates:** Not likely. **Delivery/PDI:** $1,205. **Depreciation:** Slower than average. **Insurance cost:** Higher than average. These cars are favourites with thieves—whether for radios, wheels, VW badges, or the entire car. **Parts supply/cost:** Not hard to find. Parts and service are much more expensive than average. **Annual maintenance cost:** Higher than average. **Warranty:** Bumper-to-bumper 2 years/40,000 km; powertrain 5 years/80,000 km; rust perforation 12 years/unlimited km. **Supplementary warranty:** A must-have. Maintenance costs are higher than average once the warranty expires. **Highway/city fuel economy:** *1.8L:* 7–11.4L/100 km; *2.8L:* 7.5–13.3L/100 km; *2.8L Synchro:* 8.2–13.7L/100 km.

Quality/Reliability/Safety

Pro: Warranty performance: Acceptable, until the faulty coil packs came to light. VW staffers in the States (there's no office in Canada) have been fairly sensitive to Passat complaints. Still, I'm bothered by a company that doesn't feel its customers merit a regional office. Hell, Suzuki and Hyundai managed it.

Con: Quality control: German build quality has been seriously overrated. Can you believe the company can't make distortion-free windshields for a $30,000–$45,000 car? **Reliability:** Predicted below-average reliability during the first three years, judging by NHTSA database entries. The tranny complaints are all the more annoying because VW had serious Passat transmission problems in the early '90s and assured me then that the defects had been remedied at the factory. **Owner-reported problems:** An incredible number of automatic transmission malfunctions, breakdowns, and early replacements. Transmission vibration and noise; premature CV joint failure; oil pan is easily punctured due to low ground clearance; and CV boot wears out prematurely. Some electrical and fuel system glitches cause chronic stalling; ignition coils still seem to be a primary cause of stalling and loss of engine power. Gas pedal remained stuck to the floor. Brakes don't grab as well when vehicle is cold, premature brake wear, and noisy braking. Poor fit and finish highlights: sunroof rattles; driver's seat memory feature fails; front spoiler and rear trim fall off; distorted windshields; and heated seats that are a pain in the...well, you know. Seat belt latch doesn't hold the metal tongue. **Service bulletin problems:** Updates for the ECM and TCM modules. **NHTSA safety complaints/safety:** Several reports that fire ignited in the engine compartment; excess raw fuel flows out of the exhaust system; hard starts and chronic stalling; Check Engine light comes on intermittently and then engine shuts down; while cruising, vehicle speeds up, when brakes applied, it slows down until foot is taken off the brake, when it surges again—one owner describes it this way:

> Dangerous situation. When accelerating from a complete stop vehicle does one of three things: 1) gasps for gas and goes nowhere (dangerous when making left-hand turns); 2) tries to move ahead like a carbureted car with vapour lock; or 3) performs normally.

Vehicle runs out of fuel despite the fuel gauge showing a quarter tank of gas; braking doesn't disengage cruise control; hesitation, long delays when accelerating; automatic transmission suddenly drops out of gear; airbags failed to deploy or deploy for no reason. One VW employee told U.S. federal investigators he was fired for complaining about the airbag hazard:

> The driver's side head airbag (air curtain) of a 2003 Volkswagen Passat W8 sedan deployed spontaneously while I was driving the car.... A few minutes later, when the car was stopped, the steering wheel airbag deployed spontaneously.... I suffered a permanent wrist injury and am suffering from post-traumatic stress syndrome.... The incident, which

> happened during a test drive, was reported to the management of the VW dealership for which I was working and to VW of America by the management.... I sent a detailed incident report to both VW of America [VWOA] and VW AG (Germany) CEOs.... A VW customer service representative admonished me to not make a fuss about the incident, saying it was a once-in-a-lifetime occurrence unlikely to happen again, and advised me that my letter to the VWOA CEO had been received but that he had not read it.... The owner of the dealership advised me in a special meeting that he had been contacted by a VWOA management representative, who was unhappy with the complaints I had made to the customer service representative about the lack of follow-up by VW and the means I was considering of alerting VW owners about the incident.... I left the meeting believing my job and future in the car business, at least selling VW products, would be in jeopardy if I took any affirmative action such as filing this complaint. A promise that I would be informed by VWOA of the reason for the spontaneous deployment has not been fulfilled and I have not been contacted by any VW representative in almost four months.... Recently, I was fired because of an incident I consider related directly to the incident.... I am filing this report because I have just learned that another person has, reportedly, experienced the spontaneous deployment of an airbag in a VW car and I am now confident that what I experienced was not a freak occurrence....

Airbag warning light stays lit; many complaints of windshield distortion (there's an accordion effect where letters and objects expand and contract as they pass by); passenger window suddenly exploded just after being rolled up; windshield wipers cut out; plastic engine nose shield fell off; rear tire failure damaged the fuel-filler neck, causing a fuel leak; and super-heated seats.

Road Performance

Pro: Acceleration/torque: Decent acceleration with the base 1.8L turbocharged engine, unless saddled with an automatic transmission. Better acceleration times with more torque can be wrung from the V6 (0–100 km/h: 7.9 seconds). Diesel power is relatively quiet. **Transmission:** Great performance with the manual gearbox. Smooth and quiet shifting with the automatic gearbox. The 4Motion full-time all-wheel drive shifts effortlessly into gear. **Emergency handling:** Better than average; impressive, with the AWD system; no turbo lag. **Steering:** Quick, precise, and predictable. **Routine handling:** Suspension is both firm and comfortable; precise handling outclasses most of the competition.

Con: The 4-cylinder engine power is seriously compromised by the automatic transmission. **Braking:** Less than impressive, with some brake fade after successive stops.

Comfort/Convenience

Pro: Standard equipment: This is VW's most luxurious car, so it comes fairly well appointed. Heated outside mirror. Blue instrumentation lighting is a nice touch. **Driving position:** Very good. Comfortable seating and good fore and aft visibility. Fairly good dashboard layout. **Controls and displays:** User-friendly instrument panel. **Climate control:** Efficient and quiet operation. Includes a dust and pollen filter. **Entry/exit:** Relatively easy front and rear access. **Interior space/comfort F/R:** Spacious interior seats four in comfort, as long as the rear passengers aren't too tall. All seats are supportive and comfortable. **Cargo space:** Average for the sedan. Many small storage spaces. **Trunk/liftover:** Large and accessible trunk has an innovative latch and low liftover. **Quietness:** Additional sound-deadening material keeps interior noise to a minimum.

Con: The styling of the rear pillars compromises rear visibility. Cheap-looking cupholders and ugly chrome VW logo on the steering wheel. Rear seatbacks don't fold for added storage and rear seat isn't wide enough for three-abreast seating. The 4X4 option reduces trunk space by a third.

Passat

List Price (firm)	**Residual Values** (months)			
	24	**36**	**48**	**60**
Passat GLS: $29,550 (21%)	$22,000	$18,000	$14,000	$11,500
GLS V6: $33,050 (22%)	$24,500	$21,500	$16,500	$13,500
GLX V6 Wagon: $42,905 (24%)	$30,500	$26,500	$21,500	$17,000

Technical Data

Powertrain (front-drive/AWD)
Engines: 1.8L 4-cyl. (170 hp)
• 2.0L TD 4-cyl. (134 hp)
• 2.8L V6 (190 hp)
• 4.0L V6 (270 hp)
Transmissions: 5-speed man.
• 6-speed man.
• 5-speed auto.

Dimension/Capacity
Passengers: 2/3
Height/length/width: 57.4/184.1/68.5 in.
Headroom F/R: 39.7/37.8 in.
Legroom F/R: 41.5/35.3 in.
Wheelbase: 106.4 in.
Turning circle: 38 ft.
Cargo volume: 15 cu. ft.
Tow limit: 1,000 lb.
Fuel tank: 47L/prem.
Weight: 3,250 lb.

Safety Features/Crashworthiness

	Std.	Opt.
Anti-lock brakes	■	—
Seat belt pretensioners F/R	■	—
Side airbags	■	—
Traction control	■	—
Head restraints F/R	**	**
Visibility F/R	*****	**

Crash protection (front) D/P	*****	*****
Crash protection (side) F/R	****	****
Crash protection (offset)	*****	—

Volvo

Volvo has always distinguished itself from the rest of the automotive pack through its much-vaunted standard safety features, crashworthiness, and engineering that emphasized function over style (in fact, I've always seen it as the quintessential NDP car, but that's another story). But unfortunately, these noteworthy features were eclipsed by bland styling, ponderous highway performance, inconsistent quality control that compromised long-term reliability and drove up ownership costs, and chancy servicing by a small dealer network (hell, my NDP analogy isn't all that off-base). Furthermore, Asian carmakers have successfully encroached upon Volvo territory by bringing out new products that are as safe and comfortable to drive, with greater reliability thrown in.

Volvo has changed with the times. While the mainstream U.S. automakers go retro with creased, less rounded styling and dashes of chrome, Volvo has dumped its boxy station wagons and rediscovered rounded edges, all-wheel drive, and high-performance powertrains and handling. The automaker's curvy, AWD V70 XC is the latest example of a mindset change that is already in full swing with the company's R-designated S60, S70, S80, and V70, the C70 coupe, and smaller 40 Series cars.

Ford is counting on Volvo to garnish its empty corporate wallet with much-needed profits, but Volvo doesn't have the new products that will bring in that extra money. In fact, the company is going through one of its worst selling periods in over a decade. One gets the impression that Ford will milk the company for all it can get until it unloads the company in a "three-for-one" with Jaguar and Land Rover.

Furthermore, Ford's involvement may mean some sharing of components—a good idea for Jaguar and Rover, a bad idea for Volvo. Such a scheme will likely make Fords safer and more reliable and Volvos less so (I doubt Ford's biodegradable engine head gaskets or failure-prone automatic transmission and front suspension coils would have ever found their way onto a Volvo assembly line.)

Volvo quality control decline is worrisome (see NHTSA's online database). A perusal of Volvo owner comments shows serious safety- and performance-related factory defects that weren't seen in Volvo before it was being acquired by Ford.

Coincidence? I don't think so.

S40, V40 ★★★

RATING: Average. Volvo's smallest car. **Strong points:** Average acceleration; lots of innovative safety features; dual rear integrated child booster seats; good braking, when brakes are functioning normally (see NHTSA's online safety complaints/safety); exceptional steering and handling; easily accessed and functional instruments and controls; plenty of front room and comfortable front seating; large trunk has a low liftover; "Good" offset crash and head restraint rating; and average reliability. **Weak points:** Acceleration inferior to the Japanese competition; annoying turbo lag; suspension dampened on the soft side; some wandering in panic stops; engine exhaust boom and growl at high revs; rear seating only for two adults; when front seats are pushed back, forget about footroom and easy entry/exit; visibility obstructed by rear head restraints; trunk utility hampered by its small opening; serious safety-related complaints recorded by government investigators; premium fuel required for turbo engines; and an unacceptable $250 "retailer participation charge" for leases. **New for 2004:** Nothing significant. The big change occurs with the 2005 model, which will get many of the Mazda3's chassis components (not an entirely bad idea).

OVERVIEW: The S40/V40 ($31,495/$32,495, plus an unacceptably high $1,495 freight/PDI) are Volvo's latest small sedan and wagon that are built in a joint venture with Mitsubishi. Priced about $6,000 less than the V70 wagon, Volvo's entry-level model comes with a 1.8L 160-hp turbocharged 4-cylinder engine coupled to a 5-speed manual or automatic transmission. Side airbags, four-wheel disc anti-lock brakes, anti-whiplash front head restraints, AC conditioning, cruise control, and power windows are also standard.

Cost analysis/best alternatives: The 2004 model has nothing to justify a higher price; look for a steeply discounted 2003 version, instead. Other cars worth considering are the Acura TL, Audi A4, Infiniti 135, Lexus ES 300, and Toyota Camry. **Options:** Dual integrated child booster seats. **Rebates:** Look for additional price cuts in the new year as Ford scrambles to staunch Volvo's red ink and increase market share—$3,500 rebates, lots of low-interest financing, and advantageous leasing deals. **Delivery/PDI:** $1,495. **Depreciation:** Slower than average. **Insurance cost:** Higher than average. **Parts supply/cost:** Parts aren't hard to find, but may be much more expensive than average. **Annual maintenance cost:** Predicted to be higher than average due to small dealer network. **Warranty:** Bumper-to-bumper 2 years/40,000 km; powertrain 5 years/80,000 km; rust perforation 12 years/unlimited km. **Supplementary warranty:** A must-have in view of the car's unproven history and flurry of consumer complaints. **Highway/city fuel economy:** 6.8–10.5L/100 km.

Service bulletin problems: Moisture in side-impact sensors; transmission clunking; and troubleshooting tips to detect and remedy oxygen sensor faults. **NHTSA safety complaints/safety:** *S40:* Under-hood electrical fire; cracked fuel regulator pump spilled fuel onto hot engine and spread fumes into the interior; sudden, unintended acceleration when vehicle put into Drive; chronic stalling attributed to faulty idle control valve and air mass meter; complete loss of braking; brake pedal hard to depress, reducing brake effectiveness; brakes don't stop vehicle in a reasonable distance; brake pedal is too close to the gas pedal; brake pedal snapped, went to the floor while going downhill; when applying the brakes in cold weather, pedal won't depress, causing extended stopping distance (dealer confirmed vacuum pump motor was defective); vehicle pulls to the left when accelerating or coming to a stop; premature replacement of the front and rear rotors and pads; repeated automatic transmission failures that often begin with slippage from Second to Third gear; airbag light stays lit; faulty forward/backward seat adjustment; and noisy engine and sunroof. *V40:* More complaints that the pedal won't depress, causing total brake loss or extended stopping distance and premature wearout of the front brake pads (around 20,000 km).

S60 ★★★

RATING: Average. Fewer owner complaints, but performance is outclassed by the competition. **Strong points:** Fair acceleration; exceptional handling and braking; good array of user-friendly instruments and controls; comfortable front seating; very good head restraints, offset, front and side crashworthiness scores; and predicted better-than-average reliability. **Weak points:** Lots of turbo throttle hesitation and torque steer pulling when accelerating; imprecise manual shifter; ride is a bit jarring with some tire thump (worse with the T5) when passing over uneven pavement; rear visibility obstructed by high parcel shelf, obtrusive head restraints, and descending roofline; rear room adequate only for two adults and rear legroom disappears when front seats are pushed back only halfway; narrow trunk with a small opening; and turbo engines require premium fuel. **New for 2004:** The S60R, a 300-hp, high-performance AWD spin-off.

OVERVIEW: A sporty mid-range sedan ($36,495, plus an exorbitant $1,495 freight/PDI and bogus $250 leasing charge) that uses the same large car platform as the V70 and S80, but provides more interior room. A 197-hp 5-cylinder engine teamed to an all-wheel-drive powertrain has been added to the lineup, giving buyers AWD benefits without having to choose the taller, more SUV-like Cross Country. Drivers will have the choice of three inline 5-cylinder engines: a 168-hp naturally aspirated version, a 197-hp low-pressure turbo, and a 247-hp high- pressure turbo. Other cars worth considering are the Acura TL, Audi A6, Infiniti 135, and Lexus ES 30. Leftover S60s have been deeply discounted, particularly in view of Volvo's declining sales.

Service bulletin problems: Transmission slippage; silencing wheel resonance; and AWD squeaking. **NHTSA safety complaints/safety:** Right front wheel fell off; sudden stalling; ABS brakes don't work properly; and Pirelli tire blowouts.

C70 ★★★★

RATING: Above Average. **Strong points:** Strong acceleration with lots of torque; exceptional steering and handling slower-than-average depreciation; "Good" head restraint rating; and surprisingly better-than-average reliability. **Weak points:** Engine turbo lag; excessive engine noise; difficult rear-seat entry/exit; no crashworthiness test results; and a jarring suspension. **New for 2004:** Nothing significant; the coupe was dropped late last year.

OVERVIEW: This $63,995 (plus $1,495 freight/PDI) HT luxury convertible seats four comfortably. It is based on the S70 platform, and marketed primarily to high-performance enthusiasts. There are two turbocharged engines offered: a base 2.4L 190-hp inline 5-cylinder and a 2.3L 236-hp variant. Either engine can be hooked to a 5-speed manual or a 5-speed automatic transmission. Of the two engines, the 190-hp appears to offer the best response and smoothest performance.

Acceleration is impressive, despite the fact that the car feels underpowered until the turbo kicks in at around 1500 rpm—a feature that drivers will find more frustrating with a manual shifter than with an automatic. Steering and handling are first class, fit and finish above reproach, and mechanical and body components have generated the neither the number nor the kindss of complaints seen with other Volvo models.

The only things not to like are a high base price, turbo lag, tire thumping caused by the high-performance tires, excessive engine and wind noise, and power-sliding rear seats that require lots of skill and patience.

Safety features include dual driver/passenger airbags, side airbags, ABS, traction control, and a platform designed to give maximum passenger protection in a collision. IIHS has awarded the C70 its highest rating for front and rear seat head restraint protection.

Other cars worth considering are the Acura CL, and BMW 3 Series. As with other higher-end Volvos, expect deep discounting on both 2003 and 2004 models early in the new year.

Service bulletin problems: Transmission slippage and AWD squeaking. **NHTSA safety complaints/safety:** N/A.

V70, XC70, XC90 ★★★★

RATING: Above Average. **Strong points:** Acceleration (T6); practical to the extreme; good handling and braking; lots of cargo room; well-designed instruments and controls are easy to read and access; many standard safety features; and a "Good" head restraint rating. **Weak points:** Some torque steer; a jarring ride with vehicles equipped with 16- and 17-inch wheels; limited rear visibility; excessive engine, wind, and road noise; confusing navigation system controls; fuel-thirsty (turbo models); a number of safety- and performance-related factory defects reported to the U.S. government, notably chronic stalling and poor windshield visibility; head restraints obstruct rear visibility; and no crashworthiness ratings. **New for 2004:** The high-performance R version is added to the V70 lineup.

OVERVIEW: The quintessential Volvo wagon and Volvo's best-selling line, there are two models available: a base front-drive and an AWD; sporty T6 versions employ front-wheel drive. Front-drives and AWD use a 2.5L 208-hp 5-cylinder engine hooked to a 5-speed automatic. As expected with Volvo, these cars are loaded with safety and convenience features that include four-wheel disc brakes, head/chest front and side airbags, and high-tech seatbacks designed to minimize whiplash. Three powertrains are offered: a 168-hp 2.5L 5-cylinder, the aforementioned 208-hp variant, and a 236-hp 2.9 L V6 equipped with a high-pressure turbocharger.

XC70, XC90

Essentially a renamed V70 station wagon, the XC70 is a SUV wannabe that offers five-passenger seating, high ground clearance, sleek styling, and AWD versatility—all for $49,495. Its mechanical components are practically identical to those of Volvo's other sedans. The base engine is a 168-hp (non-turbo) 2.4L 5-cylinder. The turbo version has been replaced by a 208-hp 2.5L turbocharged 5-cylinder hooked to a 5-speed automatic gearbox. It is followed by the 247-hp 2.5L turbocharged T5 and a 300-hp variant found on the V70 R. XC90 models start at $54,995 and offer seven-passenger seating, along with a base turbocharged 208-hp 2.5L 5-cylinder and an optional 268-hp 2.9L V6 on the $59,995 T6 version. **New for 2004:** The 197-hp 2.4T has been dropped. Formerly known as the V70 XC, its only new features this year consist of upgraded rack and pinion steering and optional bi-xenon headlights. The XC90 returns with new alloy wheels.

Cost analysis/best alternatives: Discounted 2003s are the best choice, since they are practically identical to the 2004 version. Other cars worth considering are the Acura TL, Audi A4 or A6/allroad Quattro, Infiniti I35, or the Lexus ES 300. Also take a look at the BMW 5 Series wagon. **Options:** Integrated child safety seats and a full-sized spare tire. The turbo's traction control isn't worth the extra cost. Also, take a pass on the complex Navigation System option. **Rebates:** Look for many low-financing and discounting programs to kick in before year's end. **Delivery/PDI:** $1,495. **Depreciation:** Slower than average. **Insurance cost:** Higher than average. **Parts supply/cost:** Parts are highly dealer-dependent and moderately expensive, but they aren't hard to find because of the many V70s that were sold. XC70s may have back-ordered drivetrain components. **Annual maintenance cost:** Average; higher than average for the XC. Higher-than-average cost predicted for both models after their fourth year on the road. **Warranty:** Bumper-to-bumper 4 years/80,000 km; powertrain 4 years/80,000 km; rust perforation 8 years/unlimited km. **Supplementary warranty:** Recommended. The most frequent complaint concerns the brakes, an item excluded from most supplementary warranties. **Highway/city fuel economy:** *2.4L:* 7.8–11L/100 km; *2.4L turbo:* 8–11.8L/100 km; *XC:* N/A.

Quality/Reliability/Safety

Pro: Reliability: Fewer reliability problems reported than with Volvo's other models; still, chronic stalling complaints are worrisome. **Warranty performance:** Much better than average. Many warranty claims are settled through "goodwill" on a case-by-case basis. **Safety:** Standard side-impact airbags and standard traction control are a nice touch. Other safety features include rear head restraints; reinforced anti-roll bars; front and rear crumple zones; a practical, roof-mounted interior cargo net that protects passengers from being hit by objects stored in the rear; and rear three-point seat belts. As with Volvo's other models, IIHS has awarded the V70 its highest rating for front- and rear-seat head restraint protection.

Con: Body assembly and paint quality are better than average, but they can't match the Japanese competition. **Owner-reported problems:** Problem areas are limited to frequent brake maintenance (rotors and pads), chronic stalling, and electrical system and body faults, notably excessive windshield-dash glare. **Service bulletin problems:** Rough cold starts; long cranking time; tachometer jump; and engine warning light activated. **NHTSA safety complaints/safety:** Engine repeatedly stopped in traffic due to what one Volvo insider called a "weak" fuel pump that is said to affect many V70 and S80 models; same problem blamed on debris in the fuel line; vehicle runs out of gas, despite gauge showing sufficient fuel left in tank; electronic throttle hampered by a delay in fuel getting into the engine; many complaints that forward visibility is seriously compromised by the dash reflecting onto the windshield (cream-beige colour the worst offender); car rolled forward down an incline despite being put into Reverse; brake and gas pedal are mounted too close together; front doors will slam shut on a moderate slope; coffee spilled from cupholder onto airbag computer module, causing huge expense; seat belt crosses at the neck.

Road Performance

Pro: Acceleration/torque: Plenty of high-range power with the base engine, especially with a manual gearbox. With an automatic transmission, the normally aspirated base engine has a 0–100 km/h time of 9.5 seconds. No turbo lag. The 197-hp engine posted better-than-average acceleration times (0–100 km/h: 8.8 seconds) with plenty of torque. **Transmission:** Smooth and quiet automatic and manual gearboxes. **Emergency handling:** Better than average, with minimal body roll and good control. **Steering:** Predictable, rapid steering response. Handles sudden steering corrections very well. **Routine handling:** Nimble handling doesn't sacrifice passenger comfort. **Braking:** Braking performance is quite good (100–0 km/h: 35 m) on dry or wet pavement.

Con: Moderate torque steer when accelerating. The soft ride deteriorates progressively as the road gets rougher and passenger weight is added.

Comfort/Convenience

Pro: Standard equipment: Lots of standard safety and performance features. **Driving position:** Excellent driving position and plenty of seat adjustments to accommodate almost any size of driver. **Controls and displays:** Everything is in plain sight, accessible, and intuitive. **Climate control:** Excellent. The system operates flawlessly and is simple to comprehend. **Entry/exit:** Easy front and rear access. **Interior space/comfort F/R:** Passenger space, seating comfort, and trunk and cargo space are unmatched by the competition. **Cargo space:** The wagon's rear seatbacks fold flat and the seat cushion can be removed for additional cargo space. **Trunk/liftover:** Spacious trunk has a low liftover.

Con: Rear head restraints limit rear visibility. **Quietness:** Excessive engine noise upon acceleration, as well as some wind and tire noise at highway speeds.

V70, V70XC

List Price (firm)	**Residual Values** (months) 24	36	48	60
V70 2.4: $37,995 (22%)	$27,000	$22,000	$17,000	$13,000
XC70: $49,495 (25%)	$35,000	$30,000	$25,000	$18,000
V70 T5: $47,995 (25%)	$33,000	$28,000	$23,000	$17,000
XC90 T6: $61,995 (25%)	$43,000	$35,000	$27,000	$23,000

Technical Data

Powertrain (front-drive/AWD)
Engines: 2.4L 5-cyl. (168 hp)
• 2.5L 5-cyl. turbo (208 hp)
• 2.5L T5 5-cyl. turbo (247 hp)
• 2.5L T5 5-cyl. turbo (300 hp)
• 2.9L T6 V6 turbo (268 hp)
Transmissions: 5-speed man.
• 5-speed auto.
Dimension/Capacity
Passengers: 2/3
Height/length/width: 58.6/185.4/71 in.
Headroom F/R: 39.3/38.9 in.
Legroom F/R: 42.6/35.2 in.
Wheelbase: 108.5 in.
Turning circle: 40 ft.
Cargo volume: N/A
Tow limit: 2,000 lb.
Fuel tank: 73L/prem.
Weight: 3,366 lb.

Safety Features/Crashworthiness

	Std.	Opt.
Anti-lock brakes	■	—
Seat belt pretensioners F/R	■	—
Side airbags	■	—
Traction control	■	—
Head restraints F/R	*****	*****
Visibility F/R	*****	**

APPENDIX I

Useful Internet Sites

You can find lots of information about new cars and minivans on the Internet—but it may not be true, or complete. Automobile companies have their own self-serving websites that, nevertheless, feature details on suggested retail prices and specifications. Their sites can easily be accessed through Google's browser under the automaker's name followed by "Canada." Sometimes, the manufacturer's name followed by ".com" or ".ca" will also work. For extra fun and a more balanced presentation, put in the car model or manufacturer's name, followed by "lemon."

Consumer Protection

Automobile Consumer Coalition (*www.carhelpcanada.com*)
Founded by the former director of the Toronto Automobile Protection Association, Mohamed Bouchama, the ACC's Car Help Canada website provides many of the same services as the APA; however, it is especially effective in Ontario and Alberta.

Automobile Protection Association (*www.apa.ca*)
This Montreal-based consumer group fights for safer vehicles and has exposed many scams associated with new-vehicle sales, leasing, and repairs; for a small fee, it will send you the invoice price for most new vehicles.

BBC TV's *Top Gear* Car Reviews (*www.topgear.beeb.com*)
Britain's automotive equivalent to Canada's CBC *Marketplace*, *Top Gear* blows the whistle on the best and worst European-sold vehicles, auto products, and industry practices.

Big Class Action (*www.bigclassaction.com/automotive.html*)
This is a useful site for using a company's class action woes in U.S. jurisdictions for leverage in settling your own claim. Plus, if you decide to go the Canadian class action route, most of the legal legwork will have been done for you. Lexus transmissions and Chrysler engine head gaskets are two of many actions listed.

Canadian Automobile Association (*www.caa.ca*)
This bilingual AARP on wheels features a nice checklist for "Keeping Track of Your Own Vehicle Costs," at *www.caa.ca/e/automotive/pdf/driving-costs-03.pdf.*

Canadian Driver (*www.canadiandriver.com*)
An exceptionally well-structured and current Canadian website for new- and used-vehicle reviews, MSRPs, and consumer reports. Other car magazine websites:

- *World of Wheels* (*www.autonet.ca*)
- *Automotive News (www.automotivenews.com)*
- *Car and Driver* (*www.caranddriver.com*)
- *Motor Trend (www.motortrend.com)*
- *Road and Track (www.roadandtrack.com)*

CBC *Marketplace* (*cbc.ca/consumers/market/files/cars/index.html*)
Marketplace has been the Canadian Broadcasting Corporation's premier national consumer show for almost three decades. Its site has extensive links and in-depth reports on Chrysler paint delamination, ABS brake failures, airbag dangers, and a host of other automotive topics.

ConsumerAffairs.com
(*www.consumeraffairs.com/automotive/manufacturers.htm*)
Expecting some namby-pamby consumer affairs site? Won't find that here. A "seller beware" kind of website, where you'll find the scandals before they hit the mainstream press.

***Consumer Reports* and Consumers Union** (*www.consumerreports.org*)
It costs $4.95 a month to subscribe online, but *CR*'s database is chock full of comparison tests and in-depth stories on products and services.

Metro Credit Union (*www.metrocu.com/services/autoservice.html*)
A unique credit union that puts a car broker at its members' disposal to negotiate new- or used-vehicle sales.

Protect Yourself/Protegez-Vous (*www.protegez-vous.qc.ca*)
Quebec's French-language monthly consumer protection magazine and website is a hard-hitting critic of the auto industry. It contains dozens of test-drives and articles relating to a broad range of products and services sold in Canada.

Supreme Court of Canada (*www.lexum.umontreal.ca/csc-scc/en/rec/index.html*)
It's not enough to have a solid claim against a company or the government. Using a Supreme Court decision to support your position also helps. There were three pro-consumer judgments rendered in February 2002 that are particularly useful:

- *Bannon v. The Corporation of the City of Thunder Bay.* An injured resident missed the deadline to file a claim against Thunder Bay; however, the Supreme Court maintained that extenuating factors, such as being under the effects of medication, extended her time to file. A good case to remember next time your vehicle is damaged by a pothole or you are injured by a municipality's negligence.

- *R. v. Guinard.* An insured posted a sign on his barn claiming the Commerce Insurance Company was unfairly refusing his claim. The municipality of St-Hyacinthe, Quebec, told him to take the sign down. He refused, maintaining that he had the right to state his opinion. The Supreme Court agreed. This judgment means that consumer protests, signs, and websites that criticize the actions of corporations cannot be shut up or taken down simply because they say unpleasant things.
- *Whiten v. Pilot Insurance Co.* The insured's home burned down and the insurance company refused to pay the claim. The jury was outraged and ordered the company to pay the $345,000 claim, plus $320,000 for legal costs and $1 million in punitive damages, making it the largest punitive damage award in Canadian history. The Supreme Court maintained the jury's decision, calling Pilot "the insurer from hell." This judgment scares the dickens out of insurers, who fear that they face huge punitive damage awards if they don't pay promptly.

Auto Safety

Airbag Killer Sites (*www.plescia.org/indexair.htm*)
Anecdotal evidence of airbag injuries and fatalities in low-speed collisions, as well as a compendium of independent research studies.

Center for Auto Safety (*www.autosafety.org*)
A Ralph Nader–founded agency that provides free online info on safety- and performance-related defects on each model vehicle.

Crashtest.com (*www.crashtest.com/netindex.htm*)
A website where crash tests from around the world can be analyzed and compared.

Insurance Institute for Highway Safety (IIHS)
(*www.hwysafety.org*)
A dazzling site that's long on crash photos and graphs that show which vehicles are the most crashworthy in side and offset collisions and which head restraints work best.

The Safety Forum (*www.safetyforum.com*)
The Forum contains comprehensive news archives and links to useful sites, plus names of court-recognized experts on everything from unsafe Chrysler minivan latches to dangerous van conversions.

Strategic Safety (*www.strategicsafety.com/mainindex.html*)
Brought to you by the people who blew the whistle on dangerous Firestone tires. Provides research, investigation, analysis, and education on safety issues.

Toyota Celica Page (*celica.net/main.asp*)
An excellent site for owners' unbiased reviews and technical updates.

Transport Canada (*www.tc.gc.ca/roadsafety/Recalls/search_e.asp*)
You can access the recall database for 1970–2003 model vehicles, but, unlike NHTSA's website (see following), owner complaints aren't listed, defect investigations aren't disclosed, and service bulletin summaries aren't provided. A list of used vehicles admissible for import is available at *www.tc.gc.ca/roadsafety/importusa/impusae.htm*, or by calling the Registrar of Imported Vehicles at 1-888-848-8240.

U.S. National Highway Traffic Safety Administration (NHTSA)
(*www.nhtsa.dot.gov/cars/problems*)
This site has a comprehensive, free database covering owner complaints, recall campaigns, crashworthiness and rollover ratings, defect investigations, service bulletin summaries, and safety research papers.

Mediation/Protest

Roadfly's BMW, Mini Cooper, and Porsche Message Boards
(*www.roadfly.org/forums*)
Another site that's no butt kisser. Here you'll learn about BMW fan fires, upgrades, and performance comparisons.

Allpar: Chrysler, Plymouth, and Dodge Car Information (*www.allpar.com*)
This is an excellent website that's jam-packed with historical information, tips on fixing common problems inexpensively, and advice on how to deal with Chrysler representatives and dealer service managers.

Chrysler Products' Problem Web Page
(*www.wam.umd.edu/~gluckman/Chrysler*)
A resource for Chrysler owners who have had problems in dealing with Chrysler, including issues with peeling paint, transmission failure, the Chrysler-installed Bendix-10 ABS, and other maladies.

DaimlerChrysler Problems Web Page
(*www.daimlerchryslervehicleproblems.com*)
This is a great site for technical info and tips on getting action from DaimlerChrysler.

Ford Insider Info (*www.blueovalnews.com*)
Pssst! Don't tell Ford (they've tried twice to shut down this site). Here's where you get all the latest insider info on Ford's quality problems and plans for future models.

Neon Enthusiasts (*www.neons.org*)
Has lots of technical info and troubleshooting tips. Includes service bulletins filed according to the component failure. Chrysler's head gasket service bulletin can be downloaded to buttress your small claims court claim for a free engine repair involving any Chrysler 4-cylinder engine.

VW Lemon Page (*www.myvwlemon.com*)
A bit biased, but lively and informative discussions of VW's quality problems. Another good site is *www.suckercars.com.* It features links to many VW, Audi, and other protest sites. Incidentally, this was the first site in U.S. history to win an anti-cybersquatting lawsuit lodged by an automaker.

Information/Services

Alberta Vehicle Cost Calculator
(*www1.agric.gov.ab.ca/app24/costcalculators/vehicle/getvechimpls.jsp*)
Estimate and compare the ownership and operating costs of vehicles with variations in purchase price, options, fuel type, interest rates, or length of ownership.

ALLDATA Service Bulletins (*www.alldata.com/recalls/index.html*)
Free summaries of automotive recalls and technical service bulletins are listed by year, make, model, and engine option. You can access your vehicle's bulletins online by paying a $25 (U.S.) subscription fee.

The Auto Channel (*www.theautochannel.com*)
This website gives you useful, comprehensive information on choosing a new or used vehicle, filing a claim for compensation, or linking up with other owners.

The Auto Extremist (*www.autoextremist.com*)
Rantings and ravings from a Detroit insider.

Automobile News Groups
These news groups are compilations of email raves and gripes and cover all makes and models. They fall into four distinct areas: *rec.autos.makers.chrysler* (you can substitute any automaker's name at the end); *rec.autos.tech*; *rec.autos.driving*; and *rec.autos.misc*.

Autopedia (*autopedia.com/index.html*)
An automotive encyclopedia, Autopedia offers a compendium of automotive-related information. Its legal section and lemon law listings by state are particularly helpful.

Canadian Driver (*www.canadiandriver.com*)
An exceptionally well-structured and current Canadian website for new- and used-vehicle reviews, MSRP prices, and consumer reports prepared by Canadian journalists.

Carcostcanada.com (*www.carcostcanada.com/en/*)
Canadian shoppers can finally get new vehicle invoice prices at a reasonable cost (about $20). This agency also supplies info on rebates, incentives, transport fees, and option prices on virtually every new vehicle sold in Canada.

Carfax (*www.carfax.com*)
Use Carfax (Tel.: 1-888-422-7329) to see if a vehicle has been "scrapped," had flood damage, is stolen, or had its mileage turned back. There's a fee of $14.95 (U.S.) if the order is placed via the Internet.

Cartrackers (*www.cartrackers.com*)
Used cars, consumer advice, and environmental issues are all well covered in this site, which features a terrific auto image gallery and an excellent automotive glossary.

Kelley Blue Book and Edmunds (*www.kbb.com*; *www.edmunds.com*)
Prices and technical info is American-oriented, but you'll find good reviews of almost every vehicle sold in North America, plus an informative readers' forum.

Metric Conversion Online (*www.sciencemadesimple.net/conversions.html*)
A great place to instantly convert gallons to litres, miles to kilometres, etc.

Phil Bailey's Auto World (*www.baileycar.com*)
Phil Bailey owns his own garage and specializes in the diagnosis and repair of foreign cars, particularly British ones. He's been advising Montreal motorists for years on local radio shows and has an exceptionally well-written and comprehensive website.

Straight-Six.com (*www.straight-six.com/theCrank/index.htm*)
Okay, for you high-performance aficionados, here's a website that doesn't idolize NASCAR, Earnhardt, or the Porsche Cayenne SUV (they call it the Ca-Yawn).

Vehicle Information Centre of Canada (VICC)
(*www.vicc.com/english/MeasureUp.htm*)
Part of the Insurance Bureau of Canada. VICC's "How Cars Measure Up" compares different model years' insurance claims experience for collision, personal injury, and theft losses.

Woman Motorist (*www.womanmotorist.com/index.php/welcome*)
Not for women only. This site's news and reviews are fresh and pertinent—and yes, women do sweat the details.

Women's Garage (*www.womensgarage.com*)
Three Canadian mechanics with a combined 100 years' experience set up this site to take the mystery out of maintaining and repairing vehicles. Don't be confused by the site title—males will learn more than they'll care to admit.

Finally, here are a number of other websites that may be helpful:

- *www.bmwboard.com*
- *www.bmwnation.com*
- *www.carforums.com/forums*
- *www.consumeraffairs.com*
- *www.consumeraffairs.com/automotive/ford_transmissions.htm*
- *www.datatown.com/chrysler*
- *www.epa.gov/otaq/consumer/warr95fs.txt*
- *www.flamingfords.com*
- *everythingfordrivers.com/carforums.html*
- *www.ford-trucks.com/forums*
- *forum.freeadvice.com*
- *www.hotbimmer.net*
- *us.lexusownersclub.com*
- *forums.mbnz.org*
- *www.troublebenz.com/my_opinion/actions/links.htm*
- *www.ptcruiserclub.org*
- *www.ptcruiserlinks.com*
- *www.thesambaforums.com/forum*
- *www.vehicle-injuries.com/suv-safety-news.htm*
- *www.worktruck.com*

APPENDIX II

Good Cars for a Tight Budget

These are volatile times, with the price of cars, fuel, and insurance going through the roof; and now's not the time to invest heavily in a vehicle that's more than you need or can afford. The following new- and used-vehicle choices will help you hunker down and get reliable, safe, fuel-efficient, and inexpensive wheels that can serve as your second car, get you to school and back, or help you through that first job. Once you get on a sound financial footing, you'll get that "dream machine" you've always wanted.

First, consider the following 10 rules:

1. Delay your purchase until mid-2004, when you can double-dip automaker and dealer clearance rebates.
2. Used is always a better buy than new—little depreciation or up-front costs.
3. Buy a Japanese/American co-venture vehicle, like the Toyota/GM Matrix/Vibe.
4. Search out five- to 10-year-old Japanese compacts or Hyundai models not older than three years.
5. Buy entry-level Japanese cars or minivans from Honda, Mazda, Nissan, and Toyota.
6. Stay away from most American front-drives—more frequent failures and costlier repairs.
7. Shop for used, rear-drive, full-sized wagons or converted vans instead of American minivans.
8. Don't buy any European models—parts and servicing can be a problem, and quality control is declining.
9. Be wary of buying for fuel economy alone: Toyota Prius and Honda's Insight/Hybrid can't beat a base Civic for savings and performance.
10. Don't buy "nostalgia" cars: they aren't as good as the memories they invoke.

New Best Buys

Acura—1.8 EL
GM—Cavalier and Sunfire
Chrysler—Sebring, Avenger, and PT Cruiser
Ford—Mustang
Honda—Civic, Accord, and Odyssey
Hyundai—Accent, Elantra, Tiburon, and Santa Fe
Mazda—Protegé, Miata, and B-series pickups
Nissan—Sentra
Subaru—Forester

Suzuki—Aerio and Swift
Toyota—Echo, Corolla, Camry, Prius, and Sienna

If an independent mechanic gives you the green light and if you have the time, knowledge, and parts suppliers to do your own maintenance and repairs, you might seriously consider a Ford/Mazda Probe/MX-6, a Suzuki Sprint or Esteem, or even a recently discontinued minivan like the Chrysler Colt/Summit or Nissan Axxess.